Cambridge Monographs in African Archaeology
45
Series editor: John Alexander

The Emergence of Food Production in Ethiopia

Tertia Barnett

BAR International Series 763
1999

Published in 2016 by
BAR Publishing, Oxford

BAR International Series 763

Cambridge Monographs in African Archaeology 45
The Emergence of Food Production in Ethiopia

ISBN 978 0 86054 971 0

BAR Publishing is the trading name of British Archaeological Reports (Oxford) Ltd. British Archaeological Reports was first incorporated in 1974 to publish the BAR Series, International and British. In 1992 Hadrian Books Ltd became part of the BAR group. This volume was originally published by Archaeopress in conjunction with British Archaeological Reports (Oxford) Ltd / Hadrian Books Ltd, the Series principal publisher, in 1999. This present volume is published by BAR Publishing, 2016.

Printed in England

BAR
PUBLISHING

BAR titles are available from:

BAR Publishing
122 Banbury Rd, Oxford, OX2 7BP, UK
EMAIL info@barpublishing.com
PHONE +44 (0)1865 310431
FAX +44 (0)1865 316916
www.barpublishing.com

Preface

Food production is possibly the single most important step in human cultural development. An understanding of the processes inherent in the emergence of food production is critical to our understanding of human cultural evolution.

Ethiopia has long been considered as one of several 'centres' in which food production arose indigenously at an early date. The principal reasons behind this assertion include the genetic diversity displayed by domesticated crops, the importance of indigenous food plants, historical linguistic arguments, and early state formation in the Ethiopian Highlands. In the absence of a comprehensive database this assertion remains unsubstantiated. This volume addresses the issue of when, how and why food production emerged in Ethiopia. A broad, integrated approach is used which synthesises disparate evidence from related disciplines.

The following work is based on a review of the existing evidence, and on original ethnographic and archaeological data. These data were collected during a two month period of field work in Ethiopia between January and March 1994, and a three week visit to Nairobi in June 1996. The fieldwork in Ethiopia and Nairobi was funded by the James A Swan Fund, the Gilchrist Trust and the University of Bristol, to whom I am most grateful.

I am grateful also for the interest showed in the project by my Ethiopian colleagues in Addis Ababa, especially Dr Bahru Zewde at the Institute of Ethiopian Studies, Dr Kassaye Begashaw in the Department of Culture and Sports Affairs, Dr Solomon Wordeql at the National Museum, Dr Dawit Tadesse at the Biodiversity Institute, Dr Jean Hansen at the International Livestock Research Institute and Dr Sue Edwards at the National Herbarium. My thanks particularly goes to Ayele Tarekegn, Isaac Ejugu, Alan Fairweather and Tania Kaiser, and to Ezra Tsegaye for being in the right place at the right time. The work in Nairobi was facilitated by Dr Charles Nelson, Dr Simuyu Wandibba and Simon Katisya. I am indebted to the British Institute in Eastern Africa for aiding and abetting my relationship with Ethiopia, and to John Sutton and the British Institute staff for their support in Nairobi.

I would like thank those who have advised me during the writing of this book, particularly Mark Horton, Vanessa Straker, Mark Roberts, Caroline Malone, Stephen Brandt, David Phillipson, Roger Blench, Rudolfo Fattovich, Henry Lamb, Martin Bernal, and the Long Ashton Research Station. The support and encouragement of my family and friends has been invaluable throughout.

CONTENTS

TABLES

Chapter 2

Chapter 3

Chapter 4

Chapter 5

Chapter 6

Chapter 7

Appendix A

Appendix C

FIGURES

Chapter 1

Chapter 2

Chapter 3

Chapter 4

Chapter 5

Chapter 6

Chapter 7

Appendix B

Chapter 1.

INTRODUCTION

1.1. THE BACKGROUND: ETHIOPIA AS A CENTRE OF AGRICULTURAL DEVELOPMENT

1.1.1. Ethiopia and the Origins of Agriculture

Ethiopia has long been regarded as 'one of the world's greatest and oldest centres of domesticated seed plants' (Sauer 1952 p.76), where agriculture is believed to have developed independently following the domestication of indigenous plant resources. This theory is primarily based on four lines of reasoning:

> ***Ethiopia as a centre of plant diversity***
> Ethiopia has been identified as a centre of genetic diversity for a range of crop plants and, by implication, a centre for their domestication.
> ***Indigenous Ethiopian progenitors***
> A number of important crops in Ethiopia are derived from endemic plant species. These crops are not grown for food in other regions of the world and they are believed to be the relicts of an early, indigenous agricultural complex.
> ***Historical linguistics***
> Historical linguistic reconstruction of the Ethiopian palaeoeconomy suggest an indigenous origin for agriculture, with domesticated plants and animals present by the 7th/8th millennium BP.
> ***State formation and urbanisation***
> The emergence of organised states and urban centres in the Northern Ethiopian Highlands at an early date relative to tropical Africa implies that a developed system of food production was already established by the 3rd millennium BP.

The evidence for each of these hypotheses is discussed below, and the validity of the theory for independent agriculture is questioned. This volume is not looking for a centre of agriculture domestication in Ethiopia, but for an insight into the prehistoric subsistence economy during the Holocene. The principal aims are to determine the pattern of subsistence change leading to the emergence of food production, and the factors that describe this pattern.

The work presented here is necessarily impressionistic. A broad approach is required to overcome the absence of a coherent database and to avoid too narrow an interpretation of the evidence. This volume draws together evidence from disparate disciplines in order to construct and rationalise a crude framework of human-environment interaction in Ethiopia during the past 10 000 years. By offering a fresh reinterpretation of the existing data, this approach provides a new insight into an old problem.

Ethiopia as a Centre of Plant Diversity

Archaeological enquiry into the origins of agriculture developed during the late 19th and early 20th centuries (eg. de Candolle 1882, Roth 1887, Childe 1928). The formative stages of these enquiries were strongly influenced by the work of the Russian plant geneticist Nikolai Vavilov. Vavilov proposed that plants were domesticated in restricted geographical 'centres' which were determined by the distribution and genetic diversity of modern crop plants (Vavilov 1926). Vavilov initially identified five centres, later adding a further seven, which included the Near East, Southern America, South East Asia and Ethiopia (Vavilov 1951). Subsequent archaeological field work in the Near East and Central Mexico supported Vavilov's 'centres of origins' model for these regions (eg. Braidwood and Braidwood 1950, MacNeish 1950). These successes initiated an energetic quest for further centres of agricultural innovation, and more recent work in South East Asia has clarified that agriculture also developed there indigenously (Chang 1970, Zhimin 1989).

However Vavilov's arguments are now understood to be fundamentally erroneous due to his correlation between the diversity of domesticated crop plants and centres of agricultural origin. Vavilov's centres were all located in mountainous regions with numerous micro-environments to stimulate genetic adaptation and diversification in both local and introduced crops. Several of his centres, including Ethiopia, were subsequently revealed to be 'secondary centres' of diversity of non-endemic crop plants such as wheat and barley (Zhukovsky 1975, Zohary 1970). Ethiopia is now considered as an 'accumulation centre' (Schiemann 1951) or a 'centre of concentration' (Ward 1962) for genetic diversity, rather than the region of origin of 'Near Eastern' crops. Nevertheless, despite the flaws of Vavilov's argument, his approach has been highly influential in African archaeology and Ethiopia is still perceived as having a unique place in the emergence of food production (eg. Hawkes 1983, Phillipson 1992).

Indigenous Ethiopian Progenitors

The work of Jack Harlan has been influential in promoting a botanical awareness of African plant domestication, and suggesting probable areas of origin of African food crops (eg. Harlan 1971a, 1979, 1989a, 1989b, 1992, 1993, Harlan and Stemler 1976). Harlan (1969) argued that Ethiopia is a principal region of diversity and a probable area of domestication for a number of indigenous plant resources. The Ethiopian Highlands are effectively an enclosed ecosystem that contain many unique plant species. Several of these have become domesticated and have been major food crops for at least the past 2000 years . These include the small-grained cereal *tef*, the oil seed *noog* and the false banana *ensete*. Tef and noog are derived from endemic wild progenitors, and are not grown as plant foods outside Ethiopia. Ensete is a useful wild plant in sub-Saharan Africa, but is believed to have been originally domesticated in South West Ethiopia. These resources are thought to constitute part of an ancient food production complex in Ethiopia. Simmonds (1965) made the salient point that a plant with seed as small as tef would never have become domesticated or economically viable if large-grained alternatives such as wheat and barley were available. This argument has drawn considerable support for a model of indigenous, independent domestication.

Ethiopia and Africa - the Botanical Perspective

In challenging Vavilov's model of agricultural centricity, Harlan (1971a) has proposed that the African Sahel constitutes one of three global 'non-centres'. In contrast to the concentrated distribution of resources in the Near East, African wild food plants have a diffuse distribution, with few areas of overlap (figure 1.1). Local plants may have been domesticated a number of times in disparate locations across the Sahelian zone according to their relative habitat, but the dispersed nature of these events failed to precipitate a co-ordinated subsistence shift.

Ethiopia and West Africa are both located at the periphery of the Sahel zone, and both have a relatively concentrated array of crop progenitors. Various researchers (including Harlan 1971 and Zohary 1970) indicate that these are primary criteria for agricultural development. Harlan (1971) also stipulates the importance of geographically adjacent centres and non-centres. West Africa has been proposed as a focus for indigenous agricultural development (Coursey 1976, Shaw 1977) and a comparable role can be implicated for Ethiopia. According to recent linguistic evidence, the emergence of agriculture in West Africa was gradual and protracted (Blench 1996) following long-term, non-intensive selection of certain plant resources. Could this model also apply to Ethiopia?

Historical Linguistic Reconstruction and a Framework for Enquiry

In the 1960s, J H Greenberg established an interactive relationship between linguistic analysis and the interpretation of archaeological data (1964, 1966). Greenberg used a combination of lexicostatistics and glottochronology to quantify the relationship between African languages and thereby estimate their relative time-depths, providing a broad system of classification and chronological development. Although many of Greenberg's original conclusions still apply, this technique has now largely been replaced by a system of comparative linguistics which is based on cross-cultural comparison of certain lexical items of cultural significance.

Although the validity of applying historical linguistics to palaeoeconomic reconstruction is still contentious, it has been used successfully to correlate reconstructed palaeoeconomies of modern language groups in North Africa with the archaeological evidence within a broad time-frame (Ehret 1984, 1993, Vansina 1979, 1980, Munson 1977, Williamson 1989, Blench 1996). When treated with caution, historical linguistics can provide a valuable technique for constructing a broad sequence of cultural and economic development, particularly where the archaeological evidence is negligible (Ehret 1979, 1984, 1993, Williamson 1989, Blench 1993, 1996).

Modern Language Groups in Ethiopia

Of the four major languages spoken in Ethiopia, three derive from Afro-asiatic, one of the four language families of Africa described by Greenberg (1966), and the fourth is a Nilotic language group. The possible original location of the Afro-asiatic language phylum during the late Pleistocene has been suggested - somewhat tentatively - to have been in the savannah lands along or adjacent to the Northern Ethiopian Plateau, in the Ethiopian-Sudanese border region (Ehret 1993, Skinner 1984). Figure 1.2.a. illustrates the possible original location of three of the major language families in Africa. Figure 1.2.b. shows the modern distribution of their daughter language groups for comparison. The immense diversity of languages in Ethiopia, particularly in the South Central region, has led to the suggestion that this is a very ancient centre of linguistic differentiation (Bender 1976). Of the Ethiopian Afro-asiatic languages, Cushitic, Semitic and Omotic, Cushitic is believed to be the original language of the Holocene population of the Highlands (Appleyard 1977). Today Cushitic only exists in isolated pockets, and Semitic based languages are widely spoken across the Highlands.

Linguistics and the Conventional View

The conventional view maintains that Semitic languages were introduced into the Ethiopian Highlands during the 3rd millennium BP and that this event was concurrent with the appearance of a fully developed system of agriculture (Ullendorf 1955). This proposal was based on the observed similarity between the art, architecture, writing and religion of the Axumite civilisation of the Ethiopian Highlands and the Sabaean civilisation of Southern Arabia. However, archaeological evidence now indicates that not only do these cultural attributes develop no earlier in Southern Arabia than in Ethiopia, but that the Ethiopian 'civilisation' was rooted in Africa (Fattovich 1990, Anfray 1967b).

Furthermore. linguistic reconstructions have shown that a number of words, and by implication certain concepts pertaining to food production and agricultural technology, were already embedded in the indigenous Highland language before the introduction of Semitic (Ehret 1979). Words for wheat, barley and plough were loaned *into* Semitic *from* Cushitic, indicating an earlier knowledge and dependence among Cushitic-speaking groups.

Linguistic Reconstruction and Ethiopian Palaeoeconomy

Ehret has proposed a model for the indigenous development of food production, based on historical linguistic reconstructions from the main Ethiopian language groups (Ehret 1979). This hypothesis argues that during the Terminal Pleistocene, subsistence in the Afro-asiatic 'heartland' was based on the intensive collection of wild grasses. Basic food processing technology was already being used to make flour and a form of bread. Comparative linguistics suggest that the dispersal of the Afro-asiatic language group at the start of the Holocene was coincidental with a pre-agricultural expansion of wild plant foraging across Northern Africa and the northern fringes of the Ethiopian Highlands. Certain grasses including tef, millets and sorghum were exploited intensively using pre-adaptive technology such as the digging stick (Ehret 1979).

Comparative analysis of Cushitic languages in Ethiopia indicate that indigenous cereal grains such as millets, sorghum and tef were being cultivated by the 8th/7th millennium BP (Ehret 1976, 1979). Lexical items such as '*to cultivate*' and '*ploughshare*' suggest that a knowledge of food production may have been present, although the interpretation and perception of these words may have altered since prehistory (ploughshare could mean 'ard' or tool for breaking the soil). Domesticated cattle, sheep, goats and donkeys were apparently known over much of North East Africa, with the possible exclusion of Ethiopia (Ehret 1974, 1979). Cognate sharing in certain African language families suggest a dispersal during the early Holocene (Ehret 1983, 1993).

Cushitic speech expanded across the Ethiopian Highland region during the 7th millennium BP to include an area from the Red Sea Hills to the far South East Highlands (Ehret 1976). The spread was apparently established by the 6th millennium BP, and was followed by a diversification of the language group. This is suggestive of regional fragmentation and isolation, as would be expected with more sedentary behaviour. Ehret (1974, 1979) contends further that this move to more sedentary behaviour was associated with the emergence of a fully developed system of agriculture based on indigenous and foreign animals and plants. According to Ehret's linguistic interpretation, developed agriculture was present in the Highlands by 5000 BP.

Additional Linguistic Support for a Model of Early Agriculture

This linguistic framework has provided the foundations for archaeological models for the development of agriculture in Ethiopia. Ehret's chronological structure has received support from a number of linguists who have noted inconsistencies between Ethiopian and Southern Arabian Semitic. These discrepancies indicate a more ancient origin for Ethio-Semitic that disputes the idea of an introduction in the 3rd millennium BP (Demoz 1978, Hudson 1978).

Hudson (1978) proposed that the proto-Semitic language originally split in Ethiopia and spread from there to the Near East and Southern Arabia. Intriguingly, Ethio-Semitic has a close correspondence with the archaic branch of Semitic known as Akkadian, spoken in Ancient Mesopotamia during the 5th millennium BP. The alternative opinion is that the Semitic language was introduced to Ethiopia in its archaic form during the 4th or 5th millennium BP (Demoz 1978). As there was considerable contact between Southern Arabia and the Horn at this time, this suggestion does not seem unreasonable (Zarins and Zahrani 1985, Amirkanov 1994). Although these arguments are unsubstantiated, a number of independent linguistic analyses do agree on one point - that the date for the presence of Semitic speakers in Ethiopia must be pushed back to at least 4000 BP, and probably earlier (Tylock 1975). As the Semitic speakers apparently adopted certain aspects of food production from the indigenous population of the Highlands, this implies that a form of agriculture was being practised in Ethiopia by the 5th millennium BP.

Broad Implications for Subsistence Change

Ehret (1984) notes that although the linguistic evidence is not geographically or chronologically infallible, it is too strong to be dismissed, and it provides a reasonable framework for broad palaeoeconomic developments in Ethiopia.

Elsewhere, Renfrew (1987, 1994, 1996) has argued convincingly that the spread of languages may correspond to the diffusion of technological and economic innovations. Renfrew (1996) collates the development of agriculture with the spread of Indo-European from Anatolia through Europe. A parallel relationship is suggested for the spread of African language families and pre-agricultural developments at the end of the Pleistocene (Sutton 1974, Ehret 1983, 1993). Using a concept originally proposed by Sutton (1974), Ehret (1984, 1993) has indicated a 'parallel testimony' (1993 p.120) between archaeological and linguistic evidence in North

Africa that relates to the development and spread of cattle pastoralism and plant foraging. Additionally, Ehret (1979) suggested that grass-seed harvesting and processing technology promoted the initial dispersal of the Afro-asiatic language family across North East Africa and into the Near East, following the expansion of the open grassland habitat at the end of the last Glacial. An early transfer of incipient knowledge is also tentatively implied by the archaeological data from Africa (Wendorf and Schild 1980, Harlan 1971). Bernal (*in prep.*) goes even further by suggesting that the concept and practice of plant cultivation that emerged in the Levant around 10 000 years ago was derived from African ancestry.

An analogous situation may be evident in the Ethiopian Highlands during the 7th millennium BP, when the Cushitic language expansion may reflect the development and spread of a major subsistence innovation.

State Formation and Urbanism

The emergence of PreAxumite states in the mid 3rd millennium BP and the fluorescence of the Axumite civilisation in the early 2nd millennium BP herald the dawn of Ethiopian history (Fattovich 1990b, Michels 1988, Phillipson and Rerynolds 1996). This phenomenon is incongruously early for our current knowledge of Black Africa away from the Nile Valley, and it has caught the imagination of archaeologists and historians.

Urbanism and state formation are believed to be a natural progression from developed agricultural societies (eg. Adams 1966, Steward 1955, Trigger 1972). Intensive farming is a critical factor in supporting the dense population, the non-subsistence specialisation and the social and economic infrastructure that characterise state formation and urban settlements in the Old and New Worlds (Adams 1966). This pattern has strongly influenced our perception of urbanism, and has provided an analogue for its emergence in the Northern Ethiopian Highlands. Although at present there is no firm evidence for the economic factors responsible for the development of PreAxumite and Axumite urbanism, it is hoped that current work in and around Axum will shed light on this problem.

This volume is concerned with the pre-urban period in Ethiopia, and therefore focuses on events in the Holocene before c.2000 years BP.

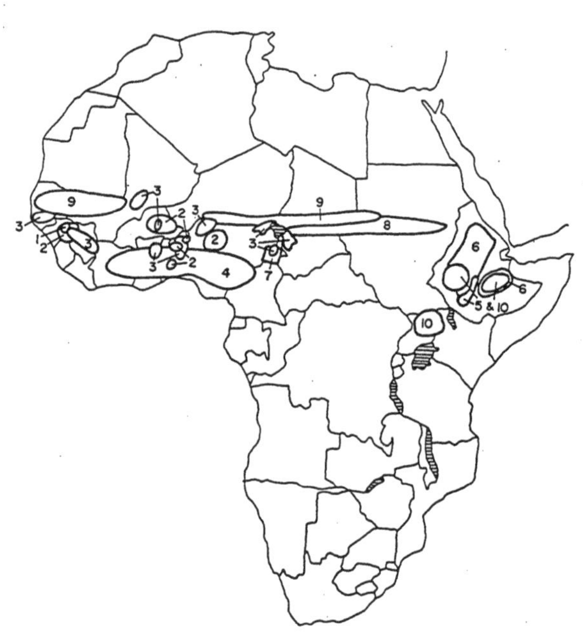

Figure 1.1. Probable areas of domestication of selected African food-plants (after Harlan 1971).
1. Brachiaria deflexa. 2. *Digitaria exilis, D.iburua.* 3. *Oryza glaberrima.* 4. *Dioscorea rotundata.* 5. *Musa ensete* and *Guizotia abyssinica.* 6. *Eragrostis tef.* 7. *Voanzeia* and *Kerstingiella.* 8. *Sorghum bicolor.* 9. *Pennisetum americanum.* 10. *Eleusine coracana.*

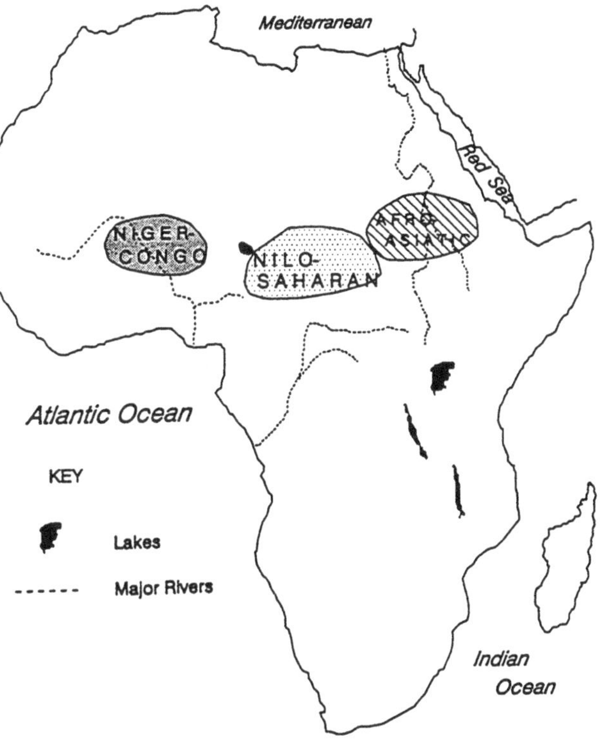

Figure 1.2.a. Suggested original location of Afro-asiatic and the major African language families in the late Pleistocene (after Blench 1993).

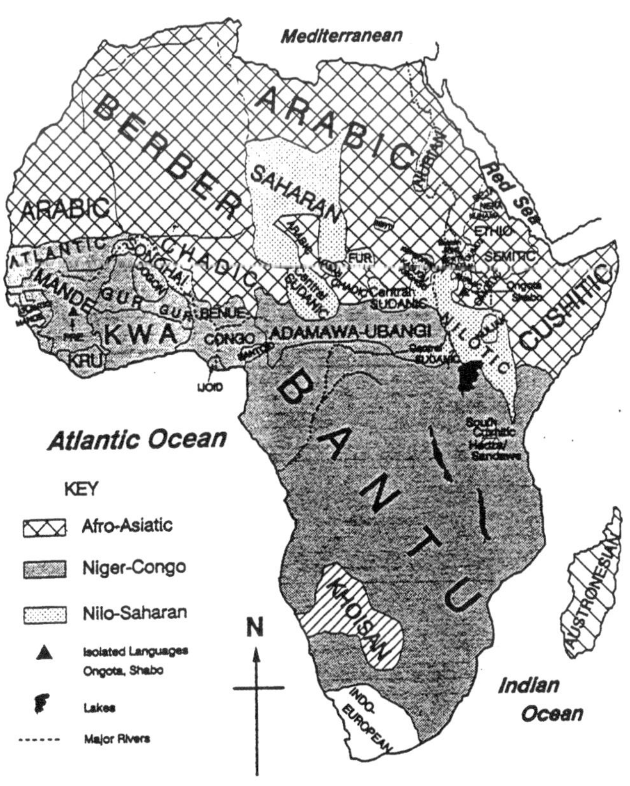

Figure 1.2.b. Modern-day distribution of African language families (after Blench 1993).

Summary

In summary, Ethiopia is considered as a centre for early, independent agricultural development for a number of reasons:

- Ethiopia was identified as one of several global centres for the independent origin of agriculture. Whilst this model has yet to be substantiated, it continues to influence our perception of the development of food production in Ethiopia.
- The presence of a number of unique, endemic food crops suggests an independent process of domestication.
- Historical linguistic reconstruction advocates a model for knowledge and exploitation of certain indigenous domesticates in Ethiopia from an early date. While the linguistic framework does not unequivocally claim that food production emerged independently, it does infer that agriculture was practised from at least the 5th millennium BP.
- The relatively early rise of states and urban centres in the Northern Highlands implies that agriculture was well established in this region by the 3rd millennium BP.

In the absence of a coherent archaeological database, this evidence structures our perception and provides the foundations for archaeological models to describe the emergence of food production in Ethiopia.

1.1.2. Traditional Views of the Origins of Agriculture in Ethiopia

The Traditional Models

The idea that agriculture arose independently has received little support in more recent years. Archaeologists and historians have devised a range of alternative models to account for how the concept of food production may have developed in the Highlands. Following Brandt (1984), five main themes can be identified:

1. Diffusion of agriculturists from Egypt up the Nile Valley and into the Ethiopian Highlands c.5th millennium BP (Simoons 1965, Doggett 1965, 1991).

This model proposed that a mixed farming economy was brought to the Ethiopian Highlands from Egypt as a result of prolonged knowledge of and contact between these regions. A date in the 5th millennium BP has been suggested for the introduction of agricultural technology in association with a 'Near Eastern package' of wheat, barley, chickpea, lentil, pea, flax and domesticated cattle and caprines Simoons (1965).

2. Diffusion of 'C-Group' pastoralists from the desert regions of the Nile Valley during the 4th millennium BP (Clark 1962, 1967, 1980).

Clark argued that cattle and caprids (meaning here sheep/goat) were introduced to the lowlands of the Horn and the fringes of the Ethiopian plateau by C-group pastoralists from the Nile Valley and the surrounding arid regions. The migration was seen as a response to increasing desiccation in these areas during the late 5th/early 4th millennium BP (Williams 1984). Clark proposed that these groups brought a knowledge of plant food production and may have introduced crops such as wheat and barley. The concept of food production then spread from the lowlands to the Ethiopian plateau, initiating the domestication of local food crops. However this explanation is invalidated by Clark's (1980) subsequent statement that indigenous plant such as tef are unlikely ever to have become domesticated if large-seeded foreign alternatives were available.

3. Migration of farming groups from the Eastern Sudan into Western Ethiopia and the Ethiopian plateau during the 4th millennium BP (Murdock 1959, Clark 1976, 1980).

Murdock (1959) suggested that tropical domesticates (sorghum, millets, cotton, root crops and domesticated animals) and an agricultural way of life were introduced to the Western fringes of the Ethiopian plateau following movements of 'Pre-Nilotic' farmers from the Nile Valley and Eastern Sudan. Contact between these groups and the indigenous population of the Highlands precipitated the domestication of local food plants, such as tef and noog in the North and ensete in the South. Clark (1976) proposed further that the Pre-Nilotes migrated into the Highlands in response to mid Holocene climatic deterioration, thus providing a mechanism for the introduction of tropical domesticates and the consequent domestication of local resources.

4. Migration of Sabaean people from Southern Arabia to the Northern Ethiopian Highlands during the mid-3rd millennium BP (Schweinfurth 1868, Steihler 1948).

The movement of agriculturists from Southern Arabia into the Ethiopian Highlands was concomitant with the introduction of a Near Eastern 'package' and plough and irrigation technology. Indigenous plants were subsequently domesticated and incorporated into the agricultural complex. This model has received little support in recent years.

5. Independent domestication of local plants such as sorghum and finger millet in the North East African 'hearth' in Eastern Sudan and the fringes of the Ethiopian plateau (Doggett 1988, 1991, Ehret 1979).

Doggett (1988) proposed that tropical food-plants were domesticated along the fringes of the Ethiopian Highlands during a period of climatic aridity when they were removed from their natural habitat to higher altitudes by groups of cultivators. Indigenous food-plants were domesticated in the Highlands as a result. Doggett (1991) also argued that temperate wild progenitors of crops such as barley were distributed in the Nile Valley as far south as Nubia during humid climatic phases in the Holocene. Deteriorating climatic conditions encouraged incipient, pre-agricultural groups to migrate with their subsistence resources into less arid regions such as the Ethiopian Highlands. The wild resources became domesticated as a result of this selection pressure.

The Inadequacies of the Traditional View

In the absence of a firm archaeological framework for the cultural chronology of Ethiopia during the Holocene, the traditional models remain subjective and, at worst, conjectural. These theories need to be rigorously tested through an interdisciplinary, regional approach.

Recent writers have modified the traditional argument for the origins of agriculture in Ethiopia, and have adopted a more conservative view. This view agrees that Near Eastern crops were introduced 'at an early date' and that domesticated animals are of 'considerable antiquity' in Ethiopia, but little is offered in the way of explanation (eg. Hawkes et al 1991, Phillipson 1992, 1993, Phillipson and Reynolds 1996). Several leading questions are implicit in this approach: How and why should Near Eastern plant domesticates be introduced to the Ethiopian Highlands 'at an early date'? From where did domesticated animals originate, and when? How did Ethiopia interact with its African context? And why should food production *not* develop indigenously in Ethiopia? These questions will be examined in the course of this publication.

Causality in the Traditional Models

Causality has often been overlooked in models for the development of food production in Ethiopia. However the effects of climate change are explicit in a number of theories. Clark (1976, 1980), Ehret (1979) and Doggett (1988, 1991) for example implicate the effects of climatic deterioration in initiating the migration of food producing groups from arid regions into the more humid Ethiopian Highlands during the 4th or 5th millennium BP.

The inadequate nature of the database means that at present these models cannot be supported or refuted. Furthermore, cultural explanations for subsistence change cannot be proposed as viable alternatives due to the paucity of cultural data. The proposition that climate change was instrumental in subsistence change in Ethiopia has not been scrutinised, and a coherent framework to describe the process has yet to be presented. This volume challenges the prevailing view for the emergence of food production in Ethiopia and explores the possible causal role of climate change in this process.

Implications and Outlook

Although these questions are no closer to being answered now than they were in the past, significant theoretical and methodological developments have been made recently within the archaeological discipline as a whole which justify a fresh reinterpretation of the existing evidence. Furthermore, a substantial body of data has accumulated in related fields, particularly those of genetics and palaeoclimate research. These advances have major implications for our current understanding of the relationship between Africa and Ethiopia, and of the nature of subsistence change in each of these areas.

1.1.3. The Questions And The Problems

The central problem then is whether or not food production emerged at an early date in Ethiopia. This problem is examined through both a cultural and economic context (chapters 2, 3, 5 and 6), and a climatic and environmental context (chapters 2 and 4) with reference to the Northern Ethiopian Highlands.

There is negligable direct evidence for subsistence change in Ethiopia during the entire Holocene period prior to the mid-3rd millennium BP, but does this absence of comprehensive archaeological data mask a real pattern of change? This question can be answered by integrating the indirect evidence from disparate fields, including plant genetics, botany, agronomics, archaeology ethnography and linguistics. This integrated approach establishes a simplified chronological and conceptual outline for the pattern of subsistence development during the Holocene, and proposes that there is substantial indirect evidence for a critical change in subsistence patterns in Ethiopia.

This second line of enquiry investigates certain factors that may be influencing human behaviour. Climate change during the Holocene has been proposed as the primary factor in controlling subsistence change in Ethiopia, and a substantial body of relevant palaeoclimatic and palaeoenvironmental data has accumulated in recent years. The extent of the changes and their impact on the environment have not been closely examined from an archaeological perspective, and their consequences for human behavioural adaptation remain virtually unresearched.

Chapter 2.

ETHIOPIA IN THE PRESENT: AN ETHNOGRAPHIC PERSPECTIVE ON THE EMERGENCE OF FOOD PRODUCTION

2.1. GEOGRAPHICAL DIVERSITY AND AN ISLAND ENVIRONMENT

2.1.1. The Physical Geography of Ethiopia

Ethiopia occupies an area roughly 1.1 million km^2 in the Horn of Africa (figure 2.1). The country is strategically located between the African interior and the Indian Ocean, although the modern political boundaries are landlocked by the recent independence of Eritrea. In addition Ethiopia acts as a corridor between Northern and Southern Africa, avoiding the swamps of the Sudd and the equatorial jungles of the Central African Republic. Ethiopia is one of the few regions between Northern Africa and northern sub-Saharan Africa that is largely free of tsetse fly, with only 66 000 km^2 of the humid South West and Blue Nile Valley affected[1] (National Atlas of Ethiopia 1981) (figure 2.2). This has important implications not only for the movement of wild game, but also for pastoral transhumant routes, both of which avoid tsetse areas.

Topographical Diversity

Ethiopia is renowned for its topographical and environmental complexity. The majority of the country comprises an uplifted volcanic plateau, or highland zone. The escarpment of this plateau rises steeply from the surrounding arid regions and the Highlands are diagonally bisected by the Rift Valley into the Northern and Southern regions. The Highlands have been massively eroded to form a series of towering mountain ranges and isolated plateaux, or *ambas*, and scoured by deep ravines and river gorges with a depth of over 1000 m. These physiographical features produce a maximum altitude range between the highlands and lowlands of 100 m below sea level to over 4500 m above sea level (asl), and a variation of over 2000 m is not uncommon within contiguous highland regions.

[1] Tsetse fly (*Glossina* sp) is a vector of the disease Trypanosomiasis (sleeping sickness). Trypanosomiasis can affect cattle, camels, donkeys, mules and horses, causing loss of productivity and death. The disease is also fatal to humans and wild bovids and equids. Tsetse fly have effectively sterilised large tracts of potential pastoral land in Africa and have precluded the development of mixed farming (Ford 1975). The 4-5% of Ethiopia affected is the most sparsely populated region of the country, and few domesticated cattle are kept (National Atlas of Ethiopia 1981).

The Highlands are cloaked in thick black and red soils, predominately alfisols, aridisols and inceptisols (Birch 1967), which are liable to cracking in drought conditions and waterlogging in the rains. The black soils are especially common in shallow valleys and enclosed plains. They are unusually fertile and can support up to twelve consecutive annual crops without fallow (Murphy 1963). Conversely the lowland regions have poorer sandy or gravely soils that are less suitable for agriculture.

Climatic Systems

Ethiopia is located at the Eastern-most end of the Sahelian zone and its climatic pattern is influenced by the annual movements of the Inter-Tropical Convergence Zone (ITCZ). Macro-scale pressure changes and monsoon flow produce a pronounced biennial alternation of dry winters and wet summers (eg. Yemane et al 1987). In addition, the Ethiopian climate is influenced by the Indian Ocean monsoon system which causes shorter, lighter rains in the spring months (March-May) (Westphal 1975). Localised precipitation and temperature vary immensely within these seasonal parameters according to elevation and micro-topography, for example annual average rainfall may vary tenfold at the same latitude with an altitude change of c.2500 m (Westphal 1975). Figure 2.3 shows the seasonal variation in annual rainfall patterns in different regions of Ethiopia, and the variation in average annual rainfall is shown in figure 2.4.

Ecological Diversity

The climatic and topographic variety in Ethiopia creates a range of ecological zones which support a wide spectrum of plant and animal life. Figure 2.5 illustrates the relationship between vegetation and altitude. The macro-biogeographical pattern of vegetation varies from the hyper-arid flora of the Afar Rift to the afro-alpine flora of the well-watered Highlands. More than 45 vegetation zones have been identified and 75% of the total vegetation cover comprises forest-savannah, woodlands, steppe and grassland (Hamilton 1982, White 1983). The vegetation of Ethiopia belongs to the palaeotropical kingdom, and with 33% endemism it constitutes an independent floristic region. Several species which also belong to the South European flora are found in the Highlands between c.1800 and 2400 m asl, and many central European herbaceous species flourish between 2400 and 3000 m (Schweinfurth 1868). Above 3000 m the vegetation of the Northern Highlands belongs predominately to the afro-montane archipelago-like regional centre of endemism, in which many of the plants species are similar to the relict flora of other East African mountains such as Mount Kilimanjaro, Mount Kenya and Mount Elgon (White 1983, Gillet 1955). High species diversity in the Northern Highlands has resulted from the recurrent contraction and expansion of

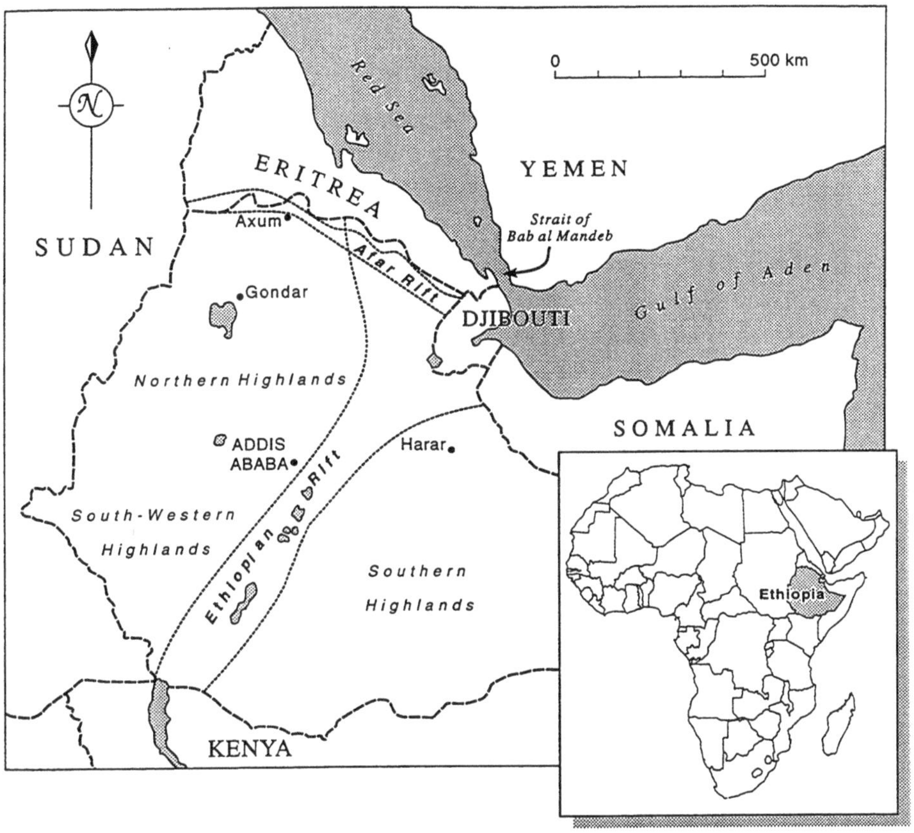

Figure 2.1. Location of Ethiopia showing major physical and political boundaries.

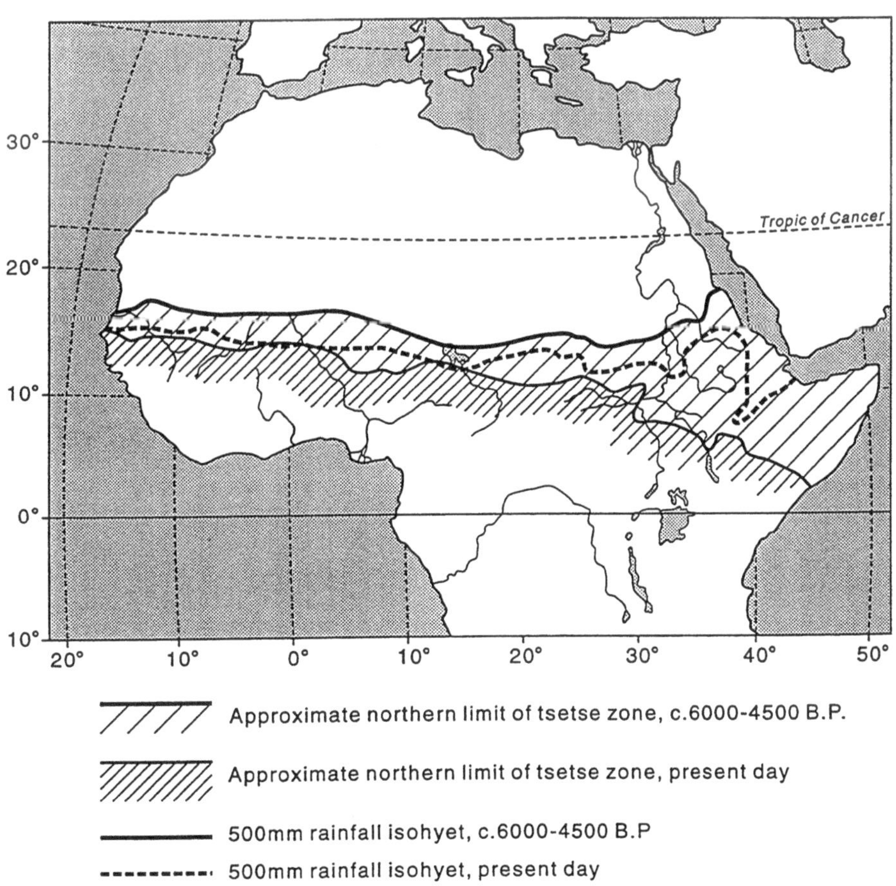

Figure 2.2. Location of past and present tsetse fly zones in Northern Africa and their relationship to the suggested 500 mm rainfall isohyet (adapted from Ford 1975).

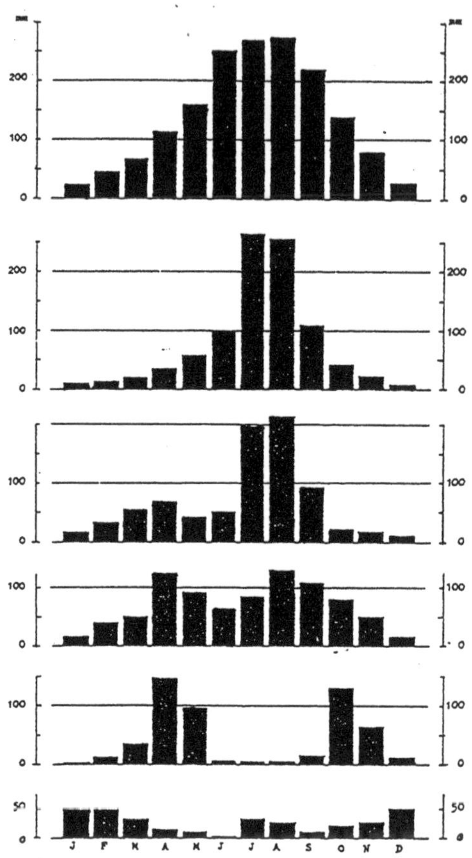

Figure 2.3. Localised seasonal rainfall pattern in Ethiopia based on monthly means recorded over 6-30 years (after Bethke 1976).

Regions from which measurements derive are (top to bottom): South West Ethiopia, North West Ethiopia, the Ethiopian escarpment, the Southern Highlands, the Rift Valley, the Red Sea coast.

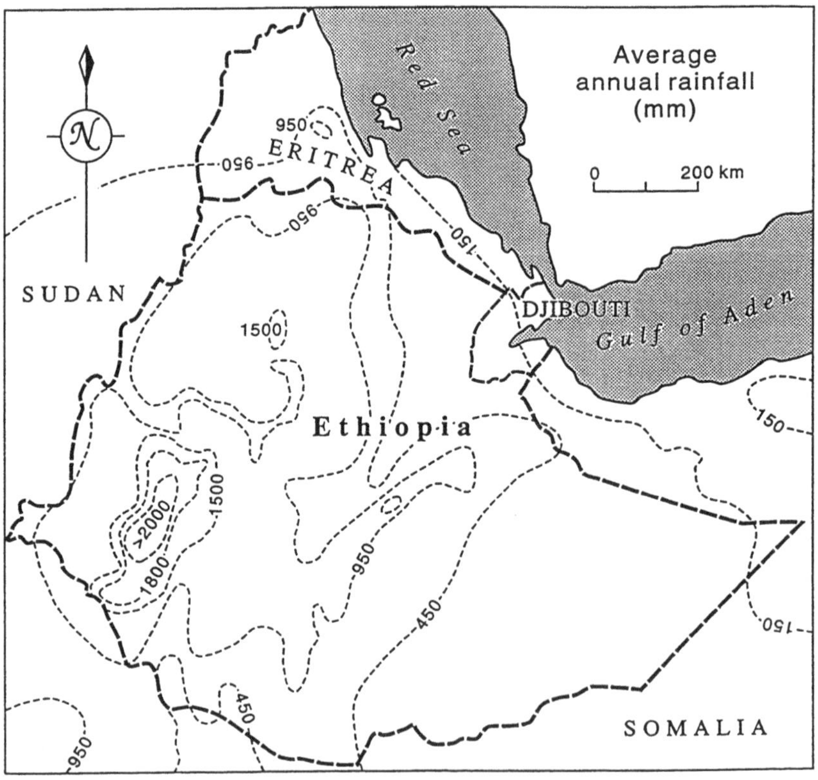

Figure 2.4. Variations in average annual rainfall in Ethiopia.

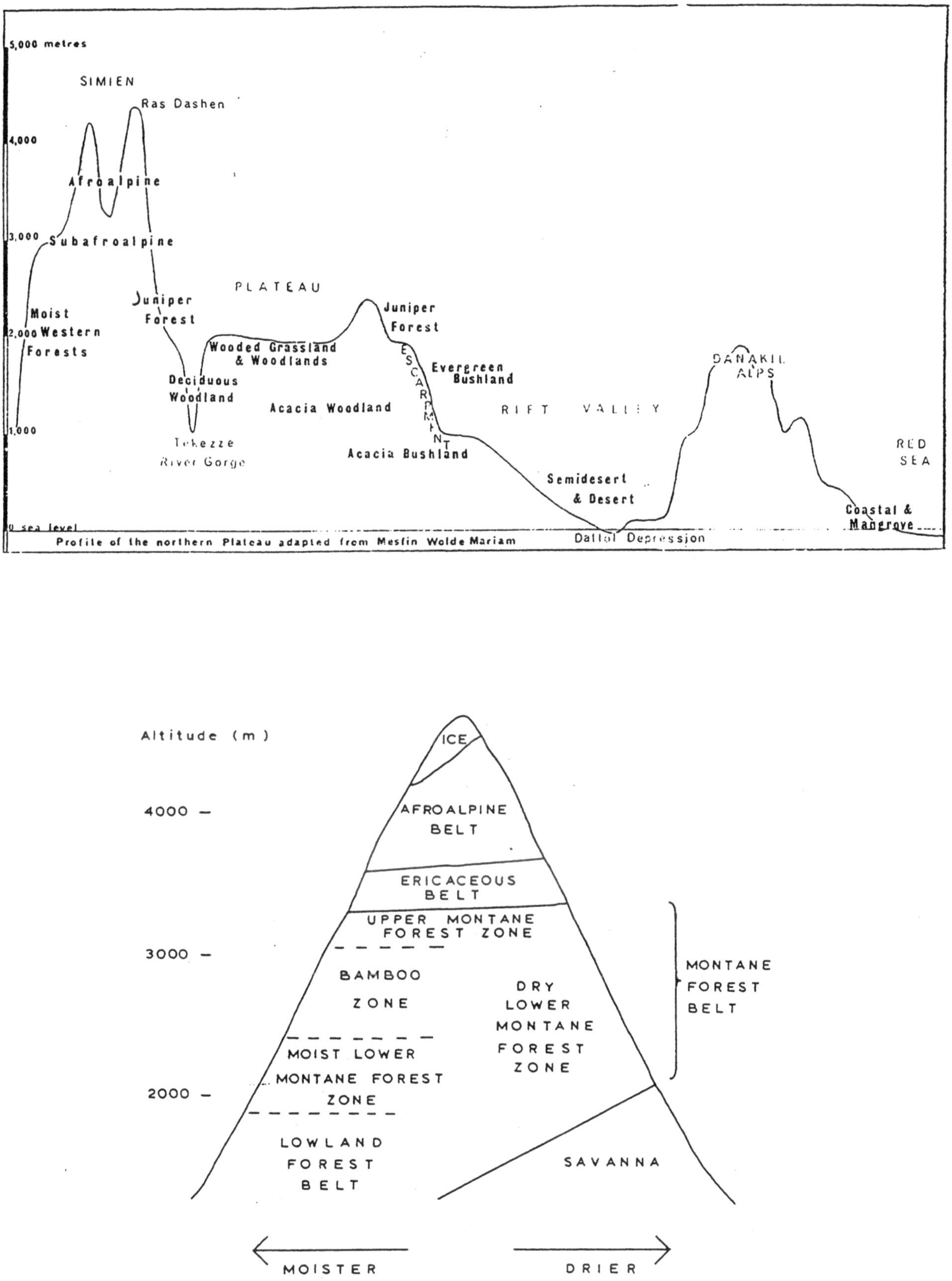

Figure 2.5. Relationship between vegetation patterns and altitude in Ethiopia.
(Top diagram after Edwards 1982, bottom diagram after Hamilton 1982).

forest refugia with climate change. Glacial periods initiated allopatric speciation of plant types within discrete refugia, and these species subsequently migrated as the forest coalesced under more favourable climatic conditions (Prance 1987).

In addition there is an important floristic divide between the Northern and Southern Ethiopian Highlands, with the Southern Ethiopian flora showing closer homology to Kenyan and Ugandan species than those of Northern Ethiopia. This implies long-term genetic separation of the two areas and isolation of the Northern Highlands.

Little of this natural diversity remains today and the lush Abyssinian Highlands of historical documentation are now largely a treeless wasteland due to the drastic effects of denudation and non-sustainable land-use. Anthropogenic forces have reduced the country to a critical condition, with one of the largest soil erosion rates in the world and a precarious level of subsistence farming (Hunri 1983). Deforestation has reduced tree cover from over 45% to 2.7% of the total land surface area (Simoons 1960). As a result 50-60% of rainfall is lost as runoff, carrying 2-3 billion tonnes of topsoil away annually (Campell 1991). Historical accounts report that the fertile volcanic soils of the Highlands have been intensively cleared and cultivated for at least the past 1000 years, and as recently as the 18th century farmers were still reaping up to three crops a year (Thomas 1938, Gibb 1929, Bruce 1790, Alvares 1881, Salt 1814). In more recent years mass deforestation of the relict forest, combined with burning and extensive overgrazing, has irreversibly destroyed much of the primary vegetation. The montane juniper (*Juniperus procera*) for example is especially popular in house construction and is also collected extensively for firewood (Mahaney 1989). Such selective exploitation causes a change in species equilibrium and leads to bush encroachment and rapid degredation of forest (Mackel et al 1989). Vegetation patterns have been altered in many areas, particularly around towns where the traditional juniper forest has been replaced by quick-growing eucalyptus trees. These provide firewood for an ever expanding population and stabilise the rapidly eroding soils. Knowledge of the original vegetation has been preserved through early botanical surveys, and through more recent documentation of the few plant refugia which persist on uncultivated land and around churches (Cufodontis 1953-72, Pichi-Sermoli 1955, Lemordant 1971, Dombrowski 1971).

Environmental Zones

Ethiopia is traditionally divided into four altitudinal zones which broadly correspond to the major vegetation bands. The origins of these divisions are unknown and they are part of an ancient verbal tradition that was established before the first written records appear in the early 2nd millennium BP, suggesting that the relative characteristics and merits of Ethiopian biogeography have long been understood by her inhabitants (Dove 1890, Huffnagel 1961, Westphal 1975).

The distinction between different zones is blurred and individual species are often common to several zones, forming a mosaic-type distribution. Some of the characteristic species from each zone are detailed in *Appendix A*, Tables A.1.1-A.1.8, and their approximate geographical distribution is shown in figure 2.6. The traditional catagorisation, which will be used throughout this volume, is described as follows:

KWOLLA

The arid or semi-arid lowland zone between 500 and 1800 m asl with low annual rainfall and temperatures in excess of 20°C, often greater than 30°C. The vegetation of the *kwolla* is typically savannah-woodland comprising xerophytic species of grass, many of which are edible or have grazing potential, eg. sorghums and millets (Chapman 1992, Renvoize et al 1992), and deciduous trees and shrubs such as acacia and baobab. As the land continues to rise towards the Highland plateau, rainfall increases in direct proportion to altitude due to the prevailing westerly monsoon influence, and the vegetation becomes richer and denser. In the upper *kwolla*, between 1000 and 1800 m, the vegetation of the lower slopes of the escarpment is typically open deciduous woodland. Numerous tall trees, between 7 and 10 m height, provide light, discontinuous cover over short, *Hyparrhenia*-dominated grassland which is interspersed with thickets of bamboo.

WOINA DEGA

The temperate zone between 1800 and 2400 m asl, with mean annual temperatures of 16-20°C and moderate rainfall. It is the most verdant region of Ethiopia comprising a mosaic of montane savannah and evergreen thicket and shrub. Soil and climatic studies indicate that 42-48% of Ethiopia and Eritrea were originally under forest, much of which would have been distributed in this zone. The forest is dominated by *Gymnosperm* sp such as *Juniperus* and *Podocarpus* which occur ubiquitously with wild olives (*Olea* sp) and a variety of shrubs. Localised pedological and topographic conditions produce variations on this general theme, for example in the Lake Tana basin, as detailed in *Appendix A*, Table A.1.4. Today only sporadic relicts of primary vegetation are preserved above 2000 m, particularly

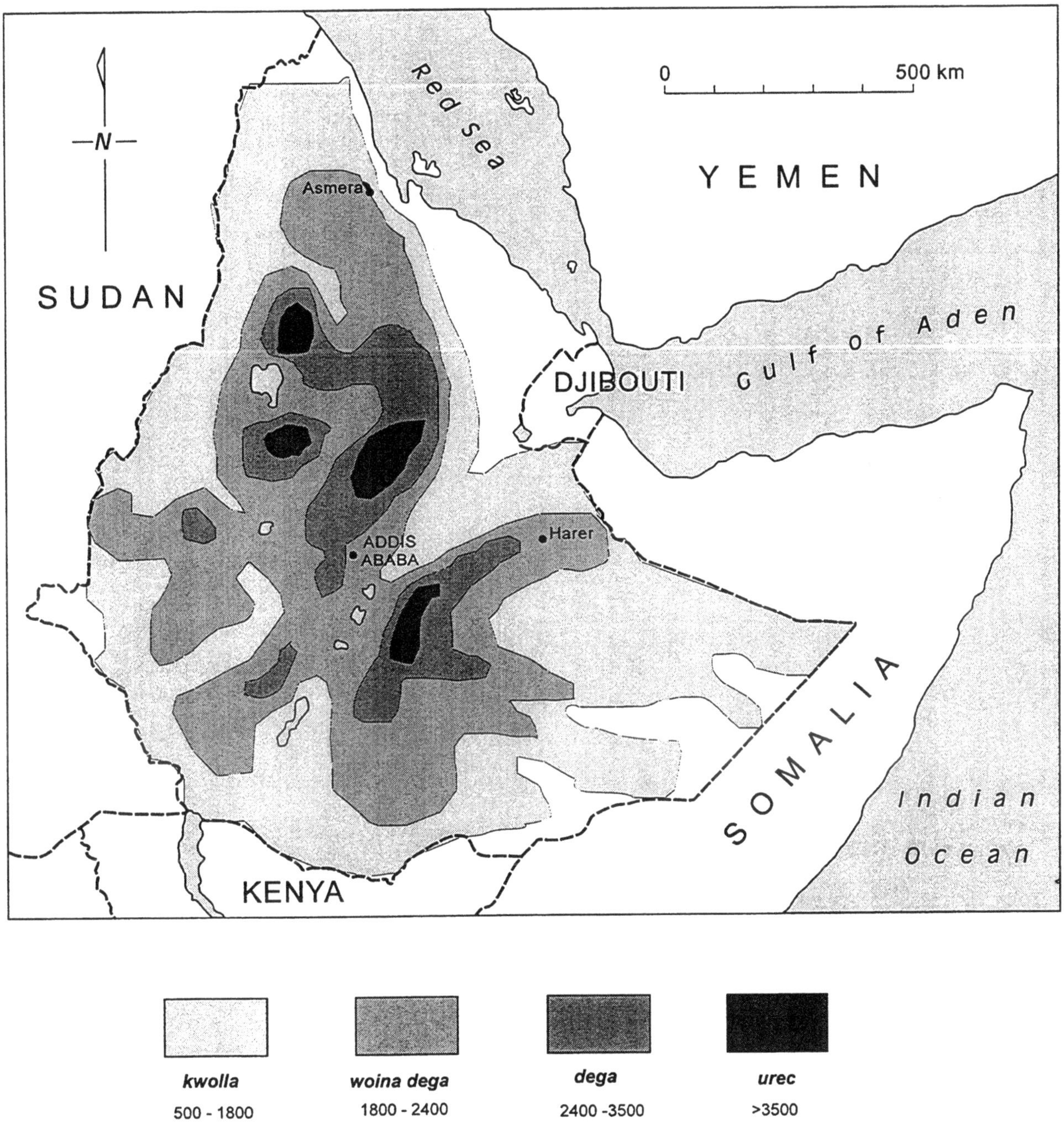

Figure 2.6. Relief map of Ethiopia illustrating the traditional ecological zones

in the mountains east of Lake Tana and on the lake islands.

DEGA

The cool zone, between 2400 and 3400 m asl with moderate rainfall and temperatures ranging between 10 and 16°C. The upper *woina dega* and lower *dega* zones, between 2200 and 3000 m, support a belt of rich montane forest dominated by dry evergreen forest, characterised by *Juniperus, Podocarpus* and *Hagenia*, which is interspersed with montane savannah of *Andropogon* and *Pennisteum* species mixed with other herbs and grasses. The richness of the vegetation decreases with altitude in the lower *dega.* At even higher elevations, the density of Ericaceous species increases and forms thickets in some undisturbed areas.

UREC

The high mountain zone above c.3500 m asl where mean annual temperatures are below 10°C, light ground frosts are common and summer rainfall is high. The upper tree-line is at c.3000 m, above which a belt of Ericaceous scrub gives way to the archipelago-like distributions of high montane vegetation of the upper *dega* and the *urec*. This comprises mosaics of high mountain scrub, high mountain steppe and afro-alpine formations. Afro-alpine vegetation mixes with the high mountain scrub and steppe at its lowest elevations, but uniquely inhabits the highest altitudes[2]. The open landscape is dotted with plants such as giant lobelia, herbs and low shrubs, with sedges, ferns and rushes in damper areas.

Summary and Implications of Geographical Diversity in Ethiopia

Ethiopia is a diverse biogeographical system. The climatic and topographical range produce a broad vegetational spectrum that is defined principally by the altitude and latitude. The region known as the *woina dega* is the most fertile ecological zone in Ethiopia. The widest botanical range is found in the ecotone between the upper *woina dega* and *dega*, between c.2200 and 2600 m asl. It is this region where indigenous agriculture is most likely to have arisen, and this is here, perhaps, that we should trace the development of food production.

2.1.2. The African Context

Much of Ethiopia comprises a cool, well-watered Highland environment above 2000 m asl. This contrasts starkly with the surrounding arid lowland regions of Sudan, Eritrea, Djibouti, Somali and Northern Kenya, making the Ethiopian Highlands unique and incongruous in their regional environmental context.

Environmental and Climatic Integration of Ethiopian to an African Context

Despite this juxataposition, Highland Ethiopia plays a vital role in the ecological equilibrium of Northern Africa. Massive rivers such as the Blue Nile, the Sobat and the Atbara drain the Ethiopian Highlands and annually replenish the fertility of the arid lowlands with tons of water and topsoil (Williams and Adamson 1980). Before a series of dams were constructed along the length of the Nile, the annual Nile flood regime was dominated by summer monsoon flood waters draining from the Ethiopian Highlands. The Highland catchment area contributed over 7 billion m^3 of water per day during August and September, compared to the average winter flow of 1 billion m^3 per day which is maintained principally by the White Nile (Hurst 1952, Grove 1978). Figure 2.7 illustrates the contribution of Ethiopian rivers to the seasonal hydrology of the Nile basin, relative to the White Nile flow. This seasonal discrepancy was responsible for the annual inundation of the Nile flood plain in Egypt and was the mainstay of Egypt's fertility and wealth. The Egyptian belief that the Ethiopian monarchs had the power to control the seasonal flood waters of the Nile was a constant source of tension between the two countries. This threat was exploited by the Ethiopians as late as the 18th century (Bruce 1790).

In addition, Ethiopia and Africa are united by their prevailing meteorological patterns. The seasonal monsoon in Africa is dictated by annual movements of the ITCZ across the entire intertropical zone (Fein and Stephens 1987). This seasonal pattern is not consistent, but varies considerably from year to year around an annual mean. Although the nature of these variations is determined by localised geographical and ecological conditions, the mechanisms responsible for them operate uniformly across the tropical zone. Human groups inhabiting the intertropical zone are constantly exposed to climatic instability. Changes in the climatic system in this region alter the environmental context in which human groups subsist. These alterations demand some form of adaptive behaviour, in relation to the degree of intensity of the change.

2 The flora is distinct enough to warrant a separate phytogeographical afro-alpine catagorisation (Hedberg 1961). This type extends across all the high altitude zones of the East African mountains and reaches its Northern-most limit in the Simien mountains of Northern Ethiopia. During colder climatic periods the afro-alpine vegetation colonised lower altitudes and was more widespread than at present, allowing the individual refugia of the East African mountains to merge (Hamilton 1982, Coe 1989).

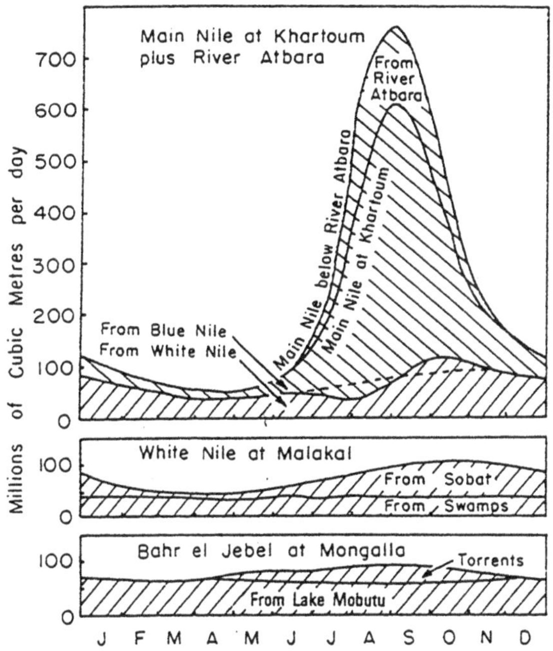

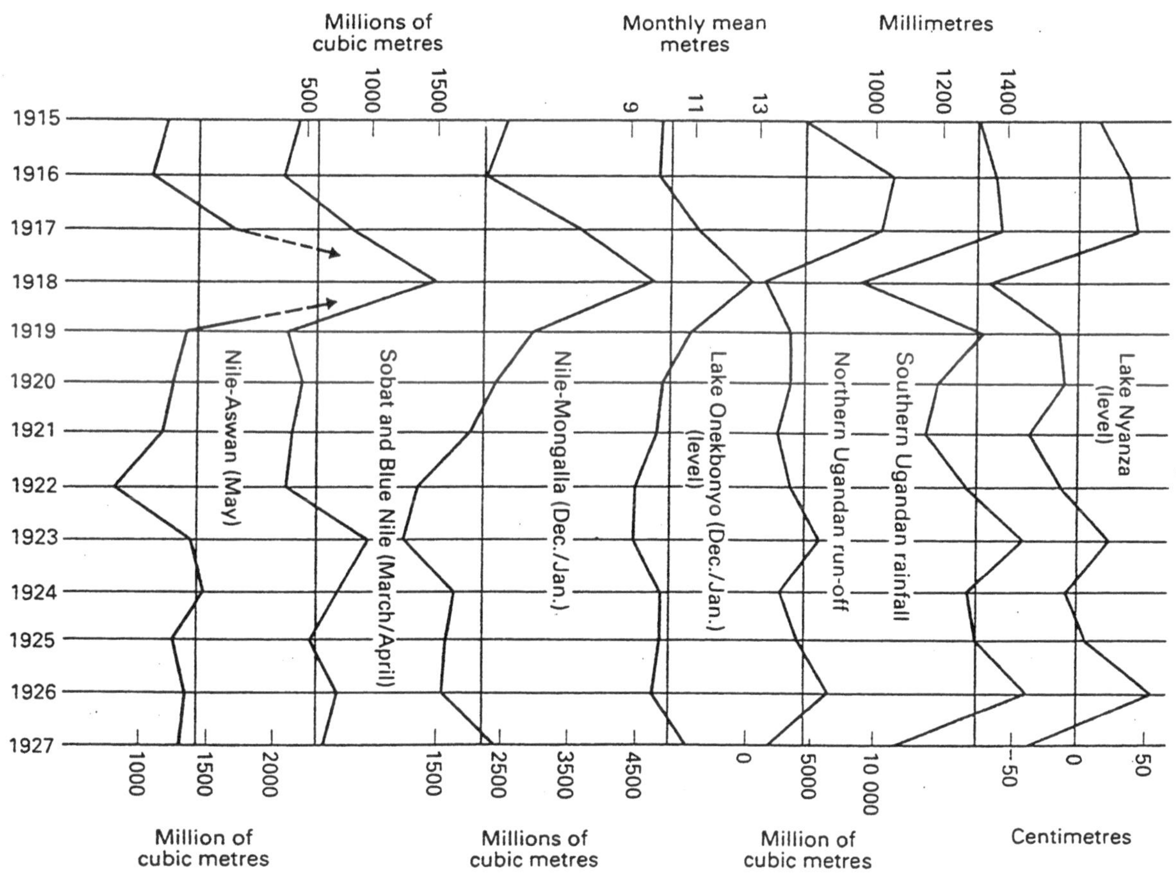

Figure 2.7. The relative seasonal contributions of Ethiopian drainage channels to the annual Nile flow (after Herring 1979).

Ethiopia as a Unique, Integrated System

The distinct biogeographical character of Ethiopia makes it unique within its regional African context. Despite this disparity, the Ethiopian Highlands are integrated to the African continent by two dynamic variables: the rivers and the climatic regime. The rivers draining the seasonal monsoon rains from the Highland catchment area are essential for sustaining life in the arid regions of Northern Africa. This seasonal flow has been responsible for nurturing the great civilisations of the Nile Valley. In addition, Highland Ethiopia is united with tropical Africa through its seasonal climatic regime. Variations in this annual pattern have been contemporaneous across the continent.

Ethiopia cannot be treated as an autonomous entity, but must be viewed as an integral component of its African context. This perspective is essential in a discussion of human development. The juxtaposition between the arid Africa lowlands and the lush Ethiopian Highlands would have had significant implications to the movement of human groups and their patterns of subsistence. Furthermore, synchronised climatic oscillations across the intertropical region would have necessitated broadly contemporaneous patterns of human adaptation during prehistory.

2.1.3. Cultural Geography and Linguistic Diversity in Ethiopia

Environmental diversity and geographical isolation in Ethiopia have promoted ethnic heterogeneity to form a cultural and linguistic mosaic. Ethiopia is one of the most varied single-family language areas of the world, with over 70 language groups and 200 dialects (Hess 1970). Attempts to categorise the Ethiopian population have concentrated specifically on cultural-linguistic factors - characteristics which correlate broadly with religious, economic, genetic and ecological patterns (Ullendorf 1960, Levine 1974). Four major groups can be identified using this system: Semitic, Cushitic, Nilotic and Omotic. The relative distribution of these groups is shown in figure 2.8.

The Semitic Cultural-Linguistic Group

The majority of Highland Ethiopians today speak Semitic-based languages such as the official language, Amharic. Ethio-Semitic languages are more diverse than in any other modern day Semitic speaking area, suggesting that the original language group dispersed a substantial time ago (Bender 1971). Semitic-speaking groups are generally sedentary agriculturalists, which may imply considerable antiquity for the spread of agriculture around the Highlands. Despite their linguistic diversity, Semitic-speaking groups display a surprising degree of cultural uniformity.

The Cushitic Cultural-Linguistic Group

Cushitic is the most geographically widespread and diverse group in Ethiopia. It currently occupies an immense area from Southern Tigray to Harar in the east, west to the tributaries of the Nile, and south to the Tana river in Kenya. Although originally confined to the arid lowlands, many Cushitic-speaking groups expanded into the Highlands during the 15th and 16th centuries (Hassen 1990). Cushitic-speaking groups are pastoralists by tradition, but the majority now practise some degree of agriculture according to their cultural and environmental context (Gamst 1969, Westphal 1975). For example, the Sidama live in a particularly rich environment in the South Western Highlands where they practice a form of agro-pastoralism (Ullendorf 1960).

The Cushitic group can be further sub-classified into Eastern and Central Cushitic speakers (Levine 1974). Eastern Cushites occupy the Eastern, Southern and Western mountain slopes and lowland plains and the Rift Valley. This comprises the largest Cushitic-speaking group, the Oromo, which is composed of several large tribal groups such as the Borana, the Mati, the Sidama and the Arussi. These groups occupy South Western regions, the Southern Rift Valley and the slopes of the Rift Valley escarpment (Hallpike 1970, Wilding 1985b).

The Central Cushites, or Agau, occupy the Northern Highlands where they survive as isolated and disparate groups such as the Qemant, the Kwara, the Khamir, the Khamta and the Falashas (Ullendorf 1960, Gamst 1969). Unlike the majority of Cushitic groups, the Agau are sedentary agriculturalists who practise plough agriculture (Gamst 1969). These groups are believed to be relicts of the original Cushitic-speaking population of the Highlands that have consequently become culturally and linguistically diversified.

Nilotic Cultural-Linguistic Group

Nilotic-speakers comprise only 3-5% of the total population of Ethiopia (Hess 1970). This group comprises several small tribes, such as the Meka, the Mao, the Gunza, the Bari and the Beni-Shangul, whose negroid features make them ethnically distinct from Cushitic and Semitic groups. The Nilotic group is believed to have spread into the Western foothills from the Nile Valley region (Murdock 1959). This group is geographically confined to the Western edge of the Ethiopian plateau adjacent to the Sudan where they practice various forms of primitive agriculture. The severe tropical climate and the high frequency of disease makes this area unappealing to Highland Ethiopians (Cheeseman 1936).

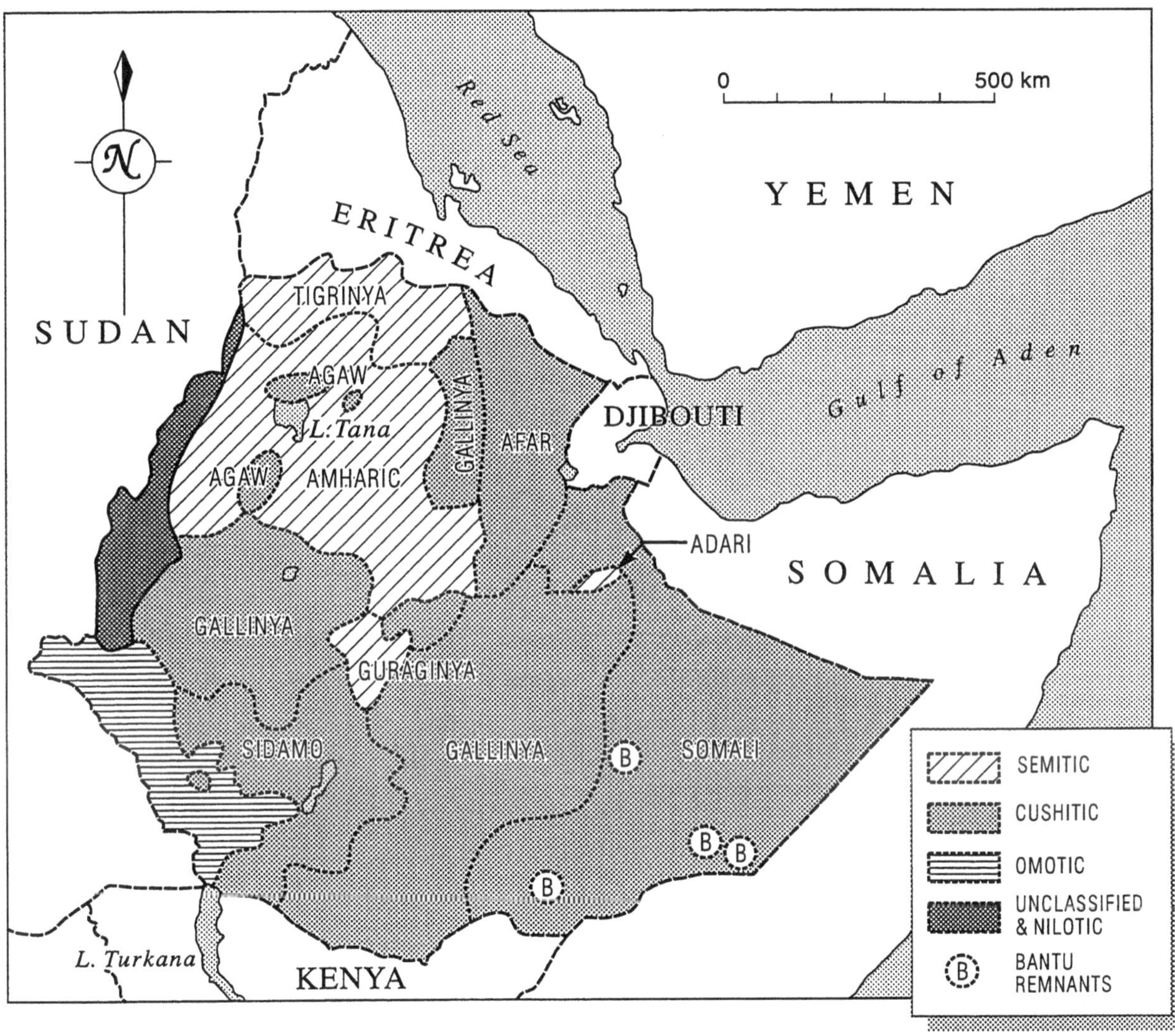

Figure 2.8. Distribution map of major Ethiopian languages (adapted from Hess 1970).

Omotic Cultural-Linguistic Group
Omotic is now recognised as a separate branch of the Afro-Asiatic language family that is unique to a small group from South West Ethiopia (Bender 1975). Omotic-speakers are the most diverse of Ethiopian core-groups, comprising over 50 sub-groups, each with a distinct language and culture. The Omotic-speakers are generally sedentary agriculturalists who exploit a particularly rich tropical environment (Westphal 1975).

The Implications of Cultural Diversity in Ethiopia
Ethiopia is a cultural melting pot where diverse ethnic groups mingle and evolve. Far from being isolated and introspective, Ethiopia benefits from dynamic cultural interaction and exchange. Many cultural groups derive from the surrounding regions of Africa, illustrating further the intimate relationship between Ethiopia and Africa.

2.2. STABILITY AND INSTABILITY IN SUBSISTENCE PATTERNS

Food Producing Systems in Ethiopia
Harlan (1969 p.313) observed that there is in Ethiopia 'a survival of an entire agricultural system little changed from prehistoric times. Ancient methods of tillage, sowing, reaping, threshing, winnowing, dehulling and processing for consumption, all have been preserved , as have the use and attitudes of these people toward their ancient crops.' The traditional food producing systems of Ethiopia hold valuable clues as to their origins. By examining what crops are grown today, their cultural significance, what resources are used to supplement them and what technology is employed, we can build a picture of how modern food producing systems have emerged.

2.2.1. Major Subsistence Patterns
Simoons (1958) identified three major food production systems in Ethiopia, outlined below, which correspond to the broad environmental and ethnic divisions outlined in Section 2.1. The predominant characteristics of these are given in Table 2.1 and their approximate spatial distribution is shown in figure 2.9.

mixed seed agriculture	settled agriculture in well-watered areas with moderate temperatures in the Central, Southern and Northern Highlands.
root and tuber agriculture	settled agriculture in high rainfall, moderate temperature areas of the South West Highlands.
pastoralism	transhumant food production in lower, more arid regions of the Rift Valley and around the fringes of the Highland plateau.

Agricultural Diversity and Agroclimatology
Ensete (*Ensete edulis*), or false banana, is cultivated almost exclusively in high rainfall areas of the South West Highlands between 1600 and 3000 m asl (Westphal 1975). The starchy pseudostem and corms of ensete are the staple resource in an economy based principally on tropical roots and tubers. These resources are supplemented by stock herding and limited seed crop cultivation. Ensete cultivation can support a dense population of around 175 people/km^2 (Huffnagel 1961) and there is less risk of soil exhaustion than with mixed seed agriculture as the crop is manured frequently. Ensete cultivation uses a technology based on wooden implements such as a two-pronged fork, hoes and digging sticks, similar to hoe-vegeculture systems in other regions of Africa (Huffnagel 1961).

In contrast to the vegeculture complex of South West Ethiopia, roots and tubers are excluded from the mixed seed complex[3] of the Northern, Central and Southern Highlands, and vegetative propagation is virtually unknown. This form of agriculture depends on seed crops and is almost totally devoid of fruits and green vegetables. The majority of soft fruits grown in the Highlands, such as banana, orange and pineapple, are for tourists and the expatriate community.

[3] With the exception of the recently introduced potato.

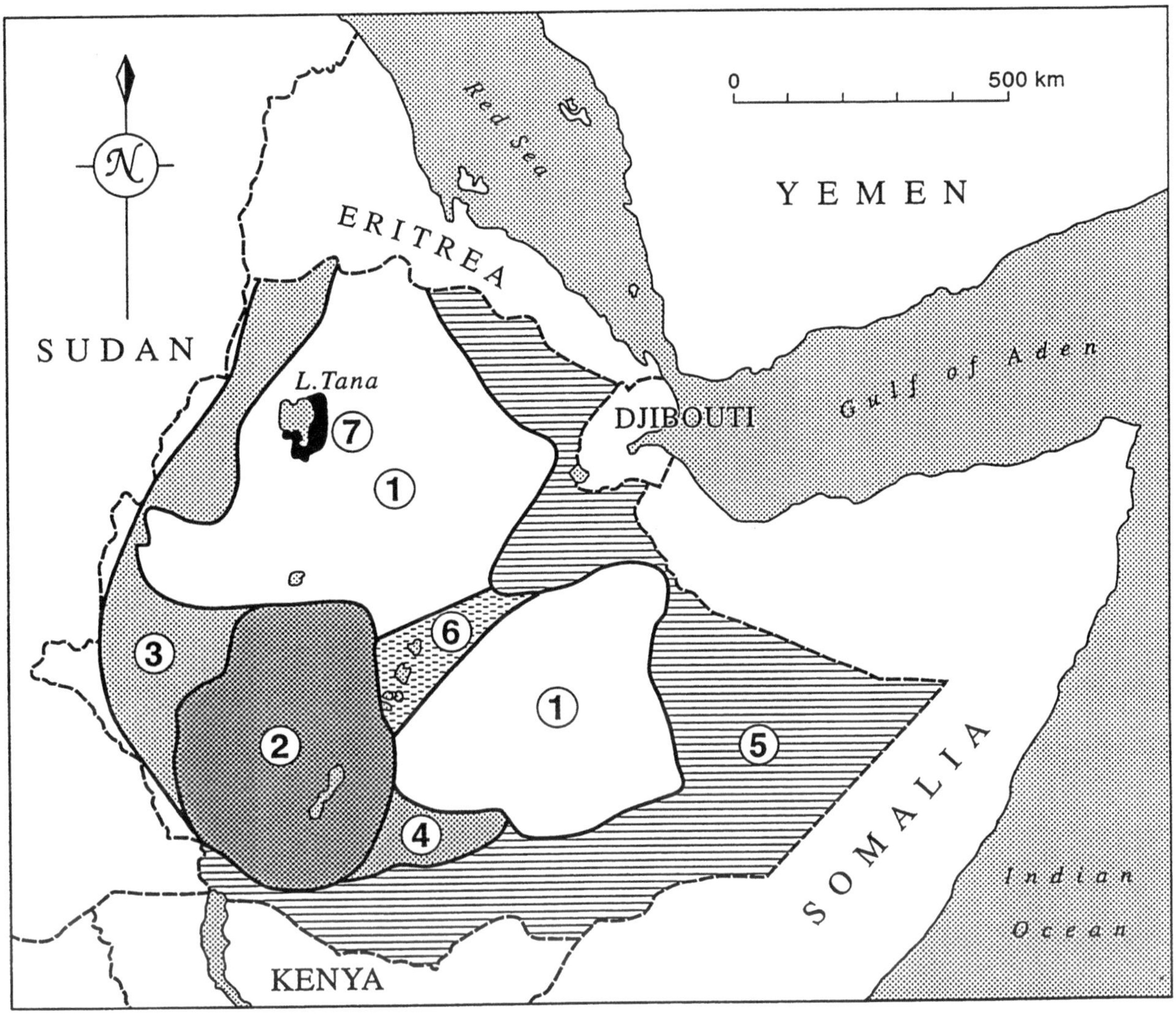

1 - Mixed seed agriculture
2 - Ensete and tuber vegeculture
3 - Shifting cultivation and horticulture
4 - Forests and woodlands
5 - Pastoralism
6 - Agropastoralism
7 - Hunting / fishing / foraging

Figure 2.9. Distribution of traditional farming systems in Ethiopia.

Table 2.1. Characteristics of the Three Major Food Producing Systems in Ethiopia

Mixed Seed Agriculture	Root and Tuber Agriculture	Pastoralism
Northern and South Eastern Highlands.	Humid South Western Highlands.	Arid and semi-arid lowland regions in the Rift Valley and around the fringes of the Highland plateau.
Elevation: min.1600-max.3000 m asl	*Elevation:* min.1600-max.3000 m asl	*Elevation:* c.0-1600 m asl
Major Plant Crops (*see also Table 2.2.*) barley wheat tef sorghum maize finger millet noog sesame linseed flax castor bean chick pea lentil horse bean haricot bean	***Major Plant Crops*** ensete taro (cocoyam) yam potato Galla potato sweet potato	***Major Plant Crops*** barley (until recently among the Oromo)
Minor Plant Crops cayenne pepper potato	***Minor Plant Crops*** cabbage coffee chat sorghum barley maize wheat tef beans	***Minor Plant Crops*** sorghum ensete millet maize cotton
Domestic Animals cattle sheep/goats mules/horses/donkeys chickens	***Domestic Animals*** variable herding of cattle sheep/goats	***Domestic Animals*** cattle sheep/goat camel horses/mules (more recent introduction)
Technology and Techniques broadcast sowing plough/hoe/digging stick sickle harvesting crop rotation/fallow threshing/winnowing limited cattle transhumance	***Technology and Techniques*** hoe digging sticks manuring vegetative reproduction	***Technology and Techniques*** transhumance or nomadism limited, primitive agricultural technology
Principal Ethnic Groups Semitic-speaking Amhara Cushitic-speaking Agau	***Principal Ethnic Groups*** Semitic-speaking Guraghe Cushitic-speaking Sidamo, Wollamo and Kaffa Omotic-speakers	***Principal Ethnic Groups*** Cushitic-speaking groups (Oromo, Somali, Afar)

As each individual plant species within each complex has a distinct agroclimatological range, their contribution to the subsistence economy varies proportionally according to localised environmental and climatic conditions, which are determined principally by elevation (Getahun 1980). Table 2.2 shows the altitude range of crops in the mixed seed complex. The majority of crops in this complex flourish in the *woina dega* zone between 1800 and 2400 m asl. This is consequently the most populous region of Ethiopia and contains 85% of the total subsistence farming population (Westphal 1975). The range of crops that can be grown at high altitudes diminishes rapidly above 2400-2500 m, and barley is the only successful crop above c.2700 m asl. Barley is therefore the staple crop in high altitude environments, and the only major crop above c.3200 m in the Simien and Bale mountains. Below c.1800 m the main Highland cereals of tef, wheat and barley are replaced by tropical seed crops including millets, sorghum and maize.

Table 2.2. Altitude Range of Crops used in Mixed Seed Agriculture (adapted from Westphal 1975)

Maximum altitude range (m asl)	Crop plant	Botanical name
cereals		
1600 - 4000	barley	*Hordeum vulgare*
1600 - 2500	wheat	*Triticum* spp
1500 - 2400	tef	*Eragrostis tef*
<500 - 2400	sorghum	*Sorghum* spp
<500 - 2200	maize	*Zea mays*
500 - 2000	finger millet	*Eleusine coracana*
pulses and legumes		
1800 - 3200	horse bean	*Vicia faba*
1700 - 2700	pea	*Pisum sativum*
1600 - 2700	lentil	*Lens esculenta*
1600 - 2300	chick pea	*Cicer arietinum*
1600 - 2100	hyacinth bean	*Dolichos lablab*
<1400 - 2100	common/field bean	*Phaseolus vulgaris*
<1200 - 2000	cow pea	*Vigna esculenta*
	grass pea	*Lathyrus sativus*
<1000 - 2000	pigeon pea	*Cajanus cajun*
500 - 2000		
oil seeds		
1600 - 2700	linseed	*Linum usticitissimum*
1600 - 2300	safflower	*Carthaus tinctorius*
1500 - 2400	noog	*Guizotia abyssinica*
<1500 - 2100	castor	*Ricinus communis*
<1000 - 1500	sesame	*Sesamum indicum*
vegetables		
<1700 - 2700	cabbage	*Brassica juncea*
1600 - 2100	onion	*Allium cepa*
1600 - 2100	garlic	*Allium satium*
<1200 - 1800	okra	*Hibiscus esculentus*

The broken topography of the Ethiopian Highlands creates a range of contiguous microenvironments. Areas that are unsuitable for arable farming, for example on steeper, broken land, are used for grazing livestock. This pattern of differential land-use reduces competition between arable and pasture requirements, and enhances the output potential of the land (Getahun 1984, 1990).

Pastoral Adaptation

'Pure' pastoralism is virtually unknown in Ethiopia, except among a few small groups such as the nomadic Afar who keep goats, camels and a few cattle in the arid, inhospitable environment of the Afar Rift (Trimingham 1952). The majority of pastoral groups now practise some form of agriculture and sedentism. As a result of wars and migration many pastoralists currently occupy agriculturally viable areas and have adopted plant crops to varying degrees, often depending on localised ecological factors and availability. Groups with similar tribal affinities may practise different subsistence activities depending on their environment. For example the Mati of the Oromo who inhabit the Sidamo region of the Southern Highlands, range from pastoralists to sedentary agriculturalists, yet they all perceive themselves as pastoralists (Westphal 1975). Similarly the North Guji and the Arussi are both branches of the pastoral Oromo. The North Guji occupy *kwolla* regions of the Rift Valley around Lake Abaya, where they grow crops such as maize, taro and cabbage and use plough agriculture (Haberland 1963), whilst the Arussi live at higher altitudes in the Rift Valley and Western Highlands and use plough agriculture to grow crops such as barley (Haberland 1963).

Livestock in Ethiopia

Caprines (sheep/goat) are kept in most regions of Ethiopia, and are particularly valuable in mountainous regions that are unsuitable for cattle. Cattle are the predominant livestock resource among both agricultural and pastoral groups, while pastoral groups in more arid areas keep browse herds of camels. The majority of cows in Ethiopia today belong to a breed of thoracic humped cattle. In addition, relict populations of small, short-horned cattle survive in isolated regions of Southern Ethiopia (Epstein 1971, Blench 1993). The implications and relative antiquity of cattle breeds will be discussed in greater detail in Chapters 3 and 6.

2.2.2. Minor Subsistence Patterns

Shifting Cultivators

Several minority subsistence regimes persist in addition to the main agricultural and pastoral complexes of modern-day Ethiopia. A number of groups in the Western and South Western fringes of the Highlands practice a form of shifting cultivation (Knutson 1969). For example the Gumuz who live in the *kwolla* region of Western Ethiopia subsist on seed crops such as sorghum and cash crops such as cotton. These are grown on plots in the forest that are shifted every 1-2 years. Variations on this pattern are found among groups such as the Berta who also keep some cattle and have recently adopted the ox-plough, the Kaffa who also grow ensete, and the Sidamo who cultivate crops such as tef, tobacco and maize in the moist, evergreen forest of South West Ethiopia (Westphal 1975). True swidden agriculture is unknown, as these groups all live in permanent or semi-permanent settlements and posses a form of land-ownership that is unknown among shifting cultivators.

Other groups, such as the Kumfel, practice a conscious form of crop rotation that makes them more intermediary between swidden and sedentary agriculture. Seed crops of sorghum, tef and finger millet are grown on plots cleared in the forest. The plots are left fallow for the second year after clearing, which enables the land to be exploited for up to eight years before the soil is exhausted (Simoons 1960) - a similar form of crop rotation is known among shifting cultivators in Uganda (Simoons 1965). This primitive form of crop rotation contrasts with the complicated and variable system practised by sedentary farmers across the Highlands.

Shifting cultivators in Western Ethiopia keep few domestic animals as the areas they inhabit are infested with animal pests and diseases such as 'horse sickness' (*nagma*) and trypanosomiasis (sleeping sickness) (Krug 1971). These diseases tend to be confined to the humid lowland regions of Western Ethiopia bordering the Sudan, although they are also known in certain regions of the plateau, including the Blue Nile and Takezze river gorges, and Dek Island on Lake Tana (Cheeseman 1936, Ford 1975). Just as foraging groups in Ethiopia may have developed from a less primitive economy, so shifting cultivation may be an adaptation to an adverse environment as opposed to a relict of an earlier, more primitive subsistence.

Hunters and Fishers

Until two decades ago, small hunting-foraging groups were known in the Highlands. These included the Way'to from the Lake Tana area and the Fugu from Central Ethiopia (Simoons 1960). Despite ethnic pressures, both of these groups persisted by exploiting resources such as fish and hippopotamus that were considered taboo among the Highlanders, therefore adapting to a restricted, under-stressed ecological niche. Hunting groups are also known in Southern Ethiopia in the Abaya area of the Rift Valley and north of the Omo river (Stanley 1966).

It is unclear whether these groups are relicts of an indigenous 'aboriginal' culture or whether they were originally food producers who have adapted their subsistence patterns in response to socio-economic pressures. Although hunting is now a marginal subsistence strategy in Ethiopia, Stanley (1966) has argued that it is deeply entrenched in modern ideology and ritual, and the distinction between hunting and food production is poorly defined.

Oromo Agro-pastoralists

The Oromo originally inhabited the Bale Highlands of Southern Ethiopia. Today they are spread across the Southern and South Western provinces of Ethiopia, where they comprise the largest ethnic group in the country (Haberland 1963, Hassen 1990). The Oromo practise a wide range of subsistence strategies. Although they are now predominantly pastoralists, they originally practised a unique form of agriculture. This primitive system of agriculture was abandoned by the majority of groups, such as the Borana, who moved to more arid environments where they now rely entirely on foraged resources and cattle (Wilding 1985a, 1985b). It is however preserved by the Mati of the Sidamo Highlands of Southern Ethiopia. The principal crop of the Mati is barley, which is grown using a simple, non-intensive method. Following a long period of fallow, the vegetation is cut back and the plots are immediately sown with grain. Ploughs and digging sticks are not used, but the Mati do use hoes. Until recently the crop would be harvested by pulling the stalks out and cutting the ears off with a knife. The Mati traditionally roast the whole grain. They do not grind their cereal or make beer from it (Haberland 1963). The barley crop is supplemented by cattle and foraged resources, but no other plant crops are grown. The Mati grow no ensete or tubers, although they inhabit an ensete-growing area. Barley is the only plant food for which the Oromo have their own word, all other crop names are loaned from Semitic

(Haberland 1963). This suggests that the Oromo have a very ancient knowledge of food production based on barley cultivation. This system of food production is now virtually eclipsed by a more recent form of intensive agriculture.

Konso Subsistence

The Konso are a small group of settled farmers who inhabit an area of the upper *kwolla* and lower *woina dega* in South West Ethiopia. The region is surrounded by pastoral groups with whom the Konso have a symbiotic relationship. Konso agriculture is remarkable for the immense range of crops grown, including cereals, oil seeds, pulses, legumes, tubers and vegetable. It can be described as an amalgam of Ethiopian agriculture, as both Highland-type seed crops and South West-type tubers are grown. Sorghum (*bicolor*) is the major staple crop of the Konso. As this displays a high diversity and an 'unusual assemblage' containing 24 types of sorghum, it has been suggested as a very ancient crop in the region (Harlan and Stemler 1976, Doggett 1988, Engels and Goettsch 1991).

The Konso practice a complex strategy of intercropping and intensive land-use (Goettsch et al 1984). Greater land potential is harnessed through terracing and crop irrigation, and permanent cultivation is maintained through manuring and crop rotation (Hallpike 1970). In addition Konso agriculture incorporates a number of unique features to enhance land-use efficiency, such as using cabbage trees to reduce soil erosion and channelling animal and human movements away from the crops through a network of walkways (Hallpike 1972).

Konso subsistence displays several exclusive elements that distinguish it from mixed seed agriculture. It is unclear whether this complex, organised system is a relict of a more widespread prehistoric agriculture, or whether it has evolved independently possibly through isolation from other agricultural groups.

Summary - a Balanced Economic Spectrum

Ethiopian and African Agriculture

Ethiopia is unusual in the context of African agriculture. Temperate crops, such as cereals, pulses and oil seeds are important subsistence resources in association with indigenous tropical African food plants. The majority of food crops grown in Ethiopia have a wide agro-climatological range and are adapted to an intricate and disparate environment (Getahun 1980, 1990). Ecological adaptations tend to override ethnic distinctions, causing numerous pastoral groups to adopt a form of agriculture. This has given rise to a spectrum of subsistence systems in which the boundaries within and between each systems are not clear-cut, as they are for instance in areas of West Africa. This subsistence mosaic is determined by environmental and geographical factors, by the agro-climatological character of the individual crops and by cultural preferences (Getahun 1980).

Diversity is found both in the type of resources exploited and also in the level of complexity and intensity of the subsistence system. An entire subsistence range is represented in Ethiopia, from hunter and forager, through various levels of shifting cultivation, to intensive and specialised economies based on sedentary plough agriculture and pastoralism.

Despite this efficient ecological adaptation, there is a fundamental divide between different food producing groups. In particular, competition for land is a potent, progressive source of tension between sedentary agriculturalists and mobile or semi-sedentary pastoralists, and a deep-seated prejudice remains between them. This tension is ameliorated by a system of trade and exchange between these groups. For example the pastoral Borana trade around 40% of their livestock to agriculturalists in exchange for grain (Wilding 1985a). Modern pastoral groups in the Afar Rift obtain materials such as pottery and metal from Highlanders in exchange for animal products such as hides (Brandt 1982, Clark and Williams 1984). This system may have operated over several millennia to provide extra resource security for economically specialised groups, and to buffer tensions between competing subsistence groups.

The nature of the food producing systems in Ethiopia suggests that they have evolved through a complex and protracted series of cultural interactions. While cultural distinctions generally determine the economic ideology (i.e. pastoralist/agriculturalist) and imply culturally specific origins, these distinctions are confused by the need for differential adaptation to the prevailing environmental context.

The Ethnographic Contribution

All these systems contain threads of their prehistoric origins. Two of the major subsistence strategies - root and tuber agriculture and pastoralism - are typically African. Similar strategies form the basis of subsistence farming and food production across the whole of Northern Africa. Conversely, mixed seed agriculture is incongruous in this context. It is not characteristic of tropical Africa, and its origins are uncertain.

The following discussion therefore focuses on the ethnography of the mixed seed complex of the Highlands, and attempts to identify the nature of its development.

2.2.3. Agricultural Techniques and Technology in Mixed Seed Agriculture

The traditional mixed seed complex of the Highlands is comparable to the prehistoric agricultural regime of the Near East, with certain common agricultural methodologies. This has led to suggestions that agricultural technology, and plough technology in particular, was introduced from the Near East, either via Southern Arabia in the 3rd millennium BP (Stiehler 1948) or via Egypt after the 4th millennium BP (Simoons 1965). Stiehler (1948) proposed that Highland agriculture had two distinct historical origins. One was purely African and was represented by hoe and digging stick technology. The other was less primitive and included 'foreign' elements such as plough and irrigation technology. The 'foreign' element is superimposed over an ancient, indigenous subsistence system, and can be considered as a later introduction. This dualism has implications for determining the processes involved in the development of food production in Ethiopia and identifying their potential origins. Indivdual elements of the Highland agricultural economy are explored in greater depth below.

Agricultural Techniques

Ploughs are used widely in the Highlands. Arable land is prepared for sowing by breaking the topsoil with a plough and digging sticks or hoes. Seed is broadcast onto the prepared field, which is then reploughed or trampled by cattle to bury the seed. Knowledge of alternative propagative methods such as vegetative reproduction, is scarce in the Highlands and farmers have even been known to broadcast seed potatoes (Simoons 1960). Usually only one type of seed crop is sown in a particular field, but sometimes two or even three types may be planted, harvested and processed together. Common combinations include wheat and barley, green peas and horse beans, sorghum and finger millet, and lupine with tef, flaxseed or barley (Simoons 1960, Westphal 1975).

Following germination the young crops are weeded by hand or with hoes, often with the entire family or community participating (Huffnagel 1961). The mature crop is harvested by hand using metal sickles. Use of stone sickles is not reported in either the Highlands or the Rift Valley, although this may be a reflection of the limited ethnographic research in many regions. The crop is threshed by oxen on a clay threshing floor in the fields, and winnowed using a range of wooden and basketry implements (Westphal 1975).

Agricultural Technology

Two types of plough are used in Ethiopia. An Arab-type seed-drill plough is used in the Red Sea coastal area, whereas in the Highlands a simple soil-breaking plough (*marasha)* is used (Hopfen and Biesalski 1953, Huffnagel 1961). Although iron plough-shares are used across the Highlands now, their wooden predecessors were not replaced in many regions until the 19th century (Salt 1814, Wylde 1901). The *marasha* design is used throughout the Highlands and, although simple, it allows the land to be farmed more intensively than with implements from other regions of sub-Saharan Africa (Liban 1955, Pankhurst 1968). This design is characteristic of the Northern hemisphere from Sudan to Sinkiang after the 5th millennium BP (Allen 1972). Simoons (1965) has proposed that the Ethiopian plough is derived from a Dynastic Egyptian prototype, which was in turn derived from Eurasia.

Pure ploughing does not exist in the Highlands. Ploughing is invariably accompanied by a wooden hoe or digging stick to help break and till the soil (Huffnagel 1961). Hoes and digging sticks alone are used where the ground is too steep or the family too poor to afford a plough (Simoons 1960). Simoons (1958) notes the incredible variety of digging sticks in the Highlands and suggests that certain of these, particularly the double-pronged digging stick used by the Guraghe, are based on an Egyptian design.

Hoes and digging sticks are characteristic elements of African agriculture throughout the continent. Their presence in the Ethiopian Highlands indicates that the agricultural system of the Highlands is rooted in Africa, and may have developed indigenously. Plough agriculture is very successful in the Highlands and it is unlikely that the more primitive technology would have been introduced **after** the plough. This implies that plough technology was assimilated into an established prehistoric subsistence system that used an effective form of primitive technology.

Land Use Patterns

The Highland soils are farmed intensively using a complex, regionally varied system of intercropping and rotation of cereal, oil and legume crops (Huffnagel 1961, Westphal 1975, Getahun 1990). The land may be cultivated for up to nine years before fallowing depending on the localised crop complex and soil type (Simoons 1960, Westphal 1975). Fertile soils, such as black soils in areas of the North West Highlands, are rarely, if ever, fallowed, whereas the poorer red soils are fallowed at least every third year (Murphy 1963). Land is sometimes only fallowed if the farmer owns enough fields, although land ownership is retained throughout the fallow period, even though this may last several years (Simoons 1960). The agricultural yield is very variable. On good soils a 60-fold yield can be achieved in a good year, falling to as little as 6-fold in a bad year (Pankhurst 1968). A peasant farmer with a pair of oxen could harvest a minimum of 3000 kilos of grain a year from his modest land (Girad 1873, Pankhurst 1968).

Although this system indicates an intimate knowledge of agronomy and the properties of a wide range of plant species, it has been described as a form of shifting cultivation because it exhibits a number of primitive elements in common with swidden agriculturalists (Simoons 1960). For example, sedentary farmers use a system of slash and burn to clear their fields from primary woodland or after a long period of fallow. Land under cultivation is never manured, but soil burning (*guai*) is a well established mechanism for increasing soil fertility on a short-term basis by releasing more minerals into the soil (Huffnagel 1961). However this practice has longer-term detrimental effects by destroying soil nitrogen and reducing the agricultural longevity (Birch 1967, Murphy 1963). Use of fire to promote productivity is well documented among present day foraging cultivators, such as the Australian aborigines (Yen 1989).

Basic irrigation and field terracing are practised in the more denuded areas of the Northern Highlands especially Tigray, Simien, Yeju and Awasa, where they may be of considerable antiquity, possibly dating back to the Pre-Axumite period in 3rd millennium BP Ethiopia (Michels 1988, Munro-Hay 1989). Although the annual agricultural cycle is adapted to take full advantage of seasonal variation in rainfall, this no longer meets the increased resource demands made in recent years by a growing population and intensive, non-sustainable land-use. Irrigation technology is now being applied more widely by NGOs through localised, community-based projects.

An Integrated Agricultural System or Disparate Elements?

Although livestock form an essential part of the modern agricultural economy of the Highlands, they are poorly integrated into the subsistence pattern. Hides serve a variety of functions and meat is important on non-fasting days, but otherwise secondary animal products are not used widely. Milk and butter are considered culturally marginal and are rarely consumed, with many people believing them to have detrimental properties. Cotton garments are preferred traditionally over woollen ones (Simoons 1960). Animal dung is never deliberately used for manuring fields, despite the beneficial effects this would have. Instead it is dried and used as a poor fuel substitute in denuded areas (Huffnagel 1961). Although oxen are important in ploughing and processing plant crops, digging sticks are of equal importance and may replace animal traction in certain areas.

The Highland agricultural system can be described as existing in a state of ecological equilibrium. Livestock provide a predictable food resource and a means for intensifying agricultural productivity without severely reducing the land available for plant agriculture. The majority of primary and secondary animal products of importance to the Highlanders, such as meat, horn and hides, can be obtained from wild fauna. These observations suggest that livestock may not have been an integral component of the original subsistence system in the Highlands. Instead, a primitive system of plant cultivation technology may have developed in the Highlands *before* the introduction of domesticated animals.

Summary: Pre-Adaptive Behaviour and the Ethnographic Record

Ethnographic evidence provides a valuable insight into how and when the traditional farming system of the Highlands may have developed. Land use patterns and agricultural technology in the Ethiopian Highlands indicate the co-existence of two integrated subsistence regimes. The evidence suggests that one of these is foreign and can be characterised by an intensive system of land use that harnesses the potential of domesticated animals and a broad range of food-plants. The other is derived from localised developments and uses a less intensive, more primitive system of plant cultivation.

The ethnographic record preserves relicts of an earlier system of resource exploitation that predates the use of draught animals and livestock. As the plough has been important for at least the past 2000 years in the Highlands (de Contensen 1963) and domesticated animals have been known in Ethiopia for the past 3500 years (Brandt and Carder 1987), we can speculate that a pre-adaptive system of plant cultivation was being practised in the Ethiopian Highlands by the start of the 4th millennium, and possibly considerably earlier.

2.2.4. Natural Resources and Indigenous Knowledge

The Role of Wild Plants in Ethiopian Culture and Economy

A brief glance at some of the 3000 or so indigenous species that are used today in the Ethiopian Highlands illustrates the versatility and proliferation of the available resources. Some of these are detailed in *Appendix A*, Tables A.1.1-A.1.8. A fuller description can be found in Edwards (1991), Westphal (1975), Getahun (1974), Wilding (1984), and in the references given in *Appendix A*. Apart from the high mountain flora of the *urec*, each vegetational zone offers a rich array of edible flora. Many more species are exploited for their pharmaceutical or cosmetic efficacy, for their fibre or oil, for their aromatic properties or for craft and construction. Wild plants are also used extensively in religious ceremonies, for magic and for medicine, and they form an integral part of indigenous Ethiopian oral tradition (Getahun 1974, Westphal 1975, Wilson and Gebre Mariam 1979). Foraged plants constitute an essential, primary source of nourishment among specialised pastoral groups with a low intake of meat and domesticated plant foods. For example, the Borana use at least 72 different edible plants and 28 medicinal plants, and a further 154 species are exploited for stock browsing, ritual, poison, fumigation and aromatics, cosmetics, construction and utensils (Wilding 1984). Even today wild plants are widely utilised by sedentary farming societies in Ethiopia, and in the 1970s over 80% of the population still depended on traditional herbal medicines (Getahun 1974).

Foraging as a Dietary Supplement

The Highland Peasant's Diet

The basic diet in the Northern Highlands comprises a spicy stew of meat or legumes (*wot*) flavoured with a hot sauce (*berberre)* made from a mixture of spices such as chilli peppers, coriander and garlic. Although several of these ingredients are recent imports (see *Appendix A*, Table A.2.1) this sauce is ubiquitous in Ethiopian cuisine, and may often replace the *wot*. Vegetable oil, particularly from *gomenzer*[4], noog, flax or sesame, is used in the preparation of the *wot* and for cooking the Ethiopian staple food, *injera*. *Injera* is a flat pancake-like bread made of fermented tef dough which is cooked on a lidded griddle, or *mitab*.

Wild Resources as Nutritional Supplements

The consumption of animal protein in rural areas is very low and may average well below 7g per day, compared with the required daily intake of 20g (Miller and Makonnen 1965). The Ethiopian Orthodox Church further stipulates that no animal products should be consumed during the 150 annual fasting days. Cow's milk is not abundant and is rarely consumed by adults, whilst goat or sheep milk is only drunk by very poor people or for medicinal purposes (Simoons 1960). In many areas there are preventative traditions or taboos associated with the intake of animal protein - for example shepherds believe that drinking cow's milk will make their teeth fall out (Selinus et al 1971).

Nutritional surveys have shown that the local domesticated plant component of the Ethiopian diet alone is deficient in certain vital vitamins and minerals including Vitamin A, Ascorbic Acid, Calcium, Riboflavin and Niacin (Agren and Gibson 1969, Selinus et al 1971). Food processing methods tend further to denature or destroy valuable biochemically active compounds - for example vitamins are destroyed when chilli peppers or garlic are pounded to a powder to make *berberre* (Simoons 1960), and certain proteins are denatured when foods are cooked at high temperatures and for extended periods of time (Stryer 1995). Young children are especially susceptible to deficiency-related conditions, although the entire community is at risk. Essential nutritional compounds not supplied by domesticated resources are obtained from wild plants.

The basic, low protein diet of *wot* and *injera* is supplemented by a range of foraged roots, berries, seeds, fruits, leaves and stems of plants (Getahun 1974). Children and young men consume the most wild resources, although certain sectors of the society, such as monks, nuns and hermits, subsist entirely on wild plant resources in rural areas of Ethiopia (Getahun 1974). The intake of wild plants by all group members increases radically under drought or stress conditions. The initial reaction to famine conditions is to alter the subsistence pattern and reduce the consumption of domesticated foods whilst increasing the consumption of wild foods (Webb et al 1992). Only under extreme circumstances will a family or community be forced to migrate, which emphasises the importance of the tie of the land among Ethiopian farmers.

Foraging or Farming?

In contemporary Ethiopia, 85% of the population depends on basic subsistence farming, yet the boundary between wild and domesticated plant resources can be very indistinct, and the division between farming and foraging is very poorly defined (Westphal 1975). For example 33 root and tuber species are used in Ethiopia, but only 11 of these are grown as crops (Getahun 1973). Many wild plants are left to grow when fields are cleared of crops because they have recognised valuable properties (Huffnagel 1961). Although there are no indigenous cultivated fruits, wild fruits such as rosehips (*Rosa abyssinica*), *wanza* (*Cordia africana*) and figs (*Ficus sycamorus*), are readily consumed by the peasant community, especially by children (Selinus et al 1971). Another wild fruit, *gaba* (*Zizyphus spinachristi*), is even collected, dried and sold in local markets to be eaten as a snack (Selinus et al 1971). Wild resources are exploited in a strict perennial cycle that complements the seasonal availability of cultivated resources so, for example, large quantities of wild fruits are consumed at the height of the dry season (February-May) when their juices and sugars are most appreciated, whereas starchy roots and tubers are consumed during the rainy season when carbohydrates from domesticated foods are least available (Getahun 1973).

Certain plants, such as the medicinal plant *indahula* that grows wild in the forest fringes, are transplanted close to settlements whilst others, such as black mustard (*Brassica nigra*) and Ethiopian cabbage (*Brassica juncea*), may be sown purposefully or left to grow wild (Dombrowski 1971). Plants like indigo (*Indigofera argenta*), which is used as a dye, are still exploited from the wild although they have been cultivated in surrounding areas such as Egypt since antiquity

[4] Ethiopian mustard (*Brassica carinata*).

(Greenway 1945), and the wild progenitor of finger millet, *Eleusine africana*, is used for making baskets in areas where the domesticated form *Eleusine coracana* is grown as a food crop (Dombrowski 1971). This evidence suggests very ancient and intimate knowledge of the wild plant. Other wild grasses and sedges such as *Pennisetum schimperi*, *Cyperus fischerianus* and *Chloris* spp are often transplanted or sown close to houses so that their fibres can be easily gathered and utilised for making baskets, which in turn are used for a variety of activities connected with food processing and production, such as winnowing, containment and storage (Simoons 1960).

Many wild plants require complex and protracted preparation processes to detoxify them or activate them. For example, a number of plants are commonly used as stupefiers for catching fish, including the seeds of the tree *Milletia ferruginea*, the leaves of the shrub *Maesa lanceolata*, the bark of the tree *Lantana trifoliata* and the fruit of the plant *Solanum marginatum*. These plants are dried in the sun for two days before being crushed in a mortar and mixed with other natural stupefying reagents (Simoons 1960). The mixture is then rubbed in the water and left for several hours before the dazed fish rise to the surface and can be picked out by hand.

Implications for a Preadaptive Economy

This complex exploitation of natural resources by Ethiopian farmers is the product of a long-term relationship. This relationship has not been usurped by the introduction of agricultural technology and foreign domesticates. Pankhurst (1966) has suggested that dietary deficiency and continual economic insecurity amongst the agricultural peasantry of Ethiopia account for the persistence of wild plant foods in the diet. However, this explanation cannot entirely account for the range of exploited plants and the obvious cultural importance with which they are imbued.

The continued importance of foraged resources to Ethiopian farmers suggests that indigenous wild plant foods formed the basis of the original subsistence economy. This palaeoeconomy was pre-adapted in such a way as to be able to assimilate incoming technology and resources, and to remain of primary cultural and economic importance despite the potential of introduced resources and technology.

The greatest variety of useful and edible resources is to be found in the undifferentiated evergreen thicket and montane woodland of the *woina dega* and lower *dega* between 2000 and 2600 m asl (*Appendix A*, Table A.1.2-A.1.4). This includes a range of palatable, digestible grasses[5], as well as wild figs, African olives, asparagus, leafy Brassica and a selection of oil-bearing seeds, medicinal plants, spices and timber. This is also the most agronomically productive and the most comfortable region of Ethiopia in which to live and consequently it is the most densely populated zone. As traditional use of wild plant resources probably reflects an ancient knowledge, the *woina dega/dega* ecotone would also have been the most attractive for prehistoric foraging.

An Ecological Model and an Ethnographic Argument

Indigenous agriculture is a product of the natural environment and it must be considered in ecological terms. The domesticated resources on which it is based must derive from this ecosystem and must be sufficient to maintain human communities without the addition of foreign elements. A traditional farming system must reflect the agricultural potential of the natural, wild ecosystem and the extent to which this potential has been harnessed by the indigenous population.

Harris (1976) proposed that the relationship between traditional agricultural systems and the origins of agriculture could be determined in the absence of direct archaeological data by looking at the spatial and temporal relationship of the components of the systems. The pattern that emerged from this study could then be validated ethnographically to assess whether that agricultural system developed indigenously. Harris applied this approach successfully to the different agricultural systems of West Africa, but it can equally well be applied to the mixed seed complex of the Ethiopian Highlands. The following questions should be asked of a traditional system:

- what domesticates are indigenous to the Highlands?
- what is their relative antiquity?
- what is the relative importance of the indigenous domesticates?
- what behavioural adaptations have been made to accommodate them?

The Relative Antiquity of Indigenous Domesticates

The mixed seed farming complex comprises a range of cereal, pulse and oil seed crop plants supplemented by a rich variety of spices and condiments. Many of the domesticated plants in this complex are introductions - some recent, some with greater antiquity, some of ambiguous source. (A full account of their relative antiquities and probable origins is given in *Appendix A*, Table A.2.1.)

---

[5] These include *Setaria* sp, *Themeda* sp, *Panicum* sp, *Sporobolus* sp, *Eragrostis* sp, *Pennisetum* sp and *Eleusine indica*.

The indigenous domesticates, noog and tef, are known to have been grown in the Northern Highlands from the early 2nd millennium BP (Dombrowski 1971, Phillipson and Reynolds 1998, Boardman *pers.comm.*). However, the earliest archaeobotanical material from the Ethiopian Highlands indicates that plants of a probable foreign origin, including barley, chick pea and horse bean, were being grown in the Highlands from the mid 3rd millennium BP (Dombrowski 1971). As these remains were recovered from a remote setting in the Highlands, far from any obvious or recorded routes of communication, the assumption is that a system of food production was well established by this date and had probably developed substantially earlier. These crops are still fundamental to the traditional farming complex of the Highlands today, and the wild plants identified in the early samples are still eaten, indicating continuity in the subsistence base in the Highlands for at least the past 2500 years.

Although there is considerable debate at present over the phylogenic background of the plants found in early contexts in Ethiopia (discussed in detail in Chapter 3) we have no definite evidence that they were indigenous. In the absence of further data, the logical deduction is that the basic components of the Highland agricultural complex were introduced. At present the direct archaeobotanical evidence implies that Ethiopian agriculture is derived from a complex of foreign domesticates combined with several indigenous edible plants and that this system was well established across the Highlands by the second half of the 3rd millennium BP.

However vital questions still remain. In particular we need to know when were foreign crops introduced, were indigenous plants domesticated before the introduction of foreign crops, and can we trust the evidence? These questions will be approached in more depth in chapters 3, 5, 6 and 7. In the meantime this section investigates the potential of the indigenous resources to form an effective subsistence base.

The Relative Importance of Indigenous Domesticates

Consider Table 2.3 below. Excluding the crops of uncertain origin, the remaining resources comprise a vegetable, two oil seeds and a starchy, high protein cereal grain. Hardly a substantial diet. In the absence of foreign crops, the indigenous subsistence economy must have included a broad range of wild plants in addition to tef, noog and cabbage. This suggestion is reflected in the modern agricultural system of the Highlands in which a large number of forage resources supplement the indigenous and non-indigenous arable crops.

This suggests that indigenous crop plants were exploited from an early date as part of a broad spectrum of wild flora, prior to the introduction of foreign domesticates. An advanced system of agriculture did not develop, but certain plants may have been exploited preferentially for various reasons, for example taste, availability, ease of harvest, nutritional properties. These resources may have been exploited intensively and possibly cultivated prior to the introduction of foreign domesticates.

Table 2.3. Indigenous and Introduced Crops in the Traditional Farming System of the Highlands

Indigenous crops	Crops of possible Ethiopian origin	Introduced crops
tef	barley	emmer wheat
	sorghum	durum wheat
	finger millet	bread wheat
		maize
	cow pea	field/common bean
	chick pea	pigeon pea
	lentil	pea
		grass pea
		hyacinth bean
noog	castor	safflower
cabbage seed	sesame	linseed
cabbage	okra	onions
		garlic

2.2.5. Summary: an Ethnographic Model

The points raised above regarding the agricultural technology, the systems of land use, the predominance of plant resources over domesticated animals, and the integration of wild plants into the farmer's diet all indicate that two strata of subsistence economy co-exist in the Ethiopian Highlands. Furthermore, these integrated strategies are relicts of two discrete systems. One is based predominately on wild plant foraging with selective cultivation of certain indigenous resources using a typically African technology. This system is focused on the diverse vegetation of the *woina dega/dega* ecotone, and implies that this is the region in which prehistoric food production may initially have emerged. The other is based on intensive farming of foreign plant domesticates using plough technology.

This hypothesis suggests that before the introduction of foreign domesticates and technology by the mid-4th millennium BP, wild plants were being exploited, possibly through an increasing scale of intensification, using a pre-adaptive economy. In this model the economic developments would have been accompanied by distinct behavioural adaptations.

This hypothesis is examined through the available data using the following framework of inquiry:

- What indigenous resources were exploited from an early date in Ethiopia?
- What evidence do we have for behavioural change relating to subsistence?
- What evidence do we have for the introduction of resources?
- Do the observed cultural developments have implications for indigenous subsistence change, or do they result from introductions?

These questions are approached in chapters 3, 5, 6 and 7 respectively.

2.3. ECONOMIC INSTABILITY AND CATASTROPHE

2.3.1. The Effects of Short-Term Climate Change

Ethiopian subsistence represents an efficient adaptation to an ecologically diverse environment. However the system exists in fragile equilibrium, which has devastating consequences on the population when it breaks down. Ethiopia is remembered by many Westerners for the severe droughts and famine in the past two decades. In 1973 and 1975 these affected over 10% of the population and caused the death of over 250 000 people and 50% of livestock in certain areas. In 1984/1985, over eight million Ethiopians were affected and over one million died (Rahmato 1988). Droughts and famine have been a recurrent phenomenon throughout Ethiopian history following their earliest documentation in the 9th century AD (Hussein 1976, Webb et al 1992, Hancock 1985). In addition Egyptian flood records dating back to the 5th millennium BP indicate rainfall variability in the Blue Nile catchment area of the Ethiopian Highlands and the catastrophic effect of low rainfall for the Egyptians, although the implications for the Ethiopian population are unknown (Bell 1970, 1971, Wood 1977, Butzer 1984). More detailed historical records from Ethiopia over the past 200 years document a series of severe droughts throughout the 19th century, particularly between 1888 and 1892 during which over 30% of the population and 90% of cattle died (Pankhurst 1968). These are not isolated events, and droughts of variable severity have occurred approximately every ten years throughout the 20th century (Wolde-Michael 1985).

Causal Mechanisms and Famine Severity

Droughts and famine are localised phenomena. Their onset, extent and severity are controlled by several interactive factors, shown diagramatically in figure 2.15. The success of the subsistence system depends ultimately on the consistency of annual climatic cycles within certain parameters. Failure of the rains causes drought conditions and crop failure, but does not automatically result in famine. The onset of famine can be exacerbated by pests or diseases such as locust and rinderpest, by localised population pressure, or by the marginality of the environment (Hussein 1976). For example the droughts at the end of the 19th century were accompanied by rinderpest epidemics which caused massive death of livestock and amplified the effects of crop failure (Pankhurst 1968). This effect was most severe among sedentary agriculturalists in highly populated regions of the Northern Highlands. Currently over 88% of the population is concentrated in the *woina dega* and lower *dega* between c.1600-2600 m asl, where the density can exceed 200 people/km^2, making this a particularly vulnerable area (Gryseels and Anderson 1983). Pastoral groups are also particularly vulnerable because they tend to occupy more marginal environments and have a specialised, high risk subsistence strategy. For example in the 1970s the pastoralists in the arid Afar Rift Valley were the hardest hit sector of the population, with the death of 25-30% of their total number (Bondestam 1974).

The threat of recurrent drought restricts the development of highly specialised economies in the absence of a 'fallback' mechanism, such as trade and exchange, to buffer the severity of the response.

The potential for monsoon failure as a cause of famine is ultimately determined by the socio-economic context. Garcia and Escdero (1982) have suggested a linear relationship between economic stability and the amplitude of drought-related famine which is influenced by international trade, as shown in figure 2.10 overleaf. An efficient infrastructure and a stable economy buffer the detrimental effects of drought, whereas a weak economy and unstable political framework intensify the severity of the response. As a result the correlation between drought years and subsequent food shortages is not perfect. The famine of 1984/85 for example was the culmination of a severe drought in 1984, an economy weakened by the famine of the 1970s, the ongoing civil war and an unpopular military regime (Webb et al 1992).

2.3.2. Responses to Famine Stress

Although conditions vary locally, there are identifiable patterns of behaviour associated with the onset, development and climax of famine (Dirks 1980, Corbett 1988, D'Souza 1988). These patterns differ between subsistence groups. Under stable socio-economic conditions, both pastoral and agricultural groups tolerate a single year of drought conditions by reducing their calorific intake, eating fewer meals and using more wild resources in their diet (Webb et al 1992). If stress conditions are intensified through continued drought or through adverse amplifying factors then the response pattern of these groups diverges, as shown in figure 2.11 below.

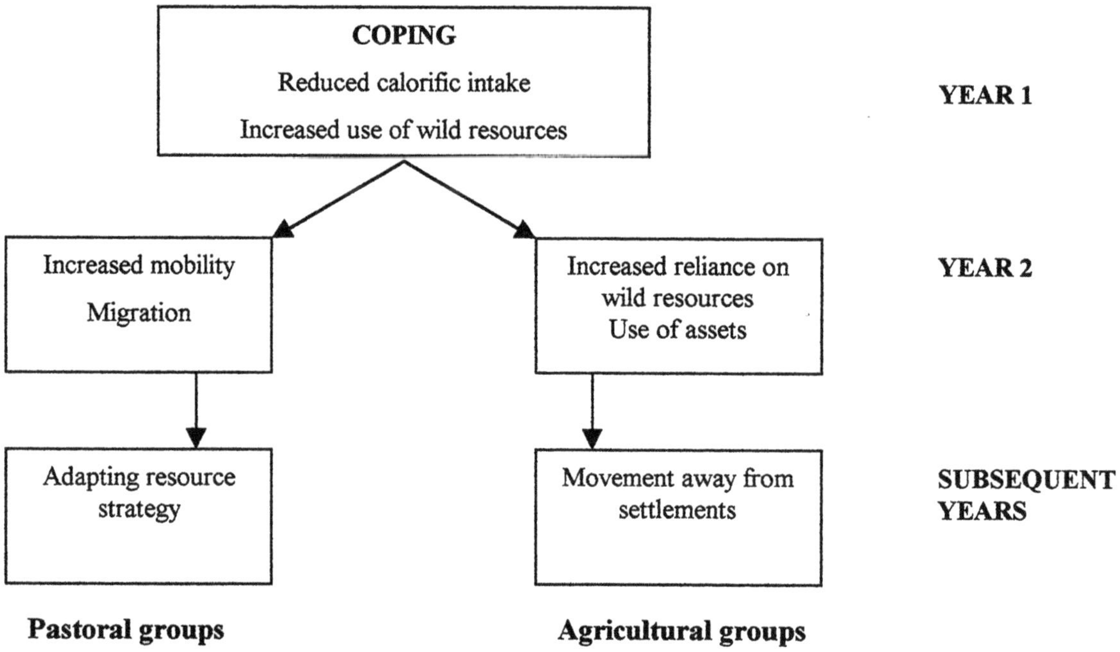

Figure 2.11. Different response patterns to stress conditions among pastoral and agricultural groups.

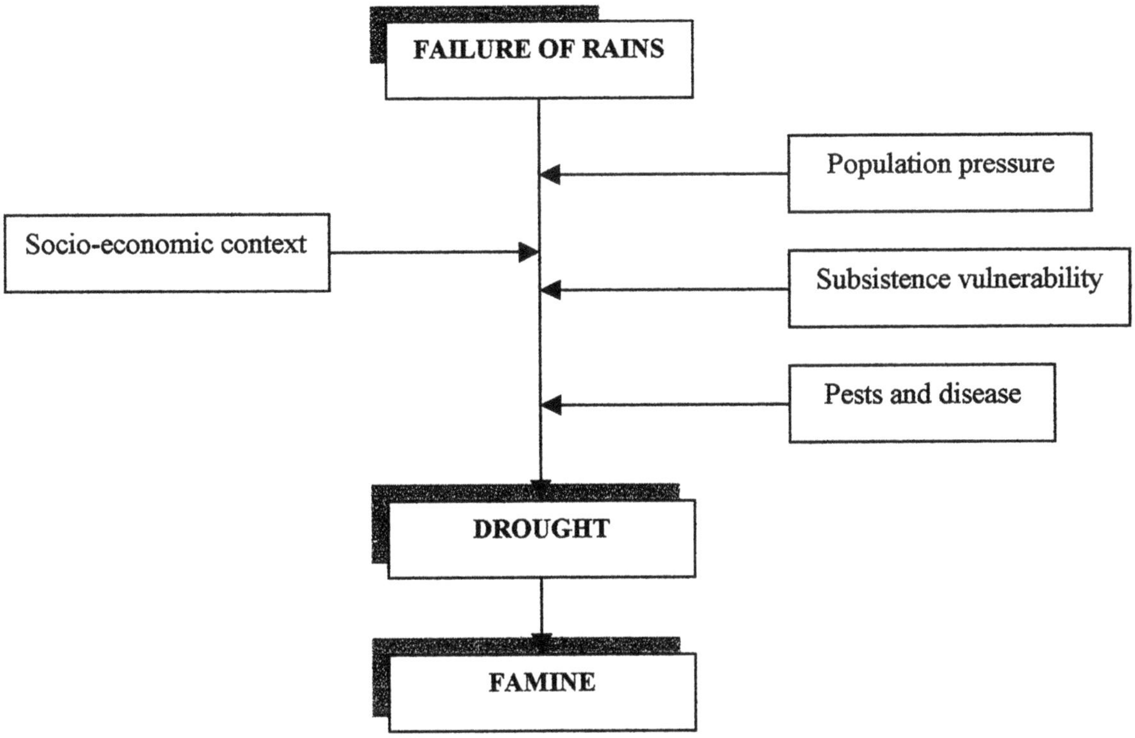

Figure 2.10.a. Simplified causal relationship between meteorological inconsistency and drought.

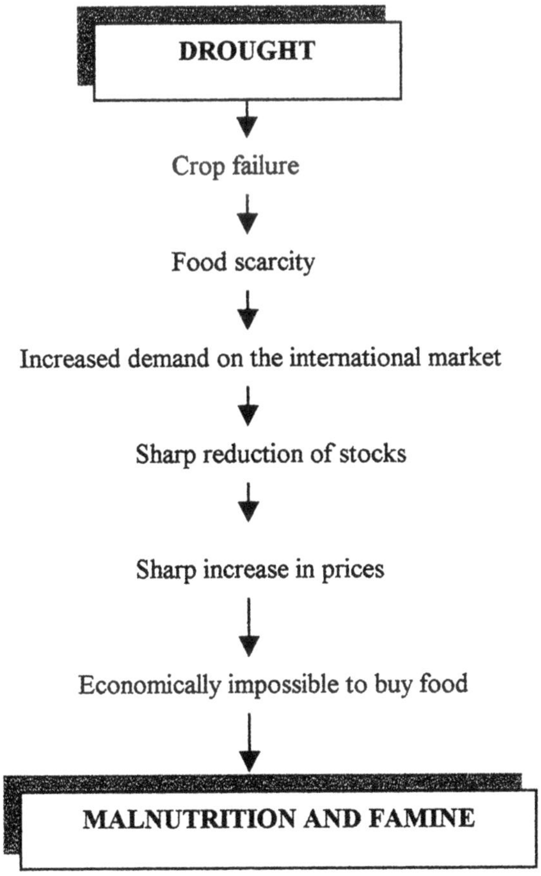

Figure 2.10.b. The linear economic relationship between drought and famine.

Pastoralists respond to continued short-term resource stress by increasing group mobility and migrating to new areas in search of grazing and water, as well as increasing reliance on animal resources (Hussein 1976). For example, stress related 'land hunger' was ultimately responsible for the northward movement of Oromo pastoralists and the southward movement of Somali pastoralists during the 15th and 16th centuries in Ethiopia (Hassen 1990), and the resulting tension between peasants and pastoralists. If this response is not possible, or if stress conditions persist, then alternative adaptations are provoked. Hogg (1980) observed that during the droughts of the 1970s, the Isiolo Borana of Northern Kenya adopted small-scale agriculture around urban centres. Other Oromo and Somali groups began limited cultivation of sorghum, millet and maize and intermarried into agricultural groups (Lewis 1982, Wilding 1985a). These responses are generally reversible if the stress conditions do not persist. In regions where agriculture is not viable, groups such as the Samburu and the Somali have adopted browse herds of camels that are more effective at exploiting the 'badlands' than grazing ruminants (Sperling 1984), or different strains of cattle, like the zebu which are more tolerant to arid conditions (Wilding 1985b). An additional response among pastoralists is to adapt the existing technology, for example to take maximise water resources through deep well watering, or to increase trade activities with agricultural groups (Wilding 1985).

Conversely, sedentary agriculturalists respond to persistent short-term stress conditions by further increasing their consumption of wild resources (roots, leaves and rodents for example) and reducing their calorific intake (Webb et al 1992). At this stage agriculturalists sell or consume their material assets, and eventually their livestock (Webb et al 1992). Only in the final stages of this escalating scale of commitment and irreversibility will agricultutalists resort to mobility and move away from their homesteads to relatives, urban centres or refugee camps (Wood 1976).

Cattle and other livestock are perceived as the most important resource, as they provide a buffer against famine. Even under extreme stress conditions, pastoralists uphold their taboos against certain foods such as fish, certain meats, or animals that have died of hunger or disease, whereas among agricultural groups food taboos break down during food shortages (Kloos 1982, Selinus 1971).

Although drought and famine are more severe in marginal environments, they recur globally throughout the tropics, and are therefore a symptom of the pronounced and variable rhythm of the monsoon. For at least the past 1000 years, climatic variability in Ethiopia has recurrently destabilised the dynamic equilibrium between human groups and their environment. As demonstrated above, adverse climatic conditions cause resource stress, which may lead to famine under specific, variable economic and environmental conditions.

Droughts, famines and epidemics are socially and economically destabilising events that interfere with, and may temporally retard, cultural development. The severity of documented famine is largely a function of over-exploitation of land, combined with socio-economic pressures. The Ethiopian economy exists as a fragile equilibrium that is constantly unbalanced by climatic instability, the parameters of which are progressively altered by anthropogenic interference. Short-term destabilisation evokes a short-term adaptive strategy, the nature and scale of which varies according to the specific cultural-ecological context. Although this response may be profound, it is also dynamic and reversible.

Analogous conditions are unlikely to have applied in prehistory, yet the impact of recurrent climatic oscillations would have been equally as destabilising. This section has discussed the implications of tropical climate change on Ethiopian food producers in order to illustrate the impact of relatively small-scale changes on human subsistence, and the response which this evokes.

Chapter 3.

CROPS AND CATTLE: THE ORIGINS OF DOMESTICATED RESOURCES IN ETHIOPIA

3.1. NORTH AFRICAN RESOURCES AND THE PATTERN OF DOMESTICATION

Agriculture, based on the exploitation of morphologically domesticated plants and animals, emerged remarkably late in Africa. Whilst mixed farming economies became established globally between 10 000 and 5000 years, agriculture did not appear in tropical Africa until the 3rd millennium BP. Why should such an apparent discrepancy exist? As discussed below, the available evidence shows that a different pattern of subsistence change prevailed in Northern Africa during the Holocene to that seen in other parts of the world. This pattern was characterised by the early appearance of domesticated cattle and animals between the 10th and 5th millennia BP, contrasting with the relatively late appearance of domesticated African plants in the 3rd and 2nd millennia BP.

3.1.1. The Origins of North African Cattle

The Conventional Model

Modern African cattle are genetically diverse, giving rise to several theories over their origin. Domesticated cattle are conventionally believed to derive from three separate phases of introduction (eg. Bisschopp 1931, Epstein 1971):

- The humpless Eurasian long-horn, *Bos taurus,* is thought to have been introduced along with domesticated ovicaprids to Egypt and North Africa from the Near East during the 7th millennium BP. Domesticated animals subsequently spread up the Nile Valley, along the Mediterranean littoral and across the Sahara during the 'Neolithic Wet Phase' in the 7th millennium BP, giving rise to the rock art of the Sahara (Epstein and Mason 1984, Muzzolini 1993). Limited osteological data from North Africa suggests that cattle size decreased substantially following exposure to a more demanding environment in the Sahara (Carter and Clark 1976, Carter and Higgs 1979). However, as Grigson (1991) notes, the sample size used in this interpretation is insufficient to exclude factors such as age, sexual dimorphism and anthropological selection pressures.
- Humpless, short-horned cattle were introduced to North Africa from the Near East during the 4th millennium BP (Smith 1980, Gautier 1987, Clutton-Brock 1989).
- Humped zebu cattle, *Bos indicus*, with cervico-thoracic humps were introduced to the Horn of Africa from Southern Asia after the mid-4th millennium BP (Epstein 1971). These interbred with autochthonous long-horned breeds to produce modern populations of thoracic humped Fulani and Sanga cattle in West and East Africa respectively. Figure 3.1 shows the modern distribution and relative predominance of humpless and humped cattle in Africa.

The available evidence, however, does not entirely corroborate this model. Cattle remains from late 7th/early 6th millennium BP contexts, at sites such as Adrar Bous and Uan Muhuggiag in the Sahara, have been interpreted as belonging to short-horned breeds (Carter and Clark 1976, Smith 1984). Furthermore both long-horned and short-horned cattle are represented in prehistoric rock art in the Sahara and the Horn, and humped bovids are also occasionally represented (Clark 1954, Muzzolini 1983, Mohammed 1986, Le Quellec 1987). Significantly, comparative metrical analyses of osteological remains indicate that African cattle are morphologically distinct from Near Eastern cattle.

Genetic Research and Current Opinion

This taxonomic distinction is supported by genetic studies on protein polymorphism from African and Asian cattle (Baker and Manwell 1980). Furthermore, recent genetic work has undermined the established view of a Near Eastern Origin for African cattle. Research on mitochondrial DNA from different breeds of cattle shows that the progenitors of Asian humped zebu cattle and Eurasian *Bos taurus* diverged roughly a million years ago, and were domesticated independently (Loftus et al 1994). This work also indicates that African and Eurasian cattle became genetically distinct around 10 000 years ago. This raises a number of possibilities, which have been outlined by Grigson (1995):

- Cattle were originally domesticated in Africa, and spread to the Near East.
- Cattle were originally domesticated in the Near East and spread to Africa.
- Cattle were domesticated simultaneously around the Mediterranean basin.
- Wild cattle populations in the Near East and Africa became isolated from one another c.10 000 years BP, possibly by a period of hyperaridity.
- Native African cattle have a common ancestor, but different evolution histories in different regions.

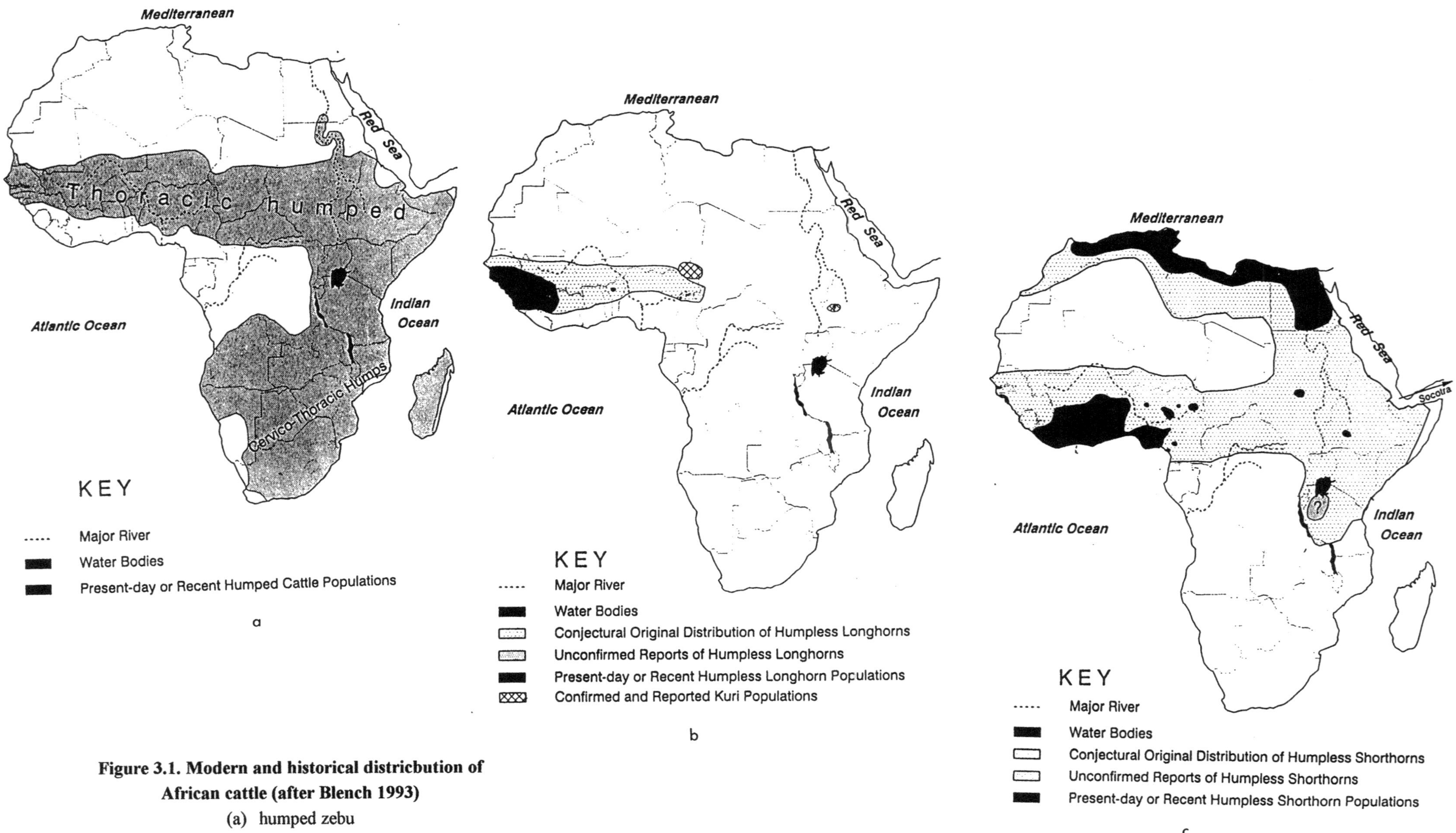

Figure 3.1. Modern and historical districbution of African cattle (after Blench 1993)

(a) humped zebu

(b) humpless long-horned cattle

(c) humpless short-horned cattle.

Until more definitive evidence can be supplied for early domesticated cattle in Northern Africa, the genetic evidence and current opinion supports a model for contemporaneous domestication of separate wild breeds in Africa and the Near East between the 8th and 6th millennia BP (Grigson *pers.comm.*).

The Evidence for Indigenous Cattle Domestication in North Africa

In the light of this recent genetic research, the question of the origin of African cattle has been entirely reviewed. A series of alternative theories has been proposed, and there is now considerable support for an indigenous African origin for domesticated cattle (eg. Gautier 1987, Grigson 1991, Hassan 1995). Remains of wild cattle *Bos primigenius* have been recovered from Palaeolithic mortuary contexts in the Nile Valley, indicating that by the late Pleistocene, cattle had a perceived cultural and ritual significance that may derive from a deeper socio-economic importance (Wendorf and Schild 1976). Wendorf et al (1984b) have suggested that cattle were domesticated within this cultural and environmental context during the early Holocene. This enabled human groups to occupy marginal areas away from the Nile Valley during the early Holocene pluvial period. The pattern of settlement across the Sahara in the late 10th-8th millennium BP was one of seasonally mobile, possibly transhumant systems characterised by relatively large, semi-permanent camps at predictable water sources, such as the Dakhla, Khargha and Siwa oases, Ti-n-Torha in the Acacus and Abu Bodras, and small, transient encampments scattered across more arid areas (Gabriel 1984, 1987). Stone architecture, storage pits, walk-in-wells, pottery, extensive flaked tool technology and heavy grinding equipment at many of the larger sites indicate that they were occupied permanently or repetitively for prolonged periods.

Domesticated *Bos* sp have been tentatively identified from faunal remains dating to the 10th/9th millennia BP at Bir Kiseiba and Nabta Playa in the Eastern Sahara, as shown in Table 3.1 (Gautier 1984, Wendorf et al 1984b). Gautier acknowledged the equivocal nature of the osteological material, but offered an ecological argument for the existence of domesticated cattle. The only other animal remains associated with the deposit were of small, desert adapted mammals, indicative of an arid environment. Large bovids would have been unable to survive in such an arid habitat without human intervention. This model is not universally accepted, although the evidence and arguments used to refute it are themselves unconvincing (Smith 1989). Ironically, the presence of domesticated bovids has been accepted elsewhere in Africa from less evidence (Carter in Brandt 1982).

Wider Implications for Cattle Domestication

If these ecological arguments are to be accepted, domesticated cattle were present in North Africa from the mid 10th millennium BP. These dates are considerably earlier than those for domesticated *Bos taurus* in the Near East (for example 7750 years BP at Catal Huyuk, Perkins 1969; 7500-7000 years BP on the Deh Luran Plain in Iran, Hole et al 1969). Little is known of when and how cattle became domesticated in the Near East. Although indigenous domestication of the wild progenitor *Bos primigenius* has been inferred from changes in osteomorphology and population structure (Collier and White 1976, Meadow 1989, Grigson 1989), the evidence is scarce and there is no guarantee that similar patterns do not result from other processes, such as genetic adaptation to different environmental conditions, or a shift in the subsistence strategy.

Grigson's size-range profile of *Bos* sp from Near Eastern sites documented a dramatic diminution during the 8th millennium BP which was interpreted as evidence of domestication (Grigson 1989). This phenomenon was only apparent at sites located in the Western Fertile Crescent, particularly in the South West region closest to North Africa. Large, wild bovids persisted in the Eastern Fertile Crescent until the 7th millennium BP. Could this evidence suggest instead that a smaller, domesticated animal was introduced to the Near East from North Africa?

Unfortunately, there has been little archaeological work in the important area between North Africa and Southern Asia. The Sinai and Negev Deserts form a land bridge which would have allowed the passage of goods and information between the two continents. Domesticated animals were apparently unknown in the Sinai in the 8th and 9th millennia BP (Bar-Yosef 1975, 1981) and the earliest substantive evidence of domesticated animals in this region dates from c.5000 years BP (Rosen 1988). Pastoral groups are tentatively attested in the Negev Desert, close to North Africa, from the early 7th millennium BP (Bar-Yosef and Phillips 1977, Rosen 1988). These cattle are morphologically closer to the Neolithic cattle of the Central Sahara than to the Near Eastern domesticates, suggesting that they originated in arid North Africa. In addition, these remains were contextually associated with polished stone axes, or 'celts', comparable to the celts which characterise pastoral assemblages in North Africa from the 7th millennium BP. These artefacts were significantly different from the stone axes that characterise 6th millennium BP assemblages of the Negev, and suggest an earlier contact with Saharan cattle keepers (Bar-Yosef and Phillips 1977).

One tentative deduction arising from this evidence is that domesticated cattle were introduced to the Near East from

North Africa by way of the Sinai and Negev, possibly as a result of increasing aridity in North Africa. At present there is no evidence to support or refute this highly controversial model, and substantive further work is needed in this area (Grigson *pers.comm.*).

The Spread of Pastoralism within Northern Africa

Banks (1984) has suggested a massive expansion of pastoralists into the desert region during the pluvial conditions of the 7th millennium BP. The appearance of ovicaprids at this time may additionally have augmented pastoral exploitation of more marginal environments. As Table 3.1 shows, the appearance of cattle was roughly synchronous throughout the Central Nile Valley, Egypt and North Africa during the mid-late 7th millennium BP. The spatial and chronological distribution of the evidence for early domesticated cattle is illustrated in figure 3.2, and the dates are given in Table 3.1.

Archaeozoological evidence from the earliest Neolithic Egyptian sites in the Fayum indicates that ovicaprids were initially a more prolific resource than cattle. Cattle increased in importance during the 6th millennium BP (Wetterstrom 1993). A similar pattern prevails along the North African coast. For example at Capelleti Cave in Northern Algeria, c.6500 BP domesticated ovicaprids comprise 89.7% of the total bone assemblage while cattle comprise only 7.3%. By c.4500 BP 24.7% of the total faunal material belongs to domesticated cattle and 70% to ovicaprids (Roubet 1978). At Ti-n-Torha in Libya, wild Barbary sheep represented 74% of the faunal remains in the early Holocene, but were gradually replaced by domesticated cattle after the 7th millennium BP (Barich 1987). Conversely, cattle appear to have been the major faunal resource in the Central Nile Valley from the late 7th millennium BP. At Kadero, for example, cattle bones comprise 75% of the domesticated fauna and over 60% of the total faunal material (Krzyzaniak 1978).

3.1.2. Plant Domestication and the Origins of African Agriculture: the Odd One Out?

Foreign Crop Plants in North Africa

One of the most striking features of African agriculture is the incredibly late appearance of indigenous African plant domesticates such as sorghum and millets in the archaeological record relative to the global evidence for plant domestication. Indigenous Near Eastern food plants such as barley, emmer and einkorn wheat, chick pea and lentil were domesticated by the 10th millennium BP (Zohary and Hopf 1993), South East Asian plants such as foxtail millet (*Setaria italica*), broomcorn millet (*Panicum miliaceum*) and rice (*Oryza sativa*) were cultivated from around 8000 years BP in the Yellow River area of China (Zhimin 1989, Chang 1970), and in Central and Southern America endemic plants such as maize (*Zea mays*), cowpea (*Phaseolus vulgaris*) and peppers (*Capsicum* sp) were being grown from the late 7th millennium BP (Pickersgill 1989).

Tropical African Crop Plants

In contrast, there is no secure evidence at present that domesticated African food plants were being exploited in Africa until the 3rd millennium BP. By this time agriculture based primarily on Near Eastern domesticates had already been practised throughout Eurasia for several millennia. This does not mean that a knowledge of horticulture was absent from Africa. A fully fledged agricultural economy based on a Near Eastern complex of emmer wheat, 2-row and 6-row hulled barley, flax, lentils and vetch, with domesticated cattle and caprines (sheep/goat) was present in Lower Egypt and the Nile Delta from at least c.6500 years BP (Caton-Thompson and Gardener 1934). The available evidence suggests that this technology then spread southwards along the fertile floodplain of the Nile Valley, reaching the Hemamieh area around after 6000 years BP and the Armant-Nagada area of Upper Egypt by 6000-5500 BP (Wetterstrom 1993). Recent, unconfirmed evidence also suggests that Near Eastern domesticates such as emmer wheat were present in Northern Sudan at their southern-most ecological limits by 5700 years BP (Reinolds *pers.comm.*). Figure 3.3 illustrates the chronology and distribution of early plant domesticates in Northern Africa.

Although agriculture was being practised from at least the 7th/6th millennium BP in the Nile Valley between 32° and 20° N, it seems to have been confined to this region and was apparently based entirely on foreign plant domesticates. As Table 3.2 shows, with the exception of an unconfirmed report of a single fragment of *Hordeum* sp from the Gash Delta dating to the 5th millennium BP (Fattovich and Piperno 1982), there is at present no evidence for the use of Near Eastern domesticates away from the Nile Valley until the 3rd millennium BP. Furthermore, there not yet any *definitive* evidence that domesticated African plants were being exploited anywhere in Africa until the 3rd millennium BP, despite rigorous attempts to prove otherwise (Shaw 1977, Phillipson 1977, Klichowska 1984, Stemler 1990, Wendorf et al 1992, Wasylikowa et al 1995, Wasylikowa and Kubiak-Martens 1995). Instead there is substantial evidence that a comprehensive range of wild plants were being exploited across the whole of North Africa until the 3rd/2nd millennium BP.

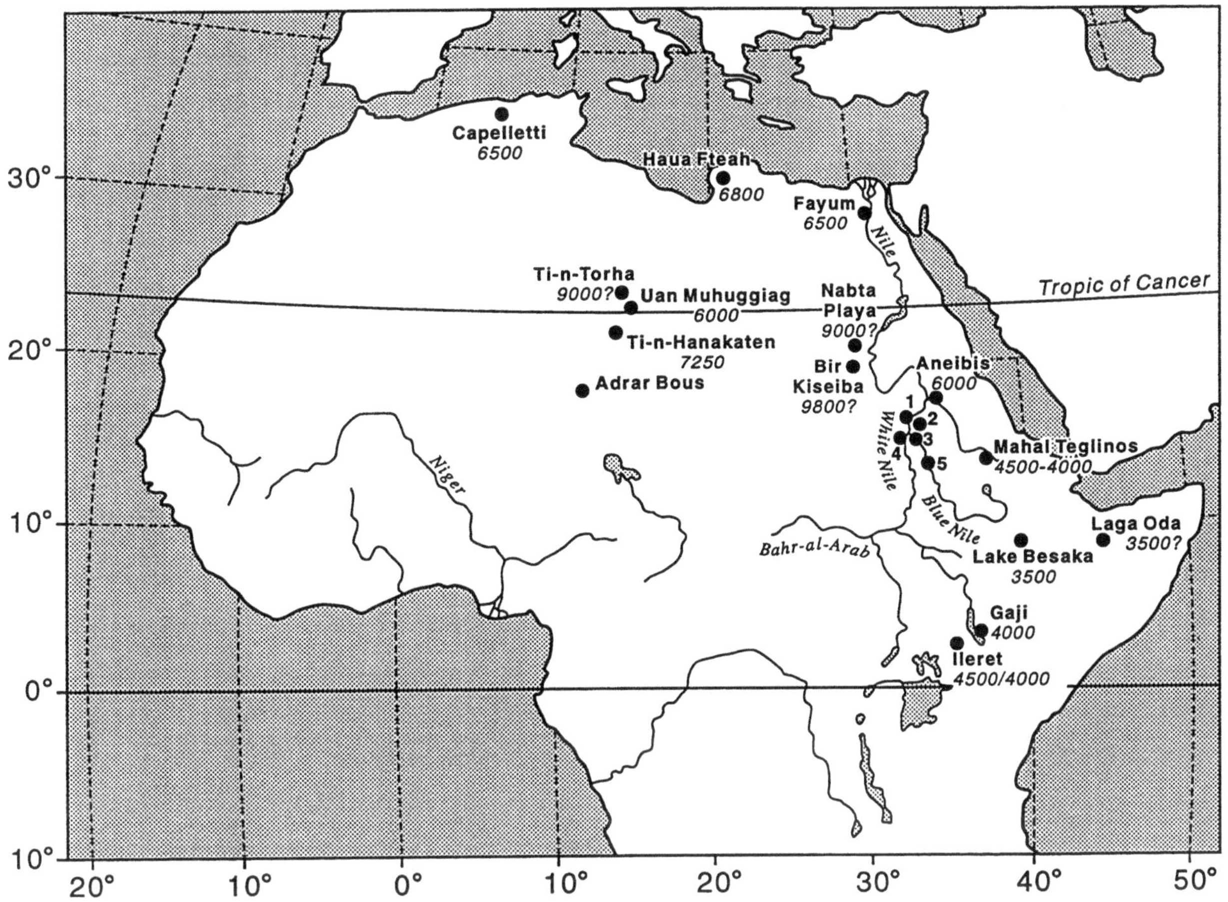

1 - **Um Direiwa** *6000 BP*
2 - **Zakiab** *6000 BP*
3 - **Kadero** *6000 BP*
4 - **Shaheinab** *5700 BP*
5 - **Rabak** *6000 BP*

Figure 3.2. Early distribution of domesticated cattle in Northern Africa

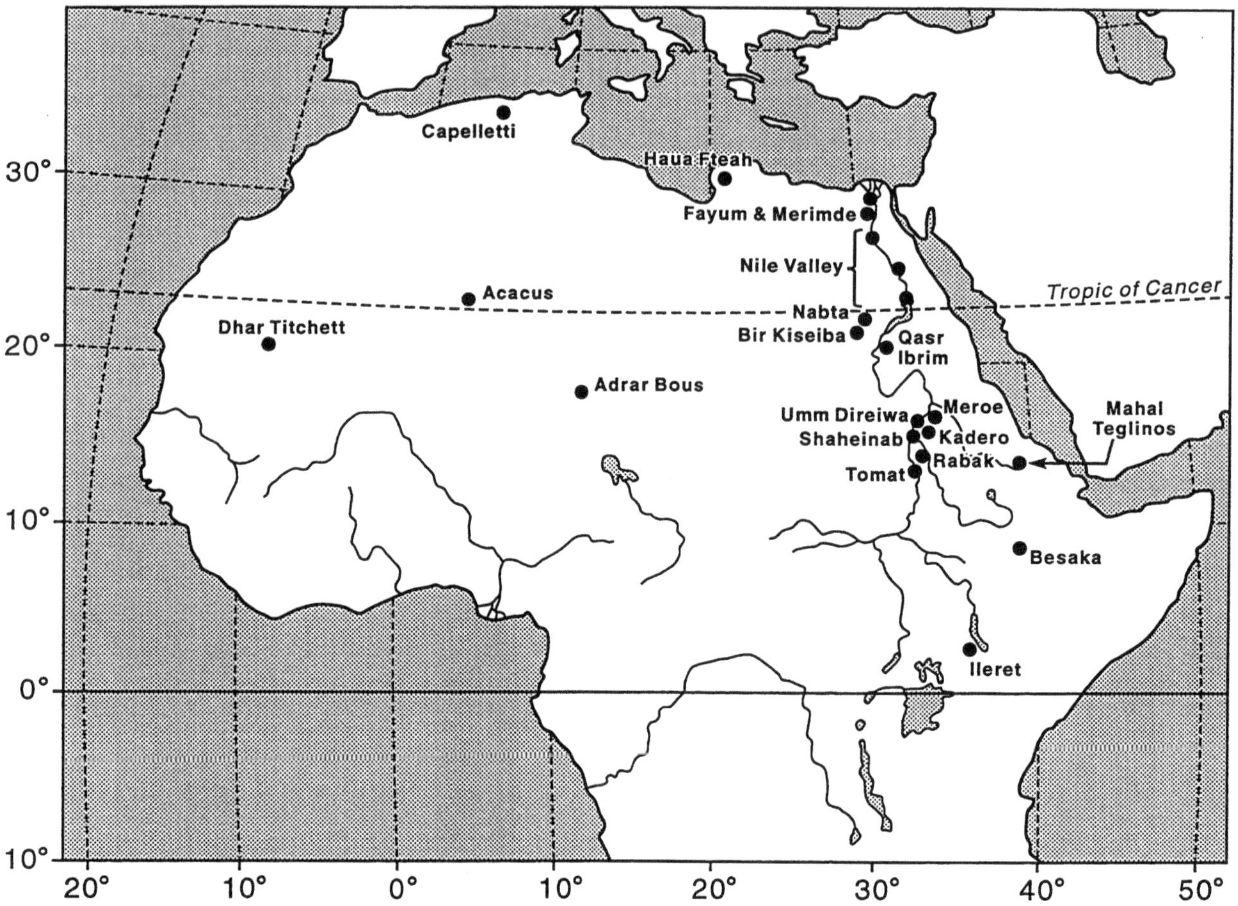

Approximate Dates (BP)

	Cattle	Sheep/ Goat	Millet	Barley	Wheat	Sorghum
Dar Titchett	3500		3100			
Adrar Bous	6000					
Acacus	7500					
Haua Fteah	6800	6800				
Capelleh	6500	6500				
Fayum & Merimde	6500	6500		6500		
Nile Valley	6500			6500		
Nabta & Bir Kiseiba						
Qasr Ibrim	9500?					2000
Umm Direiwa	6000					
Shaheinab	5700	5700				
Tomat	2000					2000
Meroe						2000
Kadero	5700					
Rabak	6000					
Besaka	3500					
Ileret	4000					
Mahal Teglinos	4500			3700?		3700?

Figure 3.3. Chronological and spatial distribution of early domesticated plants and animals in Northern Africa.

Table 3.1. Chronological and Spatial Distribution of Domesticated Cattle in North Africa

Earliest Date (BP)	Latest Date (BP)	Site and Location	Reference
9840±380?	8120±80?	***Bir Kiseiba***, Eastern Sahara	Wendorf and Schild (1980)
9360±70?	8100?	***Nabta Playa,*** Eastern Sahara	Wendorf and Schild (1980)
9080?	7990?	***Ti-n-Torha East***, Tassili, Sahara	Gautier (1982)
8960±110?	4150	***Ti-n-Hanakaten***, Tassili, Sahara	Aumassip (1984)
6800	-	***Haua Fteah***, Cyrenica, North Africa	Higgs (1967)
6530±250	5100±150	***Capelleti***, Algeria, North Africa	Roubet (1978)
c.6500	-	***Qasr el Sagha***, Fayum, Lower Egypt	Brewer (1989)
6050±90	6020±130	***Rabak***, White Nile, Southern Sudan	Haaland (1984)
6010±90	5280±80	***Um Direiwa I***, Central Nile Valley, Sudan	Haaland (1981), El Mahi (1988)
5970±80	5460±70	***Zakiab***, Central Nile Valley, Sudan	Haaland (1981)
5952±120	-	***Uan Muhuggiag,*** Acaucus, Sahara	Gautier (1982)
5810±500	5740	***Adrar Bous***, Air, Sahara	Carter and Clark (1976)
c.5750	-	***Merimde Beni Salama***, Lower Egypt	Gautier (1987)
5720±80	5650±80	***Shaheinab***, Central Nile Valley, Sudan	Hassan (1984)
5700±100	5460±70	***Kadero***, Central Nile Valley, Sudan	Krzyzaniak (1978)
5650	4750	***Ti-n-Torha North***, Tassili, Sahara	Gautier (1982)

Table 3.2. Evidence for Domesticated Crops of Near Eastern Origin in Africa before 2000 Years BP

Date (Years BP)	Site and Location	Evidence for Domesticated Crop
6500-6000	***Fayum,*** Lower Egypt (Caton-Thompson and Gardiner 1934)	emmer wheat 2-row barley, 6-row barley flax
6500-6000	***Merimde Beni Salama,*** Nile Delta (Werth 1939)	emmer wheat, free threshing wheat hulled 6-row barley lentils peas flax
c.5700	***Kadruka***, Central Nile Valley (Reinolds *pers.comm.*)	emmer wheat?
4500-3000	***Mahal Teglinos,*** Kassala, Ethiopia-Sudan border (Fattovich and Piperno 1982) (Costantini et al 1983)	barley (single fragment, unconfirmed) leguminoseae
2700-2400	***Zinchecra,*** Fezzan, Libyan Sahara (van der Veen 1995)	emmer wheat barley
c.2500	***Lalibela cave***, Ethiopian Highlands (Dombrowski 1971)	barley chick peas horse beans leguminoseae
c.2300	***Beita Giyorgis***, Ethiopia Highlands (Bard and Fattovich 1996)	wheat barley lentils

Plant Resources In African Prehistory

Plant Resources in the Prehistoric Diet

The importance of vegetable foods in prehistoric subsistence is often underestimated due to an absence of data, yet among present day hunter-gatherer groups such as the Hadza in North Tanzania, plants contribute 60-80% of the diet (Woodburn 1968). In New Guinea, plant foods are emphasised disproportionately over animal foods among both food procuring and food producing groups (Allen 1977) - for example the Tsembaga people cultivate 90 varieties of the 3 basic tubers, grow 174 local plants and exploit a further 200 wild species from a possible total range of over 650 (Rappaport 1968). The role of wild plant foods is documented ethnographically across the world - for example in Australia (McCarthy and McArthur 1960, Jones and Meehan 1989), North America (Shipek 1989) and Africa (Lee and DeVore 1976).

Even under exceptional conditions of preservation and extraction, we can only obtain a tantalising glimpse of floral abundance in the prehistoric diet. Such conditions prevail in arid regions of North Africa, allowing the recovery of charred and desiccated seeds and parenchymatous tissues that are usually archaeologically invisible, and this has allowed a glimpse of the rich dietary diversity enjoyed by the inhabitants of these sites. For example, 25 different types of local plant macrofossil were isolated from the Terminal Pleistocene site of Wadi Kubbaniya in Upper Egypt, including charred and faecal remains of parenchymatous tissue from aquatic plants (Hillman et al 1989). Forty different morphological types of fruits and seeds have been recovered from the site of E-75-6 at Nabta Playa in the Eastern Sahara (Wasylikowa et al 1995), about 25 different plant taxa were collected from Ti-n-Torha (van der Veen 1995) and 30 from Uan Muhaggiag (Wasylikowa 1992), both in South West Libya. A comparable taxonomic range occurs at late Pleistocene sites (eg. Hillman 1989), suggesting that a broad spectrum foraging economy was already well established by the end of the Pleistocene in Africa. Similar evidence of broad spectrum foraging economies comes also from contemporary sites in the Old World where, for example, 157 different plant species have been identified at Abu Hureyra and 38 at Tell Muryebit in Syria (Hillman 1975, Hillman et al 1989, Cauvin 1985).

A Stabilised Foraging Economy in Holocene North Africa

As a result of the exceptional conditions for recovery of plant remains in North Africa, several important features of the prehistoric economy have been identified:

- A wide range of wild plants were exploited during the Holocene period.

- The range of exploited plant resources was closely representative of the local vegetation, as determined by the pollen evidence (eg. Schulz 1987).

- There was negligible change in the range of resources exploited at early and late Holocene sites in similar ecological zones following the introduction of domesticated animals, as illustrated in Table 3.2 with reference to archaeological sites in the Central Nile Valley spanning a 2500 year period. Wild grasses (including *Panicum* sp, *Pennisetum* sp, *Setaria* sp, *Brachiaria* sp, *Eragrositis* sp, *Cenchrus* sp, *Sorghum* sp), African hackberry (*Celtis integrifolia*), *Zizyphus spinachristi*, date palm (*Balanites/Phoenix* sp), sedges (Cyperaceae) and wild watermelon (*Citrullus colocynthis*) were among the numerous plant resources of the Holocene prehistoric economy.

- Many of these plants are still vital wild foods today (Ahmed 1994). For example wild grasses are still harvested in Saharan and savannah regions of Africa, and are a staple for several groups such as the Northern Tuareg, the Tawaerek and the Kredas (see figure 7.1) (Harlan 1989).

Such consistency and continuity in the nature and the range of wild food plant used in North Africa throughout the Holocene implies that this assortment formed a satisfactory resource base. There was perhaps no immediate need to expand or alter this foraging economy.

The Emergence of Tropical African Domesticates

The Chronological Pattern

Morphologically domesticated tropical plants appear in North Africa from the start of the 3rd millennium BP, although the majority of domesticated African staples appear not to have been exploited until 2000 years ago. Domesticated bulrush or pearl millet (*Pennisetum typhoides*) has been identified from numerous impressions on pottery from Dhar Titchett and Dhar Oualata in Mauritania dating to c.3000-2900 years BP (Munson 1976, Holl 1985, Amblard and Pernes 1989), and to c.3000 BP from Burkina Faso (Neumann 1996). Wild *Pennisetum* (*P.fallax*) was one of several wild Saharan grasses, including *Panicum* sp, *Cencherus biflorus* and *Brachiaria deflexa*, exploited at Dhar Titchett during the 4th millennium BP. By the end of the 4th millennium the dietary importance of wild grasses at Dhar Titchett had increased, with certain grasses such as *Pennisetum* being exploited more intensively prior to the appearance of morphologically domesticated varieties (Munson 1976).

As Table 3.4 shows, the earliest positive identifications of domesticated *Sorghum bicolor* date to 2020±127 BP from Meroe in the Central Nile Valley (Stemler and Falk 1981), c.1900 years BP at Qasr Ibrim in Upper Egypt (Rowley-Conwy 1991) and 1755±60 BP from Jebel et Tomat in the Central Nile Valley (Clark and Stemler 1975). Domesticated *Sorghum bicolor* was grown widely in North Africa by the early 2nd millennium BP (Rowley-Conwy 1991). In West Africa, domesticated African rice (*Oryza glaberrima*) has been recovered from the Niger Delta region dating to c.1960 BP (McIntosh and McIntosh 1980), and there are historical references to domesticated fonio and black fonio during the early 2nd millennium BP (Purseglove 1976, Shaw 1977). Domesticated tef *(Eragrostis tef)* has been tentatively identified from c.2000 BP and 1600 BP contexts in Northern Ethiopia (Dombrowski 1971, Hansen *pers.comm.*), and the oil seed noog (*Guizoitia abyssinica*) is known from mid-2nd millennium BP contexts (Phillipson and Reynolds 1996.)

Table 3.3. Subsistence Resources Spanning the Mid-Holocene Period in Central Sudan

Date (BP)	Site Name	Livestock	Flora
c.8000	***Khartoum Hospital***	-	*Celtis integrifolia* *Elaesis guineensis*
6895±130	***Shaqadud***	-	wild grasses including: *Panicum turgidum* *Pennisetum sp* *Sorghum sp* *Sida alba* *Setaria sp* fruits and berries: *Grewia tenax* *Zizyphus spinachristi* *Solanum dubium*
6408±80	***Sorourab***	-	*Zizyphus sp* *Salix subserrata* Gramineae *Linguliflorae*
5870±130	***Islang***	cattle?	*Celtis integrifolia*
5500±70	***Kadero***	cattle caprids	wild grasses including: *Sorghum bicolor* *Eleusine coracana* *Eragrostis abyssinica* *Panicum sp* *Setaria sp* *Digitaria sp* *Andropogon sp* fruits and berries: *Celtis integrifolia* *Hyphaena thebacia* *Citrullus sp* Nymphaea
5520±130	***Nofalab***	cattle caprids	*Zizyphus sp* *Celtis integrifolia* *Sorghum verticilliflorum* Gramineae *Crotolaria sp* *Boertiluria sp*
5260±80	***Shaheinab***	caprids	*Celtis integrifolia* *Elaesis guineensis*

An Inherent Paradox

Evidence from Northern Africa suggests that indigenous wild cereals were being exploited intensively by the 6th millennium BP in the Nile Valley (Haaland 1985, Ehret 1993). A shift to a more intensive, specialised exploitation strategy is suggested by an increase in both the frequency of sites (Caneva 1988), and their size and length of occupation - for example the site of Kadero exceeds 30 000 m^2 (Krzyzaniak 1978). Haaland (1995) suggests that these changes imply the emergence of more sedentary behaviour. This proposition will be discussed in more detail in chapter 5 with reference to Ethiopia. This development in settlement behaviour was accompanied by a vast increase in the frequency of grinding stones recovered from the sites. 30 000 fragments were found

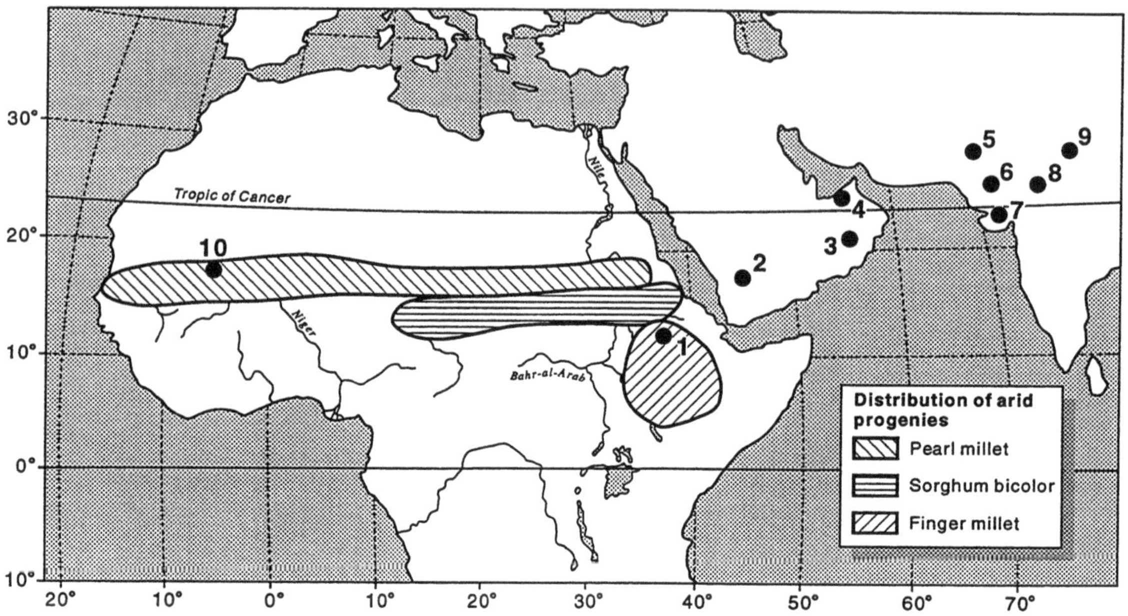

1 - Mahal Teglinos, Gash Delta
2 - Wadi Yanaim
3 - Ras al Hamra
4 - Hili, Oman
5 - Pirak, Kachi
6 - Surkotada
7 - Rojdi, Sourashtra
8 - Hallur, Karnataka
9 - Hulas, Pradesh
10 - Dhar Titchett

Modern day crop growing areas

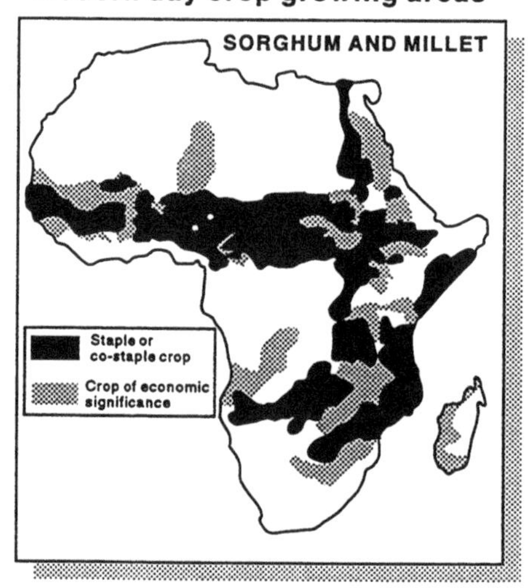

Figure 3.4. Known distribution of tropical African cereals (after Harlan 1992), and early distribution of their domesticates.

in an area of 148 m^2 at the 6th millennium BP site of Umm Direiwa, in contrast to the 100 or so fragments found at each of the similar 9th millennium sites of El Damer, Aneibis and Abu Darbein (Haaland and Magid 1991, Haaland 1995). Haaland (*in press*) argues that the repecked surfaces and fragmentary nature of these grindstones indicates a more intensive use of wild grasses during the 6th millennium BP, and proposes that particular resources such as sorghum were being selectively cultivated within their natural habitat by the 6th millennium BP. This suggestion is tentatively supported by archaeobotanical evidence, which indicates an increase in the proportional representation of certain wild grass seeds, particularly sorghum and finger millet, during the mid-Holocene (Klichowska 1984, Stemler 1990). Similar adaptations in the Near East during the early Holocene are viewed as a prerequisite for the development of agricultural societies (Harris 1977b, Rafferty 1985, Bar-Yosef and Belfer Cohen 1989). The archaeological absence of a parallel process in North Africa has been attributed to an absence of suitable selection pressures on the wild species necessary to induce genetic mutation and the emergence of morphologically domesticated plants (Stemler 1980). This concept is discussed in greater detail in chapter 7.

In addition, the absence of evidence of domesticated tropical cereals from Africa is contradicted by archaeobotanical evidence from Arabia and India. As Table 3.4 shows, domesticated sorghum, finger millet and pearl millet were known in Arabia and India from the 5th millennium BP, over 2000 years before they appeared in African contexts. This relationship is illustrated in figures 3.4 and 3.5. Despite claims to the contrary (Vishnu-Mitre 1968, 1969, Hilu et al 1976, 1979), wild progenitors for these crop plants are not indigenous to India or Arabia. Remains of the wild plants recovered from early contexts outside Africa must therefore have been introduced to those regions from Northern Africa. Furthermore, recent genetic research shows that domesticated and wild sorghum from Egyptian Nubia have identical genomes, suggesting that *S.bicolor* was domesticated indigenously in the Nile Valley around 2000 years ago (Rowley-Conwy *pers.comm.*). Haaland (*in press*) has suggested that morphologically wild African cereals were exported from the Central Nile Valley where they were being exploited intensively from the early 6th millennium BP. Once removed from their natural habitat, these plants were exposed to new selection pressures and the final phenotypic change to the domesticated form occurred. As Table 3.4 shows, with the possible exception of Mahal Teglinos, the earliest evidence for African domesticates comes from Arabia and India. Domesticated African cereals such as *Sorghum durra* and *Eleusine coracana* were subsequently introduced to Africa from Asia, at the start of the 2nd millennium BP. Haaland (*in press*) has adapted Doggett's (1988) model to explain this paradox for sorghum, as shown in figure 3.5.

Date (BP)	Savannah Belt	Egypt	Central Nile Valley	Red Sea Area	South Arabia	Indus
8th millennium	gathered	gathered	gathered ↓			
7th millennium			↓			
6th millennium	cultivated?	cultivated?	cultivated	→ →	cultivated? ↓	
5th millennium			↓		**domestic bicolor**	**domestic bicolor** ↓
4th millennium			↓		↙ ? →	
3rd millennium			↓			**domestic durra**
2nd millennium		domestic bicolor	**domestic bicolor**	↙	**domestic durra** ←	
1st millennium		**domestic durra**	**domestic durra**	↙		

Figure 3.5. Hypothetical pathway for the passage of wild, cultivated and domestic sorghum in Africa and Asia (adapted from Haaland, *in press*). shaded area - natural ecological zone of wild *Sorghum bicolor;* **bold** - palaeobotanical evidence for domesticated sorghum.

Table 3.4. Earliest Occurrences in Africa and Other Regions of Domesticated African Food Crops before 2000 Years BP

Crop	Date (BP)	Evidence from Africa	Evidence from non-African regions
Sorghum *Sorghum bicolor* from *S.arundinaceum*	5500-4800		[*unconfirmed*] Hili 8, Oman, Southern Arabia (Potts 1993, 1994)
	c.4800		Ra's al-Hamra, Oman coast, Southern Arabia (Biagi et al 1985)
	c.4500		Hili 8, Abu Dhabi (Cleuziou and Costantini 1980)
	4000-3600		Inamgaon, Maharashtra, India (Allchin and Allchin 1982)
	c.3800	[*unconfirmed*] Gash Delta, North East Africa (Costantini et al 1983)	
	c.3700		Wadi Yanaim, Central Yemen (Costantini 1990)
	3700-3000		Hulas Uttar, Pradesh, India (Saraswat 1986, 1993)
	c.3600		Pirak, Indus, India (Costantini 1979, Kajalc 1991)
	c.2000	Meroe/Qasr Ibrim, Nile Valley (Stemler and Falk 1981, Rowley-Conwy 1991)	
Finger Millet *Eleusine coracana* from *E.indica* subsp. *africana*	4500		Rojdi, Saurashtra, Western peninsula India (Weber 1991)
	c.3800		Hallur, Karnataka, India (Kajale 1974)
	c.3660		Surkotda, Gujarat Peninsula, India (Kajale 1974)
Pearl or Bulrush Millet *Pennisetum typhoides* from *P.fallax*	c.3700		Saurashtra, India (Rao 1963)
	3500-3000	Dar Tichett and Oualata, Mauritania (Holl 1985, Amblard and Pernes 1989)	
	c.3000	North East Nigeria, North Burkina Faso (Neumann 1996)	
Tef *Eragrostis tef* from *E.pilosa*?	?c.4600	[*unconfirmed*] Dassur, Egypt (*E.pilosa*?) (Unger 1866)	
	c.3400	[*unconfirmed*] Ramses, Israel (Porteres 1976)	
	c.2100		[*unconfirmed*] Hajar bin Humeid, Yemen (van Beek 1969)
	c.2000	Lalibela cave, Ethiopian Highlands (Dombrowski 1971)	

3.1.3. Cattle vs Crops in North Africa: the Implications for Ethiopia

There seems to be a chronological juxtaposition between the appearance of domesticated cattle and crop plants in North and East Africa. Domesticated cattle were widespread in North Africa by the 6th millennium BP, and in East Africa by the 4th millennium. However, foraged plants continued as the major food source throughout the period of cattle adoption, despite the presence of an established agricultural economy in the Egyptian Nile Valley from the 6th millennium BP. Although there is strong evidence for increasing intensification and specialisation of plant foraging in the Central Nile Valley by the 6th millennium BP, there appears to have been little incentive to adapt the basic subsistence strategy.

The wild progenitors of both finger millet and sorghum grow in Ethiopia. The Ethiopian domesticates of these plants are genetically diverse (Stemler et al 1977). A number of researchers have proposed that these crops were originally domesticated in Ethiopia at an early date (eg. Harlan 1969, Doggett 1988, Phillipson 1977). The geographical location of Ethiopia and the suspected antiquity of its food production implicate it as a major component in the early passage of African crops to India and Arabia, and possibly even as a region from which the wild plants or their cultivars were originally exported in the 5th millennium BP (Doggett 1988, 1991, Mehra 1963a, 1991).

The following sections investigate the evidence for domesticated cattle and domesticated plants in Ethiopia. The pattern for the appearance of domesticated resources is discussed in terms of the model outlined above for Africa, and the possibility of indigenous plant domestication is investigated.

3.2. DOMESTICATED ANIMALS IN ETHIOPIA

Cattle are the single most important resource in Ethiopia, while camels are also of great economic value especially in lowland regions where they are vital for subsistence and trade. Evidence for camel domestication in Ethiopia is more restricted than that for cattle, and this will be discussed in Section 3.2.5. There is no unequivocal evidence for the early appearance of caprines, and they will not be discussed in the body of this volume, but details of the origins of all domesticated mammals in Ethiopia are presented in *Appendix A*, Table A.3.2.

3.2.1. Current Models for the Appearance of Domesticated Cattle in Ethiopia

Cattle in Ethiopia

Cattle are vital to almost all forms of subsistence in Ethiopia. Today, over 26 million head of cattle are distributed among agricultural, pastoral and agro-pastoral groups, representing the largest population in any African country (Smith 1992). Domestic cattle in particular are integral to both pastoral and agricultural subsistence, whilst camels and equids facilitate channels of communication and trade. The appearance of domesticated cattle was therefore fundamental to the development of specialised economies and intensive crop plant exploitation.

As discussed below, domesticated cattle were being exploited by pastoral or agro-pastoral groups in the Rift Valley and along the escarpment by the mid-4th millennium BP (Clark and Williams 1978, Brandt 1982). By the 3rd millennium, domesticated cattle and livestock were an integral part of the mixed agricultural economy of PreAxumite groups in the Northern Highlands (Fattovich 1990b). Numerous rock art sites located around the fringes of the Ethiopian plateau and in Somalia attest to the extent and importance of prehistoric cattle pastoralism in Ethiopia and the Horn (eg. Graziosi 1964, Brandt and Carder 1987).

Ethiopia may have acted as a natural ecological corridor for the North-South movement of pastoral groups, due to the absence of cattle disease from large areas of the Highlands and Rift Valley, and the presence of extensive pasture land and surface water.

A similar pattern of movement from lowland regions to the Ethiopian Highlands has been repeated throughout history and prehistory during periods of stress, for example the Oromo pastoralists in the 1st millennium BP and the Beja from the Red Sea Hills and Eastern Sudan during the past 200 years (Hassen 1990, Pankhurst 1968).

Models for the Introduction of Cattle

Two speculative models currently predominate to explain the appearance of domesticated cattle in Ethiopia:

- Cattle and other livestock were introduced from the Near East via Southern Arabia, between 5000 and 2500 years BP as part of a mixed agricultural economy (Steihler 1948).
- Cattle were introduced from the Nile Valley by C-Group Nubians during the 4th millennium BP (Clark 1962, 1980). C-Group pastoralists moved into the fringes of the Highlands and subsequently migrated along the Northern edge of the Highlands, possibly confined within a

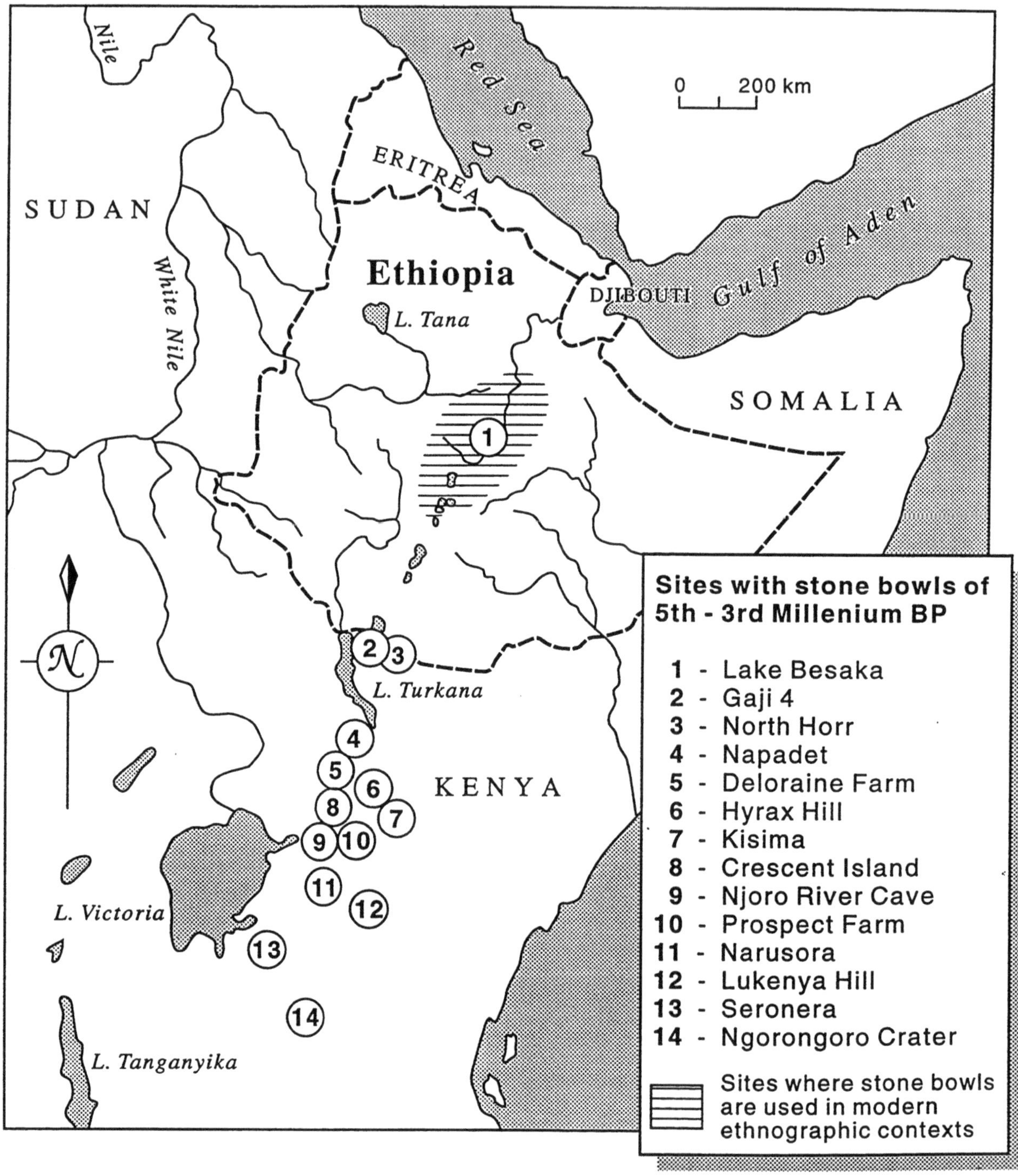

Figure 3.6. Distribution of early pastoralist sites with stone bowls in East Africa and Ethiopia.

restricted ecological zone, to the Red Sea coast and down into the Rift Valley before eventually reaching the Highland Plateau.

These models are based on tenuous evidence and they cannot be verified at present. For example, Clark's model is derived from similarities between C-Group art from the Nile Valley and the pastoral rock art from the Horn (see figure 3.10). A significant correlation has also been proposed between C-Group assemblages and artefacts from surface collections at Agordat in Eritrea (Arkell 1954). Clay figurines of domesticated cattle from Agordat have been additionally interpreted as evidence of a pastoral society (Arkell 1954, Clark 1980). The material from Agordat has been tentatively dated to the 4th millennium BP on the basis of this comparison.

However there is no evidence that the C-Group were mobile cattle-herders. Furthermore there appears to have been no mass migration to Ethiopia during the 4th millennium BP. The C-Group instead appear to have been a sedentary, agricultural population to whom cattle were one of several important resources (Adams 1977). The following sections examine the available evidence for domesticated cattle in Ethiopia and discuss possible alternative interpretations.

3.2.2. Archaeological Evidence for Cattle Domestication

The archaeological evidence for domesticated cattle in Ethiopia is extremely limited prior to the 3rd millennium, with only two ambiguous occurrences (Table 3.5):

Domesticated cattle bones were tentatively identified at Laga Oda in the Southern Highlands (Clark and Williams 1978). These came from a small sample of fragmentary material at a depth of 60-70 cm below the surface from an excavated area of $3m^2$. Only a $1m^2$ area was excavated between 70 and 140 cm, from which only two unidentifiable bone fragments were recovered (Clark and Prince 1978). The appearance of cattle bones therefore coincides with the expansion of the excavated area, and the possibility that domesticated fauna were present in earlier levels should be considered.

Domesticated *Bos* sp were also tentatively identified in the field from three teeth fragments recovered from a mid-4th millennium BP context at site FeJx3, Lake Besaka in the Afar Rift (Carter in Brandt 1982). This material has not been available for further study, and the report remains unconfirmed. Brandt observed discontinuity in the material culture patterning at FeJx3 relative to earlier 'foraging' sites in this locale, which he proposed reflected a major change in the subsistence regime and the movement of pastoral groups into the area (Brandt 1982, 1986).

This hypothesis was supported by the recovery of a shaped lava fragment, purported to belong to a stone bowl - a whole specimen being found in a disturbed context in the same area. Stone bowls are characteristic of the material assemblage of early cattle keeping groups in Eastern Africa. The earliest association of domesticated cattle and stone bowls is dated to the mid-5th millennium BP from Lake Turkana (Barthelme 1977, 1984). Their chronological and spatial distribution, shown in figure 3.6, reflects the spread of pastoralism through Eastern and Southern Africa (Leakey and Leakey 1950, Phillipson 1977a). The extension of this phenomena to Lake Besaka suggests that the Ethiopian Rift Valley was critical to the initial dispersal of domesticated cattle into East Africa.

Summary of the Archaeological Record for Domesticated Cattle in Ethiopia

As Table 3.5 shows, the indirect evidence for domesticated cattle in Ethiopia prior to the 4th millennium BP comes from undated or ambiguous contexts. Ethiopian rock art represents a valuable, but virtually untapped source of information. The available evidence for the 'iconography' of cattle domestication is discussed in greater detail below, but Ethiopian archaeology would benefit immensely from a detailed investigation of this material. The data at present can neither substantiate nor refute proposed ages for the art derived from stylistic analyses.

There is little indication of precisely ***when*** domesticated cattle became an important resource in Ethiopia, or from ***where*** they originated. The limited and uncertain nature of the data does not exclude the possibility that cattle were domesticated indigenously in Ethiopia. This proposal is discussed below, and then rejected in favour of a model for the introduction of cattle.

Table 3.5. Archaeological and Documentary Evidence for Domesticated Animal Resources in Ethiopia before the 2nd Millennium BP

Evidence	Approximate Date (Years BP)
Archaeological	
Domesticated camel tooth from Gobedra rock shelter, Northern Highlands. Possibly intrusive from a later level?	c.5th millennium (Phillipson 1977b)
Domesticated bovid remains identified from fragmented bone at Laga Oda rock shelter on the edge of the Southern Rift Valley escarpment.	c.3500 (Clark and Williams 1978)
Domesticated bovids tentatively identified from teeth at Lake Besaka in the Afar Rift. The contextual association of fragments of a possible stone bowl at this site suggests cultural links with the Stone Bowl pastoral groups of Lake Turkana.	c.3500 (Carter in Brandt 1982)
Domesticated bovids and ovicaprids tentatively identified from osteological remains at Lalibela and Natchabiet caves, east of Lake Tana in the Northern Highlands. Relatively higher frequencies of cattle remains relative to ovicaprids in earlier levels; this relationship is reversed in later contexts.	c.2500 (Dombrowski 1971)
Domesticated bovid bones from Ona Nagast, Bieta Giyorgis	c.2300 (Bard and Fattovich 1996)
Domesticated bovid bones from Gobedra	3rd millennium (Phillipson 1977b)
Documentary and Iconographic Evidence	
Rock paintings and engravings depicting domesticated long-horned and short-horned humpless bovids and ovicaprids are widespread around the fringes of the Ethiopian plateau and in Eritrea, Somalia and Djibouti.	Undated, possibly 4th millennium (Clark 1980, Brandt and Carder 1987)
Clay figurine of a domesticated bovid from Agordat in Eritrea.	Undated surface material, possibly 4th millennium (Arkell 1954)
Clay models of humpless cattle and a plough from the PreAxumite site of Hawlti near Axum, and a small stone figurine of a bull from Mahabere Dyagove in Tigray.	late 3rd millennium (de Contensen 1963, Fattovich 1990b)
Relief carvings of domesticated humpless short-horned and long-horned cattle from Punt, displayed in Queen Hatshepsut's mortuary temple at Deir el-Bahari, Egypt. Punt may include Northern Ethiopian Highlands (see chapter 6)	c.3500 (Naville 1898)
Domesticated camels from ?Ethiopia accompanied the Queen of Sheba's visit to Jerusalem, documented in the Phoenician Book of Kings. The traditional Ethiopian claim for the Queen of Sheba is contentious.	early 3rd millennium (Pankhurst 1968)
Documentary evidence for Ethiopian sheep being displayed in public on the demands of Alexander the Great. 'Ethiopia' may mean North East Africa and the Horn, as in Roman times, rather than Abyssinia.	c.3400-3300 (Rostovtzeft 1941)

3.2.3. Ambiguities in the Data

Indigenous Domestication and Problems with Progenitors

The archaeological evidence does not preclude the possibility that Ethiopian cattle were domesticated indigenously, but the absence of the wild progenitor, *Bos primigenius*, in Ethiopia makes such a proposal highly unlikely.

The modern distribution of wild progenitors is not a definitive statement of their past distribution. Wild aurochs, *Bos primigenius*, was widespread in the Northern Hemisphere during the Pleistocene and Holocene. It is described as 'a browsing, grazing ruminant that inhabited forests but could also flourish in open scrub', and a limited modern distribution of 30-60°N has been suggested (Clutton-Brock 1987 p.63).

Fluctuating environmental conditions during the Pleistocene and Holocene would have enabled the aurochs to migrate into regions to which it is currently not adapted. *Bos primigenius* is known to have had a wider Pleistocene distribution in North Africa than at present, including several Upper Pleistocene sites in North East Africa (Mohammed-Ali 1978).

Although wild bovids are well attested in Pleistocene and Holocene cultural contexts in Ethiopia the remains are usually too fragmentary to identify their specific genera. It is highly unlikely that the Holocene distribution of *Bos* would have extended as far south as Ethiopia. The possibility that cattle were indigenously domesticated can be excluded, but a number of ambiguities remain in the data that cannot be explained by a model for late introduction of domesticated cattle. These ambiguities have implications for a closer understanding of cattle domestication in Ethiopia.

Genetic and Documentary Ambiguities

The original Ethiopian gene pool has been virtually wiped out by recurrent droughts and rinderpest epidemics - for example, during the late 19th century over 90% of all Ethiopian cattle are reported to have died (Pankhurst 1968) - and the majority of cattle in Ethiopia today are descended from thoracic-humped zebu stock.

Figure 3.7. Wall paintings from Deir el Bahari temple, depicting (a) long-horned cattle and (b) short-horned cattle from the Land of Punt.

Short-Horned Cattle in Ethiopia

Although the ancestral gene pool in Ethiopia is now virtually extinct, a relict population of dwarf short-horned cattle persists sporadically in Ethiopia (Alberro and Haile-Mariam 1982). Short-horned cattle are the predominant local breed in West Africa and the North African littoral, and they occur in isolated pockets in the Sudan and East Africa (Blench 1993). The origins of the short-horn are unknown, but historical, ethnographic, iconographic and linguistic data suggests that it is an ancient African breed that originally had a much wider distribution in North East Africa, the Sahel, the North African littoral and the Socotra islands off the southern tip of Arabia (Blench 1993) (figure 3.1.c.).

Short-horned cattle are trypano-tolerant (i.e. tolerant to trypanosomiasis, the organism responsible for cattle disease or 'sleeping sickness', carried by the tsetse fly in humid regions of Africa), which suggest that they have been exposed to tropical climates for considerably longer than long-horn cattle. Conversely, long-horned cattle are better adapted to arid environments, indicating longer exposure to desert conditions.

Short-horned cattle are absent from Egyptian art until the mid-late 5th millennium BP, and earlier depictions seem to be purely of Egyptian long-horns (Epstein 1971). By the mid-4th millennium BP the Egyptians were importing 'many cattle' from the Land of Punt[1] (Breasted 1952 p.305). suggesting that the cattle from this region possessed desirable properties that distinguished them from the Egyptian breeds. Pictorial evidence from Queen Hatshepsut's temple of Dier el-Bahari shows that established breeds of long-horn and short-horn cattle were known in the Land of Punt by the mid-4th millennium BP (figure 3.7) (Naville 1898, Kitchen 1993).

Short-horned Cattle in North Africa

The origin of short-horned cattle in North Africa is intriguing. In general, body size and horn size are proportional. The short-horn may have evolved gradually from the long-horn in the absence of conscious anthropogenic selection for disproportionately large horns and large body size, as is found for example in the Kuri cattle of Chad (Epstein 1971, Grigson 1995). The evidence outlined above indicates that short-horned cattle were well established in sub-Saharan Africa before they were known in Egypt, before the 4th millennium BP. Genetic research additionally shows high frequency variation in polymorphic proteins between *Bos taurus* and African short-horned cattle, which suggests a significant period of genetic separation of early domesticated breeds of these cattle (Baker and Manwell 1980).

The Iconography of Domestication - the Paradox of Ethiopian Rock Art

Background and Distribution

The rock art of the Horn of Africa is a little known and poorly researched subject. Scattered across the Horn from the Eritrean/Ethiopia/Sudanese border at Karora, along the North and East fringes of the Ethiopian Plateau, through the semi-arid lowlands of Somalia and Djibouti, and in the fertile Highlands of South West Ethiopia around Sidamo and Yavello are numerous paintings and engravings depicting domesticated animals, humans, wild animals and a variety of geometric motifs (Clark 1954, Anfray 1967a). The sites are on exposed rock surfaces or in caves and rock shelters, often in extremely inaccessible locations. Their distribution in Ethiopia is within a defined altitude range of 1400-2000 m asl (Negash 1990) where it corresponds to a rich, open woodland environment. This is the preferred habitat of pastoralists in the absence of land competition from farmers.

The predominant iconography throughout this region is of domesticated cattle (eg. figure 3.8). Stylistically and thematically this rock art belongs to a widespread prehistoric artistic continuum spread right across Northern Africa, down into East Africa and into Southern Arabia. This distribution coincides with the appearance and spread of domesticated cattle in the early-mid Holocene. Significantly there is no record of this rock art tradition in humid regions of Africa that are suitable for agriculture rather than pastoralism.

Rock art is well documented in the Sahara and Nile Valley (Mohammed 1986, Le Quellec 1987, Muzzolini 1985, 1993, 1995), but research in East Africa, Southern Arabia and the Horn has been sporadic and site-specific (eg. Breuil 1934, Clark 1945, Drew 1952, Wright 1961, Graziozi 1964, Anfray 1967, Dekker and Selassie 1972, Chaplin 1974, Tesfaye 1979, Calegari 1996).

Style and Content in Ethiopian Rock Art.

The significance of cattle in the rock art unquestionably indicates their cultural importance. Although camels, fat-tailed sheep and goats are also depicted, they are rarer and are not represented as massive herds. Genetically diverse breeds of long-horned and short-horned cattle are represented frequently, while humped zebu cattle (*Bos indicus*) have a much lower occurrence (Clark 1945). Certain attributes of the cattle are emphasised disproportionately, including horns[2], udders, and body or horn decoration. In highly schematic paintings these may be the only discerning feature.

The Art of the Horn in an African Context

Similarities have been suggested between the art of the Horn and early styles of Saharan rock art (Clark 1954). The early Saharan rock art has been dated comparatively to the 6th and 5th millennia BP (Muzzolini 1993, 1995). Although some art of the Bubaline Naturalis and Tazina traditions in the Atlas Mountains and the Fezzan has affinities with the art of the Horn (figure 3.9) (Le Quellec 1987, Muzzolini 1995), much of the Saharan work bears little comparison. Instead, distinct stylistic and thematic similarities exist with the art of North East Sahara around Uwenat and the Dahkla Oasis in the Egyptian Western Desert (figure 3.9) (Winkler 1938, Muzzolini 1995). Significantly, this region is associated with the earliest appearance of domesticated cattle in North Africa in the 9th or 10th millennium BP (Wendorf et al 1984).

Similarities have also been proposed between the rock art of the Horn and the art of the C-Group Nubians of the 4th/5th

[1] Punt has been equated with an area of North East Africa that may include part or all of Ethiopia. A discussion of the connection between Ethiopia and Egypt with respect to Punt can be found in chapter 6.

[2] Many of the horns appear to have been deliberately mis-shapen. This practice is common among modern day pastoral groups in Africa, such as the Dinka and the Nuer, where it is a symbolic expression of ownership and spiritual association (Evans-Pritchard 1940, Brown 1990).

millennia BP (Clark 1980). However this is little more than a superficial stylistic comparison (figure 3.10). Additionally cattle are just one of many decorative motifs used in C-Group art, along with a menagerie of wild and domesticated animals. These depictions are primarily on pottery designed for widespread visual effect, rather than on inaccessible rock surfaces with exclusive access as in the Horn. Such a distinct contextual and conceptual divergence suggests that the art had a fundamentally different cultural role in Nubia and in the Horn. If direct links are to be proposed between the rock art of the Horn and the C-Group of the Nile Valley, perhaps the Horn should be considered as the precursor. A more viable explanation would be that both areas are components of a spatially and temporally diffuse tradition.

Closer parallels are evident in the art of the Horn and that of East Africa and of Southern Arabia (Cervick 1979, Leclant 1973, Chaplin 1976, Wright 1961), and interestingly, with the art of the Iberian peninsula, as Graziozi (1964) has suggested (illustrated in figure 3.11).

Establishing a Chronological Framework for Ethiopian Rock Art

The art has never been dated directly. Clark (1980) interpreted it within the context of Holocene climatic deterioration and proposed a 4th millennium BP date, on the basis of affinities with C-Group art. Rock art dating by stylistic comparison alone is notoriously inaccurate (eg. Bednarik 1996), and as noted above, there is little to support Clark's proposition.

Only a handful of painted caves or rock shelters have been investigated archaeologically in Ethiopia and the Horn. Radiocarbon dates of c.1600 and 2100 years BP from an archaeological deposit at Karin Heegan rock shelter in North East Somalia are thought to relate to the paintings (Brandt and Brook 1984). As these dates are taken from a single level in the centre of the deposit, there is no indication of its chronological span. Furthermore there is no evidence to associate the deposit with the paintings. Paintings at Laga Oda rock shelter in Ethiopia have been tentatively dated to the mid-4th millennium BP, when fragments of possibly domesticated animals appear in the cultural deposit in the rock shelter (Clark and Williams 1978). These results are also unacceptable due to the small size of the excavation and inconsistencies in the recovery. Again, there is no reason to link the deposit with the paintings.

A more viable connection has been discovered at Porc Epic Cave in the Afar Rift Valley escarpment, where used pigments have been found in association with rock paintings (Breuil 1934, Clark and Williamson 1984). Ochre pencils, pigment-stained grindstones and heat-treated pigments from a Middle Stone Age (MSA) deposit offer unequivocal evidence for a creative or decorative tradition[3]. The association between these cultural remains and the faded paintings of stylised wild animals and human figures at Porc Epic must remain conjectural at present. A radiocarbon date of c.5700 years BP was obtained on charcoal from the top of the MSA deposit, which was sealed by speleothem growth (Brandt 1986). This date was rejected by the excavators as being too young, and possibly contaminated. The paintings were consequently attributed to an undated, possibly recent, Late Stone Age (LSA) deposit that lacked evidence of pigment use (Clark and Williams 1984). As the cultural sequence for Ethiopia is poorly defined and insecurely dated, a late date for 'MSA' material is not unfeasible. A number of further 'MSA' deposits containing pigments and stained grindstones are associated with rock art sites in the Horn (Carter in Clark and Williams 1984), while certain sites depicting only wild animals may have an early date (Negash 1990).

The possibility that an artistic tradition existed in Ethiopia in the 6th millennium BP should be considered carefully. The possibility that this artistic tradition could have been employed for depicting domesticated cattle and livestock needs substantial further investigation.

Implications for Pastoralism and Domestication

The rock art of Ethiopia and the Horn is part of a widespread tradition that unquestionably communicates the significance of domesticated cattle in North and East Africa and in Southern Arabia. The Horn can be proposed as the precursor for adjacent regions. This argument suggests that the rock art tradition, and by implication, possibly domesticated cattle, reached Arabia from Africa by way of Ethiopia.

Although some of the Arabian art has been tentatively dated to the 6th millennium BP on palaeoenvironmental grounds, Cervick (1979) has proposed a more plausible chronology in the late 5th and 4th millennia BP. East African rock art is unlikely to antedate the appearance of cattle in the 3rd millennium BP. This suggests a *terminus ante quem* of at least the 5th millennium BP for pastoral rock art in Ethiopia.

The implications are that cattle pastoralism in the Horn and Ethiopia is more ancient than previously proposed. It is unclear at present whether pastoral art was adapted from an indigenous tradition, or whether it spread with domesticated cattle from North Africa. However, the close affinity between

[3] Ochre pencils and pigment-stained pestles and grindstones have been found in LSA deposits at Magosia in Somalia, where they may date as early as 15 000 BP, suggesting the existence of an early artistic tradition in the Horn that predates the appearance of domesticated animals (Clark 1954, Chaplin 1974).

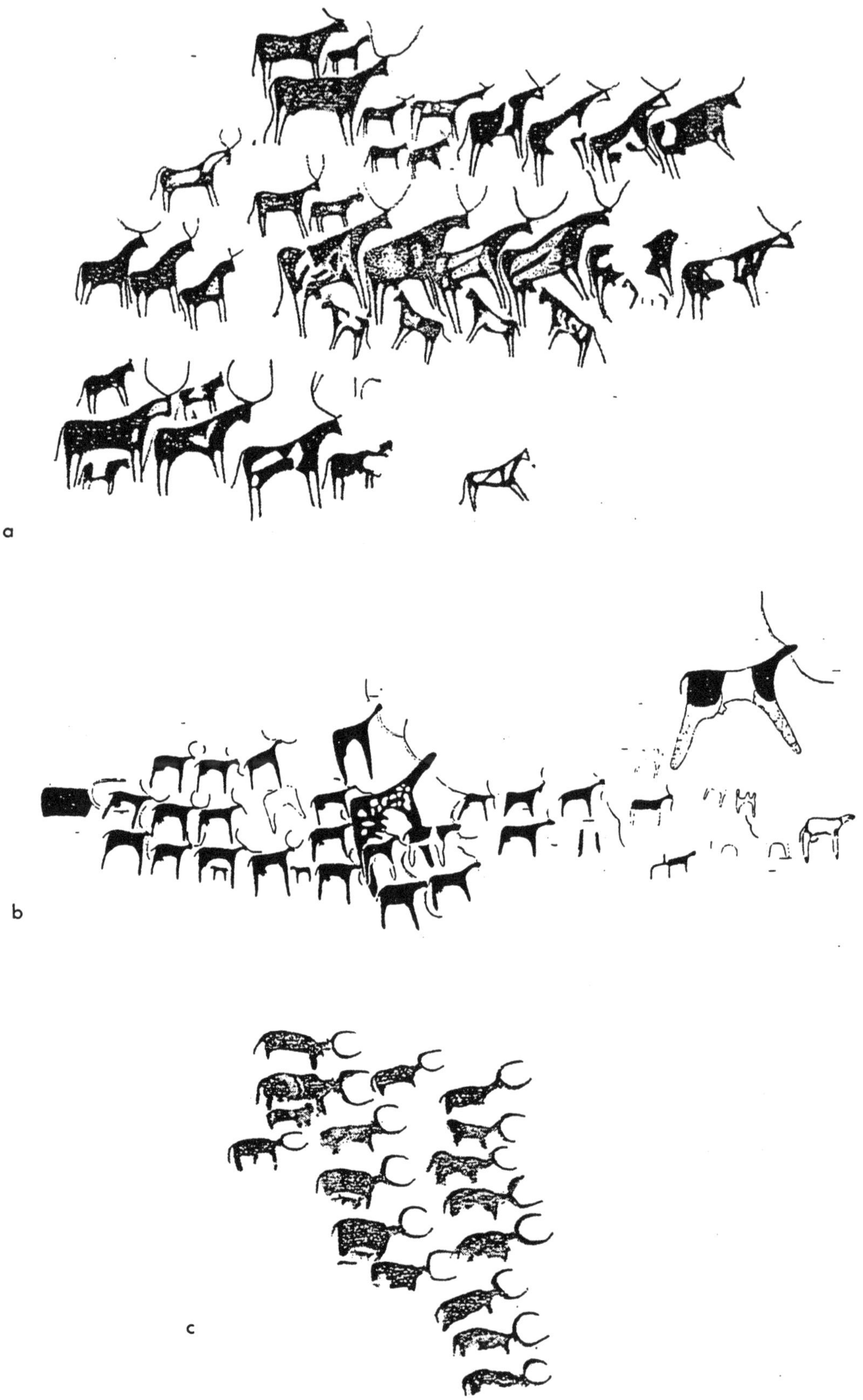

Figure 3.8. Examples of prehistoric rock art from the Horn showing herds of domesticated cattle

(a) Adi Kaieh, Eritrea (after Graziosi 1964)

(b) Karin Heegan rock shelter, Somalia (after Brandt and Carder 1987)

(c) Mai Aini, Eritrea (after Graziosi 1964).

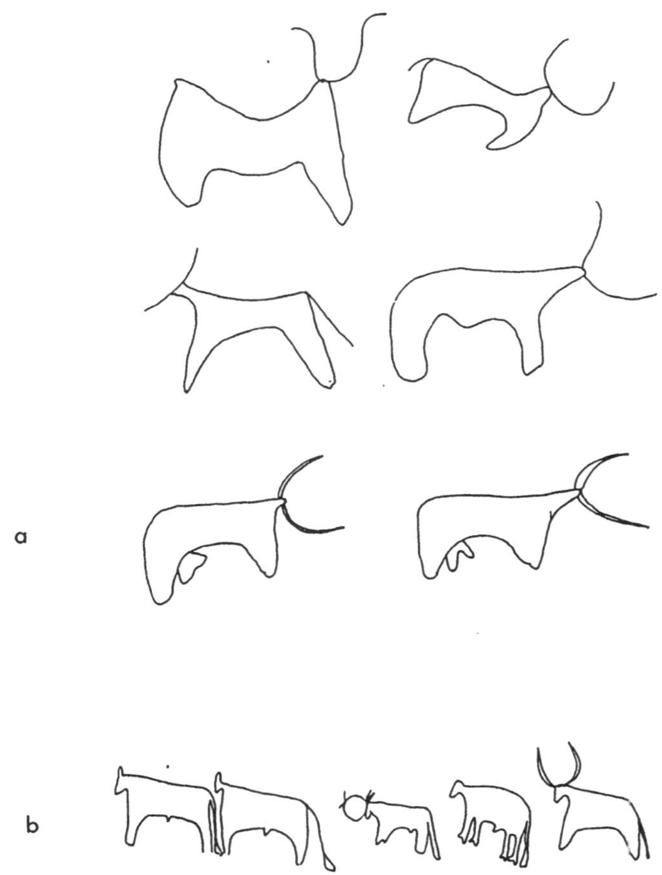

Figure 3.9. Rock art groups from:
(a) Ethiopia and the Horn
(b) Uwenat, Eastern Sahara
(c) Central and Nortehrn Sahara
(d) Butana, Eastern Sudan

Figure 3.10.
(a)-(c) Domesticated cattle on C-Group pottery (after Adams 1984, Bietak 1979)
(d) Long-horned cattle with vertical banding from the Lower Nile Valley
(e) Short-horned cattle from Sollum Ba'atti, Eritrea (after Graziosi 1964).

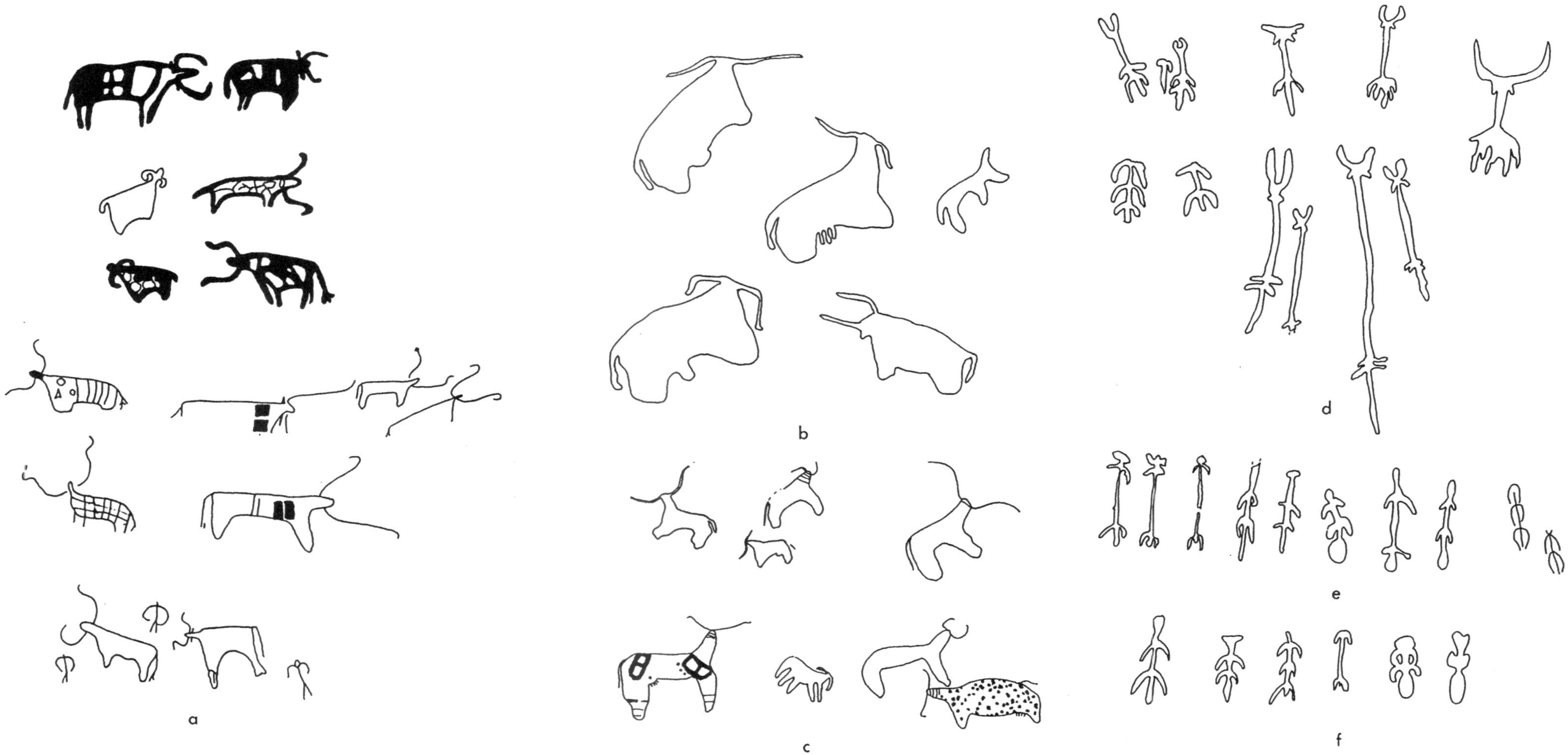

Figure 3.11. Rock art representation of domesticated cattle from:

(a) Arabia (after Khan et al 1988) (b) Mount Elgon rock shelter, Kenya (after Wright 1961) (c) Laga Oda rock shelter, Ethiopia (after Breuil 1934) (d) Lake Victoria area, East Africa (after Chaplin 1974) (e) Eritrea and Ethiopia (after Graziosi 1964) (f) Iberian Peninsula (after Graziosi 1964).

Horn art and the art of the Eastern Sahara leave little doubt that there was considerable contact between these regions in prehistory, whereas links between other regions of North Africa and the Horn are less apparent.

3.2.4. The Introduction of Domesticated Cattle

According to the archaeological evidence, cattle were brought to the Ethiopian Highlands before the mid-4th millennium BP. Iconographic and genetic evidence intimates an earlier date for this introduction. Evidence from the surrounding regions can be used to construct a clearer framework for when livestock were introduced, and from where.

Inter-Regional Archaeology

Ethiopia was intimately connected with surrounding regions throughout the Holocene. As Table 3.6 shows, cattle were present in the lowland regions immediately adjacent to Ethiopia by at least the 5th millennium BP, and in the Nile Valley by the end of the 7th millennium (see figure 3.2). Contact between cattle-herding groups in the lowlands and the Ethiopian Highlanders, possibly during the 6th and 5th millennia BP may have led to the adoption of domesticated stock.

Table 3.6. Relative Chronologies of Domesticated Cattle in Ethiopia and Surrounding Regions

Site and Location	Date (Years BP)	Reference
Gash Delta:		
Mahal Teglinos (Sudan-Ethiopian border)	mid 5th millennium BP	Geraads (1983)
Southern Arabia:		
Saar (Bahrain)	early 4th millennium BP	Nesbitt (1993)
Central Nile Valley:		
Shaheinab	c.5700	Hassan (1984)
Kadero	c.5700	Krzyzaniak (1982)
Zakiab	c.6000	Haaland (1979, 1981, 1984)
Um Direiwa	c.6000	"
White Nile:		
Rabak	c.6000	Haaland (1981)
Lake Turkana:		
GaJi4	c.4000 (4500?)	Barthelme (1977, 1984)
Ileret	c.4000	
Ethiopia:		
Lake Besaka, Afar Rift Valley	c.3500	Carter in Brandt (1982)
Laga Oda, Southern Highlands escarpment	c.3500	Clark and Williams (1978)

East Africa

From c.4500 years BP, pastoralist groups with domesticated cattle and caprines occupied the Lake Turkana Basin in Northern Kenya (Barthelme 1977, 1985). These groups were characterised by stone bowls and a range of ceramics, particularly 'Nderit' ware (Bower et al 1977, Barthelme 1984). Previously this area was inhabited by a fishing-foraging culture with a distinct ceramic tradition, reminiscent of the wavy line ware of North Africa (Robbins 1974, 1980, Phillipson 1977c, Sutton 1974, 1977). The arrival of the pastoralists in the Lake Turkana basin heralded the advent of food production in East Africa. The 'Savannah Pastoral Neolithic' (SPN) subsequently spread southwards into the Kenyan Highlands and Northern Tanzania during the late 4th millennium BP, following an apparent lull of several hundred years. By the 3rd millennium BP, Pastoral Neolithic groups occupied the East African Highlands and the lowlands between Lake Victoria in the west and Mount Kenya in the east (Ambrose 1984). The SPN diet was supplemented by fishing and by gathering of wild plants, but there is no evidence that domesticated plants were being exploited[4].

During the 3rd millennium BP a second pastoral group appeared in East Africa. This group, the Elmenteitan culture, had distinct ceramic and lithic technologies and burial practices in addition to characteristic settlement distribution patterns which occupied a separate, higher altitude ecological range from the SPN groups (Ambrose 1980). Ambrose (1984) argued that the Elmenteitan culture had a separate origin from the SPN, and that its economy was based on a caprid-dominated pastoralism rather than domesticated cattle.

A number of researchers have proposed that domesticated cattle spread to East Africa from Ethiopia, by way of Lake Turkana (eg. Clark 1980, Ambrose 1984, Brandt 1986). Although there is no certain archaeological support for this models, tentative cultural affinities can be proposed between the SPN groups of East Africa and the early pastoral groups in Ethiopia. The evidence for this can be summarised as follows:

- Stone bowls, similar to those from early SPN contexts, have been recovered from Lake Besaka in the Afar Rift Valley in Ethiopia (see figure 3.6) (Brandt 1982).
- Domesticated bovid teeth found at stone bowl sites, suggesting that the Lake Turkana Basin and the Rift Valley in Ethiopia were possibly part of a unified pastoral phenomenon.
- Similarities in rock art styles and content indicate cultural correspondence between early pastoral groups in East Africa and Ethiopia (figure 3.11).

Two additional links can be proposed:
Recent excavations in the Lake Turkana Basin have identified a complex mortuary tradition that is echoed in the 3rd and 2nd millennium civilisations of Northern Ethiopia (Costantini et al 1982, Fattovich 1988, Nelson 1995, 1996, Koch 1996). Several 'Pillar Sites' have been located around Lake Turkana, dating to c.4000 years BP. These sites, with their large, low mounds, platforms, cairns and pillars, served as cemeteries for early pastoral groups. The burials were marked by vertical stone pillars, brought from several kilometres away. Distinct parallels can be proposed between the Turkana sites and the impressive stelae fields that characterise PreAxumite and Axumite burial grounds in the Northern Ethiopian Highlands and in the Ethiopian-Sudanese border region from the mid 3rd millennium (Fattovich 1990a, Phillipson 1994).

Chapter 6 discusses in detail the evidence from Quiha rock shelter in the Ethiopian Highlands. Ceramics from this site show affinities to Pastoral Neolithic wares. Domesticated cattle remains are associated with the ceramic-bearing deposits at Quiha, suggesting possible contact between pastoral groups in the Ethiopian Highlands and the Lake Turkana region.

Southern Arabia
Domesticated bovids became widespread across the Southern Arabian peninsula during the late 5th/early 4th millennia BP. A 7th or 6th millennium BP date for their introduction has been proposed from the rock art evidence (Cervick 1979), much of which is located in areas which are now uninhabitable desert but would have been suitable for cattle pastoralism until the late 6th/early 5th millennia BP. The origin of domesticated cattle in Southern Arabia is unknown. An introduction from the north, from the Near East, is assumed but unsubstantiated.

Alternatively, the rock art evidence suggests that the Arabian peninsula had significant links with pastoral groups in Northern Africa which may have been mediated through the Horn. Significant evidence exists for contact between Ethiopia, the Red Sea coast and Southern Arabia, which dates back at least to the early Holocene (see chapter 6).This may have provided a route for the introduction of domesticated cattle from the Horn and Ethiopia to Southern Arabia, across the Strait of Bab-el Mandeb, possibly as early as the 6th millennium BP

Historical Linguistics in East Africa and Ethiopia
At present, the archaeological evidence cannot firmly identify the precise chronological and cultural relationship between Ethiopia and the surrounding regions. Historical linguistics have provided an invaluable framework through which to interpret the archaeological data.

Historical linguistic evidence demonstrates the arrival of two distinct food-producing populations in the East African Highlands before the 2nd millennium BP, one from the Ethiopian Highlands, one from the North West Nile Valley region (Bender 1982, Ehret 1974, 1984). Correlations exist with the dating and the geographic distributions of these language groups, and with the SPN and Elementeitan material from East Africa.

The assumption here is that there are broad parallels between linguistic and cultural groups (Hodder 1977, 1979, 1982).

[4] Evidence for intensive exploitation of local plants comes from Ele Bor rock shelter near the Kenyan-Ethiopian border, where numerous seeds and grindstones were recovered from a cultural deposit with domesticated livstock dating to the 5th millennum BP (Phillipson 1984). The plant remains all derive from wild species, including *Eragrostis* sp and *Sporobolus* sp, suggesting that although plant foods were an important component of the diet, they were not domesticated

Ehret claims that there was a major movement of Southern Cushitic-speaking people with domestic cattle and a knowledge of (wild) grain processing from Ethiopia into North Kenya in the 5th millennium BP. The pastoralists subsequently split and spread to Central Kenya and North Tanzania during the 3rd or early 4th millennium BP. Both the chronology and the past distribution of Southern Cushitic speakers (reconstructed from loan words) corroborate with the archaeological evidence for the SPN (Ambrose 1984). During the 3rd and 2nd millennia BP, a Southern Nilotic speaking population moved into North Kenya from the North West, possibly from the Uganda/Ethiopia/Kenya/Sudan border area, bringing domestic stock and cultivated cereals such as sorghum and finger millet. Place names, loan words and oral traditions of the Southern Nilotic speakers (Kalenjin) show that past linguistic distributions included all known Elementeitan sites (Sutton 1973, Ehret 1971, Blackburn 1974).

Linguistics and Southern Arabia

Ethio-Semitic, the language of the Northern Highlands of Ethiopia, diverges considerably from Southern Arabian (Hudson 1978), suggesting that the initial split of the language groups took place at an early date. Historical reconstructions based on lexical and glottochronology, or on educated estimates, have proposed a time-frame between the 5th and 4th millennia BP (Tylock 1975, Demoz 1978, Hudson 1978). The assertion is that a vocabulary for food production was assimilated into the proto-Ethio-Semitic language during its spread across the Ethiopian Highlands after the 5th millennium BP. Words relating to cattle pastoralism were already established in the Cushitic languages of the Highlands by this date (Bender 1976, Appleyard 1977, Ehret 1979).

By implication, domesticated cattle were not culturally significant among Semitic-speakers before the 5th millennium BP, but they were known in the Ethiopian Highlands, and may have been introduced to Southern Arabia from Ethiopia or the Horn.

Summary

The linguistic framework agrees well with the data from archaeology and rock art[5] and supports the following model:

- Domesticated cattle were introduced to Ethiopia from the arid lowland regions in the North West.
- This introduction may have occurred as early as the 6th millennium BP.
- Cattle became widespread in Ethiopia during the 5th millennium BP, and subsequently spread into Southern Arabia and the Lake Turkana basin.
- The Savannah Pastoral Neolithic group expanded into East Africa from the Ethiopian Rift and Lake Turkana basin during the late 4th millennium BP.

3.2.5. Camels in Ethiopia

The one-humped camel (*Camelus dromedarius*) is of great economic importance in arid Northern Africa, including lowland areas of Ethiopia. It is fundamental to human mobility, contact and trade in these regions, rather than human food. The camel has been an important component of desert trade in Ethiopia since pre-Islamic times (Wilding 1980)

The distribution of the wild progenitor is uncertain, and no wild camels are known from late Pleistocene or Holocene sites. An Arabian origin for the domesticate has been proposed following the recovery of dromedary remains dating to the late 9th millennium BP from Sihi on the West Arabian Coast (Grigson et al 1989). A dualistic model of introduction to Africa has been derived from the ethnographic evidence (Bulliet 1975), with one route across the Sinai to North Africa and the other across the Strait of Bab-el-Mandeb into the Horn (Wilson 1984). Camel-hair rope from Fayum in Lower Egypt indicates that camels were present in North Africa by the mid-5th millennium BP (Caton-Thompson 1934), and camel dung recovered from Qasr Ibrim in Egyptian Nubia indicates more widespread exploitation, possibly long-distance trade, by the mid-3rd millennium BP (Rowley-Conwy 1988). Camels are occasionally represented in the rock art of the Horn, but these documents remain undated (Brandt and Carter 1987).

The whole question of camel domestication in Northern Africa is confused in the light of two isolated discoveries:

At Gobedra rock shelter in the Northern Ethiopia Highlands, a single domesticated camel tooth has been dated to approximately the 5th millennium BP (Phillipson 1977b). The small size of the excavation, and the broad chronological span of the deposit from which the tooth was recovered make this age somewhat uncertain at present.

A camel molar and bone fragment were recovered from an early 5th millennium BP context at Ele Bor rock shelter, on

[5] Evidence from physical anthropology also agrees with the pattern of pastoral movement into East Africa. Two imigrant anthroplogical types can be detected in late prehistoric contexts, in addition to the indigenous population, with Sudanese or Ethiopian characteristics. The types have a mutually exclusive distribution, with the Sudanese anthropological type derived from Elementeitan deposits and the Ethiopian type occuring in SPN deposits (Rightmire 1975).

the Kenyan-Ethiopian border (Phillipson 1984). The associated faunal remains all belonged to indigenous wild animals, with the possible exception of a single domesticated caprine incisor, although these are notoriously difficult to identify accurately. While the camel remains were suggested to belong to a domesticated animal, this assertion cannot be substantiated. Instead, the predominance of wild faunal in the assemblage indicates that this was a hunting deposit.

Summary

The history of camel domestication is open to speculation. An African origin should be strongly considered, particularly if wild camels were indigenous to North Kenya and South Ethiopia in the mid-Holocene. The relationship between the fluorescence of long-distance trade and camel domestication should be considered. The role of lowland Ethiopia and the Horn may be of immense significance to this relationship, and must be investigated thoroughly.

3.3. ARGUMENTS FOR INDIGENOUS PLANT DOMESTICATION IN ETHIOPIA

3.3.1. Archaeological Evidence for Early Agriculture

Archaeobotanical evidence from Ethiopia shows that mixed farming was established in the Northern Highlands by the mid-3rd millennium BP. This evidence is summarised in Table 3.7. Remains of domesticated charred naked and hulled barley, horse beans and chick peas have been recovered from Lalibela cave to the east of Lake Tana in the Northern Highlands, dating to c.2500 BP (Dombrowski 1971). Tentative identifications of cattle and ovicaprids from later levels at this site suggest a developed agricultural tradition was practised by 2000 BP. Domesticated wheat (*Triticum aestivum*) and barley (*Hordeum vulgare*) were collected, along with cattle bones, from a deposit dating to 2335±220 years BP at Ona Nagast on Bieta Giyorgis Hill near Axum in the Northern Highlands (Bard and Fattovich 1996).

Table 3.7. Archaeological Evidence for Domesticated Plants in Ethiopia before the mid-2nd millennium BP

Crop Plant	Location	Date (Years BP)
barley (hulled and naked)	Lalibela cave	c.2500
	Ona Nagast I	c.2300
	Axum	c.1600
chick pea	Lalibela cave	c.2500
horse bean/pea	Lalibela cave	c.2500
	Axum	c.1600
tef	Lalibela cave	c.2000
	Ona Nagast III, Bieta Giyorgis	c.1500
lentil	Ona Nagast I, Bieta Giyorgis	c.2300
noog	Axum	c.1600
OTHER:		
flax	Axum	c.1600
cotton	Axum	c.1600
grape	Ona Nagast I	c.2300

By the early 2nd millennium BP, lentils (*Lens culinaris*) and grapes (*Vitis vinifera*) were also being grown at Ona Nagast, with the addition of tef (*Eragrostis tef*) by the mid-2nd millennium BP (Hansen *pers.comm.*). In addition, recent excavations at Axum have revealed that emmer and bread wheat, hulled barley, horse beans, peas, noog, flax and cotton were being grown by the mid-2nd millennium BP (Phillipson and Reynolds 1996, Boardman *pers.comm.*).

There is no archaeobotanical evidence prior to the mid-3rd millennium BP in Ethiopia, with the exception of a single seed fragment of *Amaranthus* sp[6] recovered from excavations at Gobedra rock shelter (Phillipson 1977b). The absence of

[6] Several different *Amaranthus* species are exploited in this region today in addition to the agricultural diet (Getahun 1974, Westphal 1975); *Amaranthus gracilis* is eaten as a leafy vegetable, the seeds of *A.hypochondriacus (A.hybridus)* are eaten, *A.sylvestris* is a medicinal plant, used to treat tapeworm. *A.caudatus* (Ethiopian kale) is important as a cooked vegetable when young, or its seeds are cooked as porridge or ground with tef to make injera. *Amaranthus* species are also used as famine food in the Ethiopian Highlands (Asfaw 1990). The appearance of grindstones in the associated levels at Gobedra suggests that *Amaranthus* may have been one of a number of local wild resources that were being processed, cooked and consumed at the shelter.

evidence of early domesticated plants in Ethiopia therefore cannot be considered as negative evidence that they did not exist. The potential for plant domestication in Ethiopia and the evidence to support an model of indigenous domestication are outlined below.

Table 3.8. The Ethiopian Centre of Diversity

Botanical Name	Common Name	Vavilovian Centre of Origin	Ethiopian Centre of Origin[7]
Grain crops			
Triticum durum Desf. subsp *abyssinicum* Vav.	Ethiopian hard wheat (variety of forms)	x	
T.turgidum L. gr. *abyssinicum* Stol.	cone wheat	x	
T.dicoccum Schrank	emmer wheat	x	
T.polonicum L. gr. *abyssinicum* Vav.	Polish wheat	x	
Hordeum vulgare	barley (exceptionally varied)	x	?
Sorghum bicolor	grain sorghum	x	?
Eragrostis tef (Zucc.) Trotter	tef	x	x
Eleusine coracana (L.) Gaertn.	finger millet	x	?
Pennisetum spicateum L.	pearl millet (semi-arid areas)	x	
Cicer arietinum L.	chick pea (one of centres)	x	?
Lens esculenta Moench.	lentil (one of centres)	x	?
Pisum sativum L.	pea (one of centres)	x	?
Vicia faba L.	horsebean	x	
Trigonella foenum-graecum L.	fenugreek	x	
Lathyrus sativus L.	vetch (one of centres)	x	?
Vigna sinensis Endl. var. *sinensis*	blackeye pea	x	
Dolichos lablab L.	hyacinth bean	x	
Lupinus termis Forssk.	Sicilian/Egyptian lupine (arid areas in N Ethiopia/Eritrea)	x	
Linum usitatissimum L.	flax/linseed (one of centres)	x	
Oil Plants			
Guizotia abyssinica Cass.	noog	x	x
Carthamus tinctorius L.	safflower	x	
Sesamum indicum L.	sesame (main centre)	x	?
Ricinus communis L.	castor (one of centres)	x	?
Lepidium sativum L.	cress (one of centres)	x	
Spices and Stimulants			
Coriandrum sativum L.	coriander (one of centres)	x	?
Nigella sativa L.	black cumin (one of centres)	x	
Carum copticum Benth.+Hook.	ommu (one of centres)	x	
Rhamnus prinoides l'Her.	gesho (buckthorn) - used like hops in brewing	x	x
Catha edulis Forsk.	chat	x	x
Coffea arabica L.	coffee	x	x
Vegetables			
Brassica carinata Al. Braun	Ethiopian cabbage	x	x
Allium sp	shallots	x	
Abelmoschus esculentus L.	okra (mostly wild?)	x	?
Other			
Hagenia abyssinica Willd.	kosso	x	x
Ensete edulis	false banana (ensete)	x	x

[7] Plants with wild progenitors or suspected wild progenitors in Ethiopia (Harlan 1969, Edwards 1991).

3.3.2. The Vavilovian Ideal and the Genetic Reality

Ethiopia was originally identified by Nikoli Vavilov as a global centre for plant domestication, and hence crop domestication, due to the immense diversity displayed there by all species of crop plants (Vavilov 1952). As discussed in chapter 1, Vavilov's analytical approach appears to have been oversimplified. Subsequent researchers observed that the area in which a plant was domesticated was not determined by the genetic variation of the cultivar, but by the distribution and diversity of its wild progenitor in primary habitats (Zohary 1970, Zhukovsky 1975). There is a glaring discrepancy between the range of Ethiopian domesticates identified by Vavilov and the known domesticates in Ethiopia today. This discrepancy is illustrated in Table 3.8. Details of the origins of the main Ethiopian food plants are given in *Appendix A*, Table A.2.1.

Many of the plants from the Ethiopian 'centre' have no wild progenitor in Ethiopia, even though they display immense variation both under ecologically homogenous conditions on the same plot of land *and* under ecologically diverse conditions. Ethiopia is now known to be a *secondary* centre of diversity for the majority of crop plants grown there. Important crops with wild progenitors unique to Ethiopia are restricted to tef, noog and ensete, with only tef and noog in the Northern Highlands. As these crops are still unique to Ethiopia and their earliest known domesticates come from Ethiopia, we can deduce that they *must* have been domesticated indigenously. The archaeological evidence indicates that domestication occurred before the end of the 3rd millennium BP.

Progenitors of a number of other crop plants, including sorghum, finger millet, cowpea, sesame, castor seed and okra, have a broader distribution in Northern Africa, and may have been domesticated anywhere within this region (Harlan 1971, Simmonds 1976, Blench 1991). Finally, as will be discussed later in this chapter, a minority view contends that certain crops such as chick pea, barley and lentil have, or had, a natural wild progenitor in Ethiopia and were domesticated in the Highlands at a hypothetically early date.

3.3.3. Interdisciplinary Evidence for Indigenous Domestication of Ethiopian Plant Resources

The following section concentrates on the food crops of the Northern Highlands, as described in chapter 2. In addition, two Ethiopian crop plants that are not part of the Highland complex will be mentioned below. These resources - ensete and cotton - are relevant to this discussion as they contribute to a broader perspective of the emergence of plant domesticates in Ethiopia and Africa.

Non-Highland Crop Plants

ENSETE

The false banana, *Ensete edulis*, is a major food crop in Ethiopia. It is grown almost exclusively in the South West, although there are reports that it was also cultivated in the Northern Highlands in historical times (Bruce 1790). *Ensete* plants closely resemble those of *Musa*, which include bananas and plantains. As figure 3.12 shows, the wild form, *E.ventricosum*, is widespread in sub-equatorial Africa, and may have grown further north during the mid-Holocene (Simoons 1965).

Possible representations of ensete as decorative motifs on Predynastic pottery suggest a distribution in the Egyptian Nile Valley (Laurent-Tackholm 1951). Despite the wide distribution of the progenitor, domesticated ensete is unknown elsewhere in Africa.

Musa and Ensete in Africa

The modern distribution of the crop implies that ensete was first domesticated in Ethiopia (Kells 1958). Recent interdisciplinary investigations of the domestication history of ensete reveal a remarkable similarity between the economic, magic and ritual uses of ensete and musa species across tropical Africa (Rossel 1996). *Musa* species are not indigenous to Africa, but were initially introduced from Indo-Malasia to the East African coast, from where they spread inland and across to West Africa (Rossel 1991). The chronology of this migration is unknown, but plantain phytoliths have recently been recovered from archaeological deposits in Cameroon, dating to the early 3rd millennium BP (Doutrelepont et al 1996).

The domestication histories of ensete and musa are closely linked, and Rossel (1996) argues that early cultivation and ritual association of wild ensete in sub-equatorial Africa provided a preadaptive mechanism for the subsequent assimilation of domesticated plantains. This hypothesis has immense significance for our understanding of the emergence of food production in Ethiopia, but the implications are contradictory, and more research is needed before these contradictions can be properly addressed. The current evidence raises the following questions: Was ensete domesticated in Ethiopia before the introduction of plantains (before the 3rd millennium BP)? Or was Ethiopia a backwater that was by-passed by the spread of plantains?

The possibility that ensete was domesticated in Ethiopia at an early date is tentatively supported by evidence from Egypt. Laurent-Tackholm (1951) argued that ensete was depicted on Neolithic Egyptian pottery of the Gezeran (Middle Nagada) period c.5500-5100 years BP. Simoons (1965) proposed that ensete originally grew more prolifically in the Ethiopian Highlands, particularly around Lake Tana, where it was recorded by James Bruce in the 18th century (Bruce 1790). Simoons further suggested that ensete was 'discovered' and exploited by Egyptians following the Blue Nile to its source. These plants may have been domesticated originally in Ethiopia as suggested by Vavilov (1951) and Harlan (1969), or the wild form may have been taken to Egypt and cultivated from the 5th millennium BP onwards.

COTTON

A similar dilemma is raised in the question of cotton phylogeny. The Old World species, *Gossypium herbaceum* and *G.arboreum*, now largely replaced by *G.hirsutum* from the New World (Simmonds 1976), were important before the 5th millennium BP. Unconfirmed reports of cotton seeds from Mehragrh, Pakistan, date to the 7th millennium BP (Costantini 1984), and by the early 5th millennium there is firm evidence of cotton textiles and string in the Indus Valley and Pakistan (Hutchinson 1976, Vishnu-Mittre 1977). Earliest use of cotton in Africa is attested by lint from 5th millennium BP A-Group sites in Nubia, where cotton seeds in an early stage of domestication have also been recovered (Chowdhury and Buth 1971).

The domestication history of cotton is not known, but a model of separate domestication in Asia and Africa has been suggested (Sauer 1993). Hutchinson (1947) however argued that the wild progenitors were indigenous to South Arabia and North East Africa. Wild or domesticated cotton was reaching the Indian sub-continent by the 5th millennium BP, and it may have been part of a 'package' of African plants that were being exported from Africa at this time (see Section 3.1). The discovery of impressions of finely woven cotton in late 6th millennium BP levels at Dhuweila in East Jordan indicate that cotton was being transported widely across the Old World by this time (Betts et al 1994).

Figure 3.12 Distribution of wild and domesticated Ensete in Africa (after Clark 1980).

Cotton in Ethiopia

Eight wild *Gossypium* species are known in Ethiopia, and Nicholson (1960) argued that *G.herbaceum* L.var.*acerifolium* was domesticated indigenously. Cotton has a long history of use in Ethiopia, dating to at least the PreAxumite period in the mid-3rd millennium BP (de Contenson 1981), and the recent discovery of cotton seeds at Axum will shed more light on the role of cotton in the Ethiopian economy (Boardman *pers.comm.*). At present there is no support for a model of indigenous domestication, and more work is needed to clarify this problem. The questions of when and how domesticated cotton appeared in Ethiopia have significant implications for our understanding how Ethiopia participated in prehistoric trade networks in the Old World.

Tef in Ethiopia

Background

Tef (*Eragrostis tef* (Zucc.)Trotter) is one of the major subsistence crops of the mixed seed and plough agriculture complex of the Ethiopian Highlands. This indigenous crop is pivotal in the debate on whether or not agriculture arose independently in Ethiopia. Research into the origins of tef domestication is of the utmost significance to our understanding of the timing and structure of subsistence change in Ethiopia.

Tef is grown only in Ethiopia, and although over 300 species of the *Eragrostis* genus are distributed throughout the world, tef is the only domesticate (Constanza et al 1979. Tef cultivation covers a greater area than any other arable crop in the Highlands. Between 1979/80 and 1983/4 an estimated 1.4 million hectares were under tef cultivation, compared with the second major grain staple, barley, which occupied 0.85 million hectares of arable land (Central Statistics Office 1984). The major growing regions, and areas of greatest diversity for tef today, are in the North West and Central Highlands, particularly in the *woina dega* zone at 1800-2400 m asl (Woldetatios 1973).

Agronomic Advantages

Tef is valued by Ethiopian farmers for a number of reasons:

- The plant has fewer disease and pest problems than any other crop in Ethiopia (Ebba 1969, Ketema 1991).
- Its tolerance to altitude range and variable soil quality makes it an important crop in the Rift Valley, as well as in Highland areas up to c.2800 m (Huffnagel 1961).
- Tef can germinate and become established under waterlogged conditions, which are common during the rainy season in the clayey and clay-loam vertisols that cover much of the Highland plateau.
- It can tolerate the arid conditions following the rains (Westphal 1975).
- Tef is a very nutritious cereal with high levels of iron and calcium (see Table 3.9), which may be responsible for the scarcity of anaemia in Highland Ethiopia (Constanza et al 1979). The grain also provides a good source of carbohydrate (350 kcal/100 g of grain) and protein (Rouk and Mengesha 1963, Ketema 1987).
- Stored seed remains viable for up to five years, and can be stored as food grain for nine years (Ebba 1975).

However, tef is a very labour-intensive crop. This is due in part to the small size of the seeds, which measure on average only 1-1.2 mm in length. In addition, the young plants are highly susceptible to competition from weeds of Amaranthaceae and Chenopodiaceae (Kinfe and Megenesa 1984), necessitating the preparation of a fine seed bed and repetitive hand weeding during crop growth.

Botanical and Genetic Evidence for Tef Phylogeny

Tef belongs to the *Eragrostis* family of grasses, of which 54 breeds are currently known in Ethiopia (unpublished Addis Ababa Herbarium Report on *Eragrostis* 1994). Over 35 varieties of the domesticate have so far been identified and over 2000 varieties have been collected (Ebba 1975). The progenitor is still unproven as different tef cultivars show similarities to several of the 14 endemic wild *Eragrostis* species (Constanza 1974). The closest biochemical affinity is between *E.tef* and *E.pilosa* (Bekele and Lester 1981), although there is also a possibility that *E.pilosa* is a weedy derivative of the cultigen. Like most *Eragrostis* sp in Ethiopia, *E.pilosa* favours slightly lower altitudes than tef (1400-1900 m), where it forms part of the natural scrub grassland. *E.pilosa* has a wide distribution in Northern Africa, and it is a common grass of the savannah zone, where it forms part of the Kreb complex of wild grasses[8] (Harlan 1989). The dozen or so edible Kreb grasses are a reliable forage food in savannah regions where they are harvested and exported on a commercial scale, and are considered a delicacy among the wealthier classes (see figure 7.1) (Barth 1857, Harlan 1989). Desiccated remains of similar wild grass-seed complexes recovered from sites such as Nabta Playa in Egypt indicate the importance of this resource in North Africa for at least the past 8000 years (Wasylikowa et al 1995). *E.pilosa* is also collected as a famine food in many parts of Africa, giving rise to the assumption that it may have been domesticated in times of resources stress in Ethiopia (Ebba 1975, Bekele 1978). Interestingly enough, neither *E.pilosa* nor tef are used as famine foods in Ethiopia (Webb et al 1992). No *Eragrostis*

[8] Including *Eragrostis cilianensis, E.cilisris, Brachiaria deflexa, Panicum laetum, Dactyloctenim aegyptium* and several species of *Digitaria*.

Table 3.9. Nutritional Content of some Major Seed Crops in Ethiopia (approximate % of grain content/weight)

Nutritional item	Tef	Wheat	Barley	Maize	Sorghum	Oats	Rye	Rice
Protein (%)	11.0	11.0	8.5	9.4	8.6	9.5	10.7	9.7
Fat (%)	2.6	1.9	1.5	4.4	3.8	4.8	1.7	1.8
Fibre (%)	3.5	1.9	4.5	2.2	1.9	10.3	1.9	8.8
Carbohydrate (%)	73.0	69.3	67.4	69.2	71.3	58.4	69.8	64.7
Nutritional Total	***90.1***	***84.1***	***81.9***	***87.2***	***85.6***	***83.0***	***84.1***	***85.0***
Iron (mg)	90.0	4.0	3.6	2.3	n/a	n/a	n/a	n/a
Calcium (mg)	110.0	36.0	33.0	7.0	n/a	n/a	n/a	n/a
Thiamine (mg)	0.47	0.41	0.46	0.45	n/a	n/a	n/a	n/a
Ribfloavine (mg)	0.11	0.10	0.12	0.11	n/a	n/a	n/a	n/a
Niacin (mg)	2.1	4.6	5.5	2.0	n/a	n/a	n/a	n/a

Sources:
U.S.Interdepartmental Committee on Nutrition for National Defence (1959) Nutritional Survey, Ethiopia, Section 4, Table 6.
Tareke B, Alemaya University of Agriculture, Dire Dawa, Ethiopia, published in Ketema S (1991).

species other than tef is exploited as human food in contemporary Ethiopia, although *E.pilosa* is one of several grasses gathered for livestock fodder in the Highlands.

Despite the wide distribution and exploitation of *E.pilosa* in Africa it has not been domesticated outside Ethiopia. The phylogeny of tef may be more complex than a simple evolution from *E.pilosa* and it may include a specific combination of *Eragrostis* spp (eg. *E.pilosa, E.cilianensis, E.aethiopica*) that is unique to the Ethiopian Highlands (Bekele *pers.comm.*). Harlan (1989) notes curiously that there are no hybrid swarms (i.e. populations resulting from spontaneous cross breeding between the domesticate and its progenitor) in areas of tef production, as would be expected if this were the original region of domestication, and we must conclude either that the present habitat of the plant is not the one in which it was initially exploited, or that the progenitor has been extinct for a considerable time (Tadesse 1975).

Debate on the Origins of Domesticated Tef

There is no evidence for tef in Ethiopia before c.2000 years BP, by which time mixed agriculture was already established in the Highlands. The question of whether tef was already domesticated before the introduction of foreign crops has been influenced by two principal factors:

- Domesticated tef is unknown from archaeological contexts outside Ethiopia, apart from isolated and unconfirmed reports (Porteres 1958, van Beek 1969). Domestication is therefore most likely to have occurred in Ethiopia.
- Why would tef ever have been domesticated if large grained, versatile alternatives were available (Simoons 1965, Shaw 1977, Clark 1980)?

The assertion is, then, that tef, or its wild progenitor, was a staple food resource in Highland Ethiopia *before* the cultivation of cereals such as wheat and barley. No viable alternatives have been proposed to challenge this model and, at present, theories that suggest that tef was domesticated following the introduction of wheat and barley remain unsubstantiated. The model is further strengthened by ethnobotanical and linguistic evidence, discussed below.

Ethnobotanical Evidence for Tef Ancestry

Tef has immense cultural importance in Ethiopia, and commands a high degree of status and national pride. It is preferred above all other cereals in Ethiopia, even in areas where it is not grown. Recent attempts to persuade farmers to grow more wheat and barley were unsuccessful, and there was a highly unpopular move by the military government to replace tef agriculture with maize.

The plant is deeply embedded in Ethiopian culture and economy, and all of its component parts are utilised. The grain is used to make *injera,* the staple food of all Ethiopians at all levels of society, and also to make alcoholic drinks such as beer (*t'ella*) or spirit (*katikala*); the fine hay is highly valued as cattle fodder, and is also used as a binder mixed with clay to make plaster for hut construction (Ebba 1969). Legends involving *injera* date back to the earliest written records c.2100 BP, suggesting that this use of the crop was ancient even then (Simmonds 1976). In the Northern

Highlands tef is woven into a pre-Christian legend of creation, resurrection and fertility that may have its roots deep in Ethiopian prehistory (Tekele Tsadik Mekuria, reported in Ebba 1975). This legend is strangely specific about the exact location of tef domestication, at a place called Temen Zemo, west of Axum in the Northern Highlands. The Northern Highlands have also been proposed as the birthplace of tef domestication, due to its unparalleled genetic diversity in this region (Ebba 1975), suggesting a possible merging of real and abstracted evidence.

Linguistic Evidence for Tef Ancestry

The earliest archaeobotanical evidence of apparently domesticated tef comes from Hajar Bin Humeid in the Yemen, dating to the late 3rd millennium BP (van Beek 1969). This has given rise to the hypothesis that tef was domesticated following the introduction of agricultural knowledge and technology from Southern Arabia in the mid 3rd millennium BP.

There is increasingly little support for this hypothesis. The alternative suggestion is that tef must have been exploited in Ethiopia by the end of the 3rd millennium BP, although the earliest evidence is marginally later, at the start of the 2nd millennium BP (Dombrowski 1971). Historical linguistic studies suggest an ancient origin for domesticated tef, predating Arabic influences. Ehret (1979) proposed that wild tef was exploited intensively from the Late Pleistocene, and was cultivated from the 7th millennium BP. Although the word tef has affinities with the Arabic ***tahf,*** (a name given to a similar looking wild plant that is a famine food in the Yemen) (Constanza 1974), or to the Amharic word ***teffa***, meaning lost, as the tef grain is so small that if dropped it cannot be found (Rouk and Mengesha 1963), these are both loans from a earlier Cushitic root such as ***taf,*** which is one of several similar words referring to grain food in North East Africa (Ehret 1979).

In addition, the ancient Greek word τιφη, meaning 'poor wheat', may be referring to tef from the late 3rd millennium BP (Bernal *pers.comm.*)

Summary and Implications for Domestication

Despite the numerous ambiguities, the evidence for tef phylogeny suggests the following framework:

- Tef was domesticated in the Northern Ethiopian Highlands, and now occupies a separate ecological niche from its wild ancestor(s).
- The cultural or biogeographical conditions necessary for tef domestication to occur were unique to Ethiopia.
- Tef was a staple resource in Ethiopia before the 3rd millennium BP, prior to the appearance of large-grained domesticated cereals.

A functional model can now be tentatively proposed to describe the process of tef domestication:

- Edible grasses, including several *Eragrostis* spp, were exploited as a forage resource in the Highlands from at least the early Holocene.
- 'Wild' tef evolved as a hybrid of a particular combination of *Eragrostis* spp, possibly due to unconscious anthropogenic selection, whose special qualities made it a favoured resource.
- A period of climatic hyperaridity during the 8th millennium BP caused wild grasses to migrate to slightly higher altitudes. Intensive exploitation of wild tef possibly gave rise to a primitive cultivar, but critical selection pressures were absent and the genetic evolution of the plant was suspended.
- Climatic change in the mid-Holocene caused wild *Eragrostis* grasses to retreat to lower altitudes. Certain plant species were important food resources by this stage. These may have been retained at the higher altitude by anthropogenic forces.
- The introduction of specialised agricultural technology and foreign domesticates provided a means for the more intensive exploitation of the tef cultivar within an agricultural context, and the domesticated morphology may have evolved at this stage. Despite the small size of tef, it was economically and culturally important from an early date and persisted as a major component of the agricultural economy.

Ethiopian Barley

Background

Barley is an economically important crop in Ethiopia where it is well suited to the fertile soils and temperate climate of the Highlands. It is grown extensively in most Highland areas of Ethiopia between 1800 and 2900 m asl, often on over 40% of the available arable land, and is the only crop that can be cultivated at higher altitudes up to 3600 m (Getahun 1990).

A number of diverse breeds of barley are known in the Ethiopian Highlands (Orlov 1929). Several have particularly high protein and lysine contents and many are resistant to almost all major diseases, especially rust and mildew (Engels 1991). The origins of these unusual and beneficial varieties have been the subject of considerable debate. The conventional view is that domesticated barley (*Hordeum vulgare*) was introduced to Ethiopia from Southern Arabia during the 3rd millennium BP (Schiemann 1951). Alternative theories maintain that barley was introduced from Egypt in the 5th millennium BP (Doggett 1965, 1991, Harlan 1969, Purseglove 1969, Engels 1991), or that it was domesticated indigenously from a wild progenitor (Bekele 1983, Negassa 1985). Historical linguistic models support an early date for barley cultivation (Ehret 1979). Ehret demonstrated that the root-word for barley (*sek-/geb-*) was loaned into Ethio-Semitic from the earlier Cushitic Agau languages, which implies that a knowledge of barley cultivation was already established in the Ethiopian Highlands before the introduction of Semitic languages (see Chapters 1 and 2). Bekele (1983) and Negassa (1985) go further and suggest that domesticated barley was introduced to South West Asia from Ethiopia during the early Holocene.

The Question of Barley Phylogeny

Barley (*Hordeum vulgare* subsp *vulgare*) is one of the earliest domesticated crops in the world, known from at least the 10th millennium BP in the Near East (Renfrew 1991, Zohary and Hopf 1993). Domesticated forms are generally 6-row hulled or naked (i.e. with the palea and lemma freed from the grain) or 2-row varieties. The phylogeny (genetic origin) of domesticated barleys has been contested for many years (de Candolle 1886, Vavilov 1926, Schiemann 1951, Harlan 1957, 1969, 1971, Harlan and Zohary 1966, Frost and Holm 1973, Mengesha 1975, Bekele 1983), but the predominant consensus is for a monophyletic origin from wild, weedy *Hordeum spontaneum* (*H.vulgare* L. subsp *spontaneum* (C.Koch)Thell) in the Near East (Harlan 1968, 1971, Zohary 1996, von Bothmer et al 1995).

Wild barley is known from many archaeological sites in the Near East dating back to 19 000 BP. Remains of a 2-row barley species morphologically identical to *H.spontaneum* have been preserved in an Epi-Palaeolithic context in the submerged site of Ohalo II on the shore of the Sea of Galilee (Kislev 1989). Carbonised remains of *spontaneum* occur frequently at Terminal Palaeolithic and early Neolithic sites across the Near East and Turkey between the 12th and 9th millennia BP (Helbaek 1965, Kirkbride 1966, Renfrew 1991), and both wild and primitive domesticated 2-row barleys appear at sites in the Near East from the 10th millennium BP (Zohary and Hopf 1993). Domesticated barley was a principal grain crop in the Near East during the Neolithic, and new varieties soon developed from the primitive forms, giving rise to naked and 6-row barleys from c.8000 years BP which subsequently spread throughout Eurasia between the 8th and 5th millennia (Murray 1970, Helbaek 1966, Renfrew 1991, Zohary and Hopf 1993).

Prehistoric Distribution Patterns

A model for a monophyletic origin provides a clear evolutionary profile for barley in the Near East (Staudt 1961, Bakhteyev 1971, Harlan 1971, Shao et al 1982, Holm and Frost 1983). However it does not preclude the possibility that domesticated barleys in other regions may have derived from different progenitors. There is little support at present for a polyphyletic origin, and the discovery of a wild 6-row 'progenitor', *Hordeum agriocrithon* Aberg in Tibet, India and Israel was later revealed to be a spontaneous cross pollination of wild 2-row and domesticated 6-row barley (Aberg 1948, Witcombe 1978, Kamm 1954, Murphy et al 1982). Wild barleys have a global distribution. For example a wild 2-row barley is known from China with a wide distribution in the 2800-4050 m asl altitude range (Shao 1982), and a wild 2-row barley has also been identified on the Southern slopes of the Himalayas in India, Nepal and West Afghanistan (Sakamoto *pers.comm.* in Shao et al 1990). Shao et al (1990) believe that South West Asian cultivars derived from these cold-resistant barleys. Several wild barley species grow along the North African littoral today, and isolated stands of *H.spontaneum* have also been reported from Libya and Morocco (Maire 1955, Tackholm 1956, von Bothmer et al 1995).

The possibility that *Hordeum spontanem* was domesticated independently beyond the Near East has been rejected by many scholars because of the restricted distribution of the progenitor outside this region (Harlan and Zohary 1966, Zohary 1971, von Bothmer et al 1995). This distribution is further restricted because, unlike *H.vulgare*, wild barley is sensitive to extreme cold and does not usually grow above 1500 m, preferring semi-arid conditions in steppe and desert regions with winter rainfall (Zohary and Hopf 1993). However climatic fluctuations over the past 12 000 years have regulated profound changes in the vegetation and environment, particularly in more marginal zones and in the tropics. The modern distribution of wild grasses such as *H.spontaneum* is not a valid record of their distribution in prehistory, which would have shifted with the changing climate. During the Holocene pluvial phases (c.10 000-8500 and c.6500-4500 years BP), Mediterranean conditions shifted southwards as far as the Tropic of Cancer, creating ecological conditions conducive to the proliferation of Mediterranean-type vegetation (Mohammed Ali 1982). Wright (1976) and Bernal (*in prep.*) have suggested that parts of the Maghreb and the Lower Nile Valley were a refugium for wild wheat and barley during the late Pleistocene and Early Holocene.

Plant geneticists working in different parts of the world also argue that the modern distribution of wild barley is not an infallible indicator of the place of origin (Frost and Holm

1973, Shao 1982, Bekele 1983). Progenitors can become extinct if their natural habitat is disturbed through extensive agriculture, or if the wild genome is absorbed by domesticated varieties. Frost et al (1977) argue that introgressive hybridisation is the rule rather than the exception in crop plant evolution, and in heavily farmed areas such as Ethiopia, where few primary habitats remain, the probability of the wild genome surviving intact is exceptionally low. This phenomenon is apparently responsible for the absence of a wild progenitor for domesticated maize (*Zea mays*) in South and Central America (Smith and Lester 1980, Wilkes 1984). Similarly, no wild progenitor has yet been identified for chick pea (*Cicer arietinum* L.) (Simmonds 1976).

Ethnographic and Ethnobotanical Evidence for Barley Domestication in Ethiopia

Many varieties of 2- and 6-row barley are grown in Ethiopia. 2-row barleys are most popular in the North and West; 6-row in the Eastern Highlands and Southern Ethiopia. The greatest diversity, reflecting the longest ancestry, occurs in the Northern, Central and Eastern Highlands between 2500 and 2600 m asl (Bekele 1983, Engels 1990).

Barley plays an important role in the structure of Ethiopian economy and is also an integral part of the cultural fabric. Xolocotzi (1987) proposed a positive relationship between the cultural significance of a particular plant and the relative antiquity of its exploitation in that area. The cultural significance can be qualified in terms of the extent to which the plant is incorporated into drinks, songs, tradition and legend, ritual or religious structure and ceremony. Ethnobotanical studies in Jibat and Mecha provinces of Central Ethiopia show that barley satisfies all these criteria - it is used ubiquitously for making 'dirty water' beer or *t'ella*, it is the subject of songs, the stuff of legends and is integrated into many seasonal and perennial rituals such as age-set, harvest and marriage ceremonies (Asfaw 1990). The immense cultural significance of barley in certain parts of Ethiopia overshadows its present-day economic role, and may reflect a time when barley was more important for subsistence in those regions. For example, the Oromo, an agro-pastoral group originating from the Bale Mountains on the Southern plateau, grow a few crops, including barley. Oromo legend explicitly states that barley was the original and most ancient crop grown in this area. The Oromo name for barley (*garbu*) derives from the Oromo language, in contrast to the names of other domesticated plants which are loaned from Semitic and Cushitic languages (Asfaw 1990, Ehret 1979). Barley and cattle are believed to have been created by God for sacrificial purposes, and barley is revered as the most sacred food of the Oromo (Haberland 1963). This belief persists even in areas away from the original homeland where little barley is now grown. Interestingly, many Ethiopians and plant geneticists believe the Bale mountains to be an original area for wild barley cultivation (Asfaw *pers.comm.*, Tadesse *pers.comm.*, Edwards *pers.comm.*)

Plant Genetic Evidence for an Ethiopian Progenitor

Analytical research on the *Hordeum* genus shows that, unlike most plants, barleys exhibit great variation in a group of proteins known as flavonoids that enhance adaptive flexibility and resistance to disease (Frost et al 1975). As flavonoids are coded for by a specific genetic sequence, flavonoid polymorphism can be used as an indicator of genetic variation. Frost et al (1975) used thin-layer chromatography to analyse the flavonoid protein patterns from a large sample of wild and domesticated barleys collected from around the world. Their experiments showed that domesticated Ethiopian barleys had a very distinct flavonoid sequence, the 'C pattern', that was almost unique to Ethiopia. The C pattern was also present to a limited extent in Egyptian *vulgares* but was absent from Near Eastern *vulgares* and from wild *spontaneums*. The deduction made from this data was that domesticated barley was introduced into the Highlands of Ethiopia at an early date, then exposed to a range of environmental conditions and further isolated genetically from the parent gene pool. This resulted in the formation of unique morphotype groups such as *deficiens* and *irregulare* barleys, and the immense array of land races known from Ethiopia today (Frost et al 1975).

The genetic evidence however does not support this model, but favours a separate ancestry for the following reasons:

- There is no evidence that Ethiopian crops have ever been entirely genetically isolated (see chapter 6).
- A similar level of genetic variation and diversification would be expected in other environmentally varied landscapes where barley germplasm has been removed from the parent genepool. However the flavonoid patterning in Transcaucasian and South American barleys, for example, shows little divergence from the Near Eastern pattern (Frost et al 1975).
- The rate and nature of mutation in the introduced barley would not be geographically consistent but would vary with the differential topography and selection procedures to which it was exposed, leading to high genetic diversity. Bekele (1983) showed that the distinctive Ethiopian C pattern was common to all 2-row and 6-row *deficiens* and *irregulare* barleys and was homogeneously dispersed across the Northern, Central and Eastern Highlands, with a different pattern dominating only in the South West.

- A high degree of sterility (through seed infertility) has been noted between Ethiopian barleys and barleys from Eastern Asia and Eurasia (Vavilov 1960, Harlan 1968). Independent genetic origins may be suggested by this display of genetic incompatibility. However this important trait has not been investigated further, and the results obtained should be treated with caution (Asfaw *pers.comm.*)

Summary and Discussion

The data suggest that Ethiopian and Near Eastern barleys derived from genetically distinct progenitors, or, more probably, through a divergent evolutionary pathway from a common ancestor. The Ethiopian progenitor may be an unidentified, and possibly extinct, wild subspecies of *Hordeum spontaneum*. Ethiopian barleys are fully interfertile with *H.spontaneum*, suggesting close genetic compatibility. As discussed above, the modern distribution of *H.spontaneum* is not the same as its prehistoric distribution. The ecological requirements of wild barley would have existed within Northern Africa and the Ethiopian Highlands during the Holocene.

This argument can support the tentative hypothesis, based principally on linguistic reconstructions, that wild Ethiopian barley was taken to Eurasia during the early Holocene (Ehret 1979). It became domesticated in the Near East and was then reintroduced to Ethiopia and North Africa during the 6th millennium BP (Bernal *in prep.*). A similar argument has been proposed to explain the pattern of domestication of tropical African cereals (see Chapter 6. Haaland *in press*).

Finger Millet

Background

Finger millet (*Eleusine coracana* (L.) Gaertn), or African millet, has a wide distribution across East Africa to Nigeria and south towards South West Africa and Natal (see figure 1.3) (Harlan 1969, 1992). It is a staple food crop in areas of North and Central Africa, particularly in Uganda. Finger millet is generally cultivated at higher altitudes, and grows up to 2400 m in the Ethiopian Highlands. The grain can be stored for up to 10 years without deterioration or pest damage, making this a popular fallback crop (Stemler et al 1977).

Finger Millet Phylogeny

Domesticated *Eleusine coracana* was originally thought to have derived from the pantropical weedy relative *E.indica* in India or Africa (Vishnu-Mittre 1968, Mehra 1963). More recently the progenitor of *E.coracana* has been identified as a subspecies of wild *indica*, *E.indica* subsp *africana*, which has a purely African distribution in Eastern Highland zones (Purseglove 1972, Hilu and de Wet 1976).

Debate on the Origins of Domestication

Following Vavilov, several researchers have suggested that finger millet was originally domesticated in the Ethiopian Highlands (Harlan 1969, Doggett 1970), possibly as part of a broader East African phenomenon (Harlan 1971, Hilu and de Wet 1976). The Cushitic word for finger millet, *dagussa*, was loaned into Ethio-Semitic and Arabic, possibly as early as the 5th millennium BP (Tylock 1975, Ehret 1979). Dombrowski (1971) noted that *E.africana* is used in the Ethiopian Highlands today for making baskets, indicating a very ancient knowledge and exploitation of the wild plant. Hilu et al (1976) cite biosystematic and ethnobotanical evidence to propose that finger millet was domesticated as early as 5000 years BP.

African finger millet was being exploited by the 5th millennium BP in India, where it is the earliest African domesticate to be recovered from Asian contexts (Weber 1991 and Table 3.4 above). Wild *Eleusine* has not been found in early Indian archaeological contexts, suggesting that domesticated finger millet was being exported from Africa. Ethiopia has been proposed as the centre from which the crop was originally exported (Doggett 1991). At present however there is no evidence that domesticated finger millet was grown in Ethiopia or Africa until the 2nd millennium BP.

Summary

The wild progenitor *E.africana* was present in the Ethiopian Highlands during the Holocene, and was possibly exploited from at least the 6th millennium BP as part of a broad spectrum resource base. Because of its versatility and durability, finger millet may have been more of a scarcity forage crop in Ethiopia during the Holocene rather than a staple. A number of researchers (eg. Mehra 1991, Harlan 1969, Doggett 1988, 1991) argue that *Eleusine* was domesticated in Ethiopia during the 5th or 6th millennium BP and subsequently transported to India and Arabia.

Sorghum

Background

Cultivated sorghum is one of the most important food crops in the world. Sorghum and millet together provide the staple protein and energy source in arid and semi-arid tropical regions. Wild sorghum is still an important forage crop in the African savannah (Harlan 1989). In Ethiopia, wild and domesticated sorghum grow up to 2300 m asl. Sorghum is not a major Highland crop, and is unlikely to have been domesticated in the Highland zone. However it is a traditional food at lower altitudes in the hotter *kwolla* regions, especially c.1500-1700 m asl, where it has been incorporated into certain cultural traditions, such as songs and folklore (Doggett 1991). South West Ethiopia has been implicated as a possible region of early domestication due to the diversity of forms displayed (Doggett 1991).

Sorghum Ancestry

Thirty-one species and over 150 varieties of domesticated sorghum are known. Fifteen different races have been identified, each with a broad distribution in Asia and Africa (Snowden 1936, Harlan and de Wet 1972). Domesticated sorghums are all thought to derive from wild *Sorghum bicolor* subsp *arundinaceum* (race *verticilliflorum)*. The wild progenitor *Sorghum bicolor* was confined to Africa until relatively recently (Mann et al 1983). As figure 3.15 shows, it grows particularly along the forest margins of West Africa, in Northern Uganda, Southern Sudan and Ethiopia. The main forms of domesticated sorghum in Northern Africa are race durra and type bicolor in Sudan and Ethiopia, and race guinea in West Africa. Subsequent caudatum and kaffir races developed in the Sudan, Uganda and South Africa (Doggett 1970, Harlan and Stemler 1976). Diversity among both wild and domesticated sorghums is highest in North East Africa, particularly the Sudan and Ethiopia, suggesting that domesticated sorghum has greatest antiquity in this region (Doggett 1991).

Durra and bicolor forms predominate in Ethiopia. Durra is the main form, and the best durra sorghums are found in Ethiopia. Doggett (1988) has suggested that Ethiopian durra developed by introgressive hybridisation with a local wild type, *aethiopicum*, in the drier areas of Ethiopia following an early introduction of the domesticate (Harlan 1971).

Origins of Domesticated Sorghum

There is no evidence to verify Vavilov's contention that Ethiopia was a centre for sorghum domestication (Vavilov 1952). Current opinion supports a non-centric origin for the crop plant in the broad sweep of the savannah-Sahel fringe in Northern Africa (Harlan 1971, 1989, Mann et al 1983). At present, the earliest evidence for domesticated sorghum, in the primitive form *S.bicolor*, comes not from Africa but from South East Arabia and West India dating to the early 5th millennium BP, although this evidence has yet to be rigorously tested (see Table 3.4) (Cleuziou and Costantini 1980, Allchin and Allchin 1982, Biagi et al 1985). Unconfirmed reports additionally suggest *S.bicolor* was present in Arabia even earlier, in the 6th millennium BP (Potts 1994). The significance of these findings is discussed further in chapter 6. Further unconfirmed reports also indicate that domesticated sorghum may have been present in the North Ethiopian-Sudan border region by the early 4th millennium BP (Costantini et al 1983).

Wild sorghum was therefore a major food resource in Africa by at least the 5th millennium BP (Edwards 1995). Research in the Nile Valley and North Africa has so far failed to identify domesticated sorghum before the late 3rd millennium BP (Rowley-Conwy 1991), leading to speculation that wild, cultivated sorghum may have been exported from Northern Africa to India, where it was subsequently domesticated. The processes by which bicolor sorghum reached the Indian sub-continent and was introduced into Indian Neolithic culture are not known, but impressions of African sorghums have been found in pot sherds from the Oman dating from the 5th or 6th millennium BP, indicating that an overland route may have been involved (Cleuziou and Costantini 1980, Potts 1993, 1994, Biagi et al 1985). Haaland (*in press*) has argued that sorghum was originally exported from the Central Nile Valley region, although Doggett (1988, 1991) proposed that sorghum was taken to Southern Arabia from Ethiopia, and thence to India.

This model must be reviewed in the light of recent genetic research on ancient and modern sorghums from Qasr Ibrim in Upper Egypt. While much work remains to be done, the recent results indicate that *S.bicolor* was domesticated indigenously in the Upper Nile Valley during the late 3rd millennium BP, prior to the introduction of durra sorghum, possibly from India, in the early 2nd millennium BP (Rowley-Conwy *pers.comm.*).

Additional Crop Plants Possibly Domesticated in Ethiopia

Several additional important crop plants may have been domesticated in Ethiopia. The range of plants suggested is quite astonishing. Some are endemic to Ethiopia and must have been domesticated there. In the majority of examples however, the progenitors have a wider distribution, and a number of crops are included which are generally believed to have been domesticated in the Near East. While there is currently no evidence to the contrary, the possibility of an alternative origin in Ethiopia or elsewhere should not be ignored.

This account is concentrating especially on the Northern Highlands, and plants predominately from the southern or lowland regions, such as coffee (*Coffee arabica*) and c'hat (*Catha edulis*) are not included here. The full range of Ethiopian crop plants and their origins is given in *Appendix A*, Table A.2.1.

Pulses and Legumes

COWPEA (*Vigna unguiculata* (L.) Walp)

Cowpeas are an important source of protein throughout the tropics and subtropics, where they are a very ancient crop. The cowpea was domesticated from the wild subspecies *V.dekindtinana* which grows only in the African savannah and in tropical Ethiopia (Simmonds 1976). The wild type in Ethiopia displays extreme diversity, whereas in the African savannah it is relatively homogeneous. A prevailing view is that the cowpea was domesticated in Ethiopia and subsequently spread throughout the African savannah (Westphal 1974, Faris 1965, Purseglove 1976), to be introduced to Western Asia and India by the 5th millennium BP (Purseglove 1976, Doggett 1970).

CHICKPEA (*Cicer arietinum*)

Chickpea is one of the most important pulse crops in Ethiopia, and in other tropical and subtropical regions of the world. It provides an excellent source of protein and a perfect complement to the cereal and oil-seed diet. The wild progenitor is unknown but chickpeas are believed to have been originally domesticated in the Eastern Mediterranean where the genus displays a high diversity. A related wild form, *Cicer cuneatum*, is known in Northern Ethiopia, and Ethiopia is a also centre for diversity of the domesticated form (Mekbib et al 1991). The earliest domesticates known are from Turkey dating to the 10th millennium BP (Zohary and Hopf 1993) and from the 5th millennium BP in India (Simmonds 1976). Chickpeas have been found dating to c.2500 years BP in the Ethiopian Highlands (Dombrowski 1971) in association with barley, vetch and other legumes

A genetically related wild species (*C.cuneatum*) is known from the Highlands. The crop displays great diversity in the primary agricultural zone between 1400-2900 m asl (Westphal 1974), leading Vavilov (1951) to designate Ethiopia as a centre of diversity for chickpea. Chickpea is an ancient food crop in the Highlands, but whether it was domesticated indigenously is debatable. Ethiopian varieties are infertile when crossed with Near Eastern varieties. This genetic incompatibility may reflect distinct ecological and anthropogenic selection pressures, or it may indicate a separate evolutionary pathway.

LENTIL (*Lens culinaris*)

Lentil is one of the oldest leguminous crops in the Old World, and it is also an important pulse crop in the Ethiopian Highlands. The domesticate has a wide modern distribution in the Old World, where it is generally grown together with wheat and barley. Domesticated *L.culinaris* is derived from wild *L.orientalis* which has a Near Eastern distribution (Zohary 1970). The earliest evidence of lentil cultivation coincides spatially with the distribution of the wild progenitor in South West Asia (Zohary and Hopf 1993).

Alternatively there is some evidence that lentil may have been domesticated in Ethiopia. It is generally believed to have been introduced at the same time as other foreign domesticates from the Near East during the 3rd millennium BP, although no remains have been found in archaeological contexts before the late 3rd millennium BP in Ethiopia (Purseglove 1968, Hansen *pers.comm.*). Lentil also displays great diversity, and is reported to be genetically incompatible with Near Eastern varieties (Tadesse *pers.comm.*). In addition there are recent unconfirmed reports of wild stands at Alennth about 45 km from Addis Ababa (Tadesse *pers.comm.*) which further supports an independent origin in the Highlands.

Oil Seeds

NOOG (*Guizotia abysinnica*)

Noog is the main oil seed of Ethiopia. It is endemic to the Highlands and proliferates in the *woina dega* where it was domesticated from an unknown progenitor (Seegeler 1983). Although there is no doubt that noog was domesticated in Ethiopia, there is no evidence as to when it was first grown.

ETHIOPIAN MUSTARD (*Brassica carinata*)

Ethiopian mustard is an important oil crop, as well as a leaf vegetable. It was domesticated in the Highlands, where it now grows extensively. The wild progenitor is unknown, but hybridisation between *B.nigra* and *B.oleracea* is suspected (Engels and Hawkes 1991).

CASTOR BEAN (*Ricinus communis*)

The wild plant is widespread in Eastern and Northern Africa, the Yemen and the Middle East. In Ethiopia, it grows from sea level to the Highlands and is exploited for cooking, lighting and medicine (Engels and Hawkes 1991). It was cultivated from c.6000 years BP in ancient Egypt where plant remains have been found in tombs (Purseglove 1976). The great diversity of the plant in Ethiopia has led to speculation over whether it was domesticated in the Highlands (Zeven and Zhukovsky 1975).

SESAME (*Sesame indicum*)

Sesame is one of the oldest oil seeds known and used in Africa. The plant was grown by 3300 years BP in Egypt, but was possibly introduced at a much earlier date to both Egypt and India (Greenway 1945, Darby et al 1977). The wild species are confined to Africa, with one exception, although the origins of the domesticate are disputed (Bedigan and Harlan 1986). Ethiopia has been suggested as the original region of sesame domestication due to the genetic range of the cultivar (Vavilov 1951, Purseglove 1968, Harlan 1969).

Vegetables

OKRA (*Abelmoschus esculentus*)

Okra (lady's fingers) has a doubtful origin. It is generally thought to be an African domesticate, but it does not appear to have been grown in Egypt before the arrival of the Arabs in the 3rd millennium BP (Darby et al 1977). Okra may have been domesticated in Ethiopia, at a relatively late date (Greenway 1945, Harlan 1969), but again there is no evidence to substantiate this theory.

Summary

The available genetic, archaeological, linguistic and ethnobotanical data provide a convincing argument that the ancestors of important food plants were cultivated intensively in Ethiopia by the 5th millennium BP. In addition, several domestic crops have a probable Ethiopian origin. The results of the above discussion are summarised in Table 3.10. These plants are not only the basis of Ethiopian agriculture, but most are also important tropical crops. Although there is no archaeobotanical evidence that these resources were being exploited before the 3rd millennium BP, this may reflect weaknesses in the archaeological record from Ethiopia. Section 3.1 noted that several domesticated African crop plants have been recovered from 4th and 5th millennium BP contexts in India and Arabia, several millennia before the domesticated form is attested in Africa (see Table 3.4). The majority of these resources are indigenous to Ethiopia, and may suggests that an Ethiopian plant 'package' was being exported to Asia by the 5th millennium BP.

Table 3.10. Indigenous Ethiopian Crops in the Highlands

Indigenous crops	Crops of uncertain origin in Ethiopia
tef	barley
	sorghum
	finger millet
	cow pea
	chick pea
	lentil
noog	castor
Ethiopian mustard	sesame
	okra

Ethiopian crops may have been some of the many African resources transported throughout Eurasia during the past 6000 years, and the evidence suggests that Ethiopia adheres to the general African model for indigenous plant exploitation. This evidence is in direct contradiction of the conventional model and the archaeological evidence, but is becoming accepted in other regions of Africa where similar contradictions exist (Haaland *in press*). The pattern for plant utilisation and domestication in Ethiopia parallels the general model for Africa, outlined above.

Chapter 4.

THE PALAEOENVIRONMENTAL CONTEXT

4.1. NORTH AFRICAN PALAEOCLIMATE

4.1.1. The Nature of the Evidence

Background to Palaeoclimatic Research in Africa

Current knowledge of climate change has only evolved in the last century. Over the past 40 years, the development of improved analytical and dating techniques has shown that the climate has changed massively and recurrently. For example during the Quaternary period (roughly the past 2.5 million years) the climate repeatedly swung between cold, arid glacial conditions and warm, humid interglacials. In more recent years, the implications of anthropogenic forcing on the climate and the threat of global warming have stimulated intense interest into global palaeoclimate change. This has led to extensive research in the tropics, particularly in Africa. Research has revealed that climatic oscillations in the tropics were synchronous with climate changes in temperate regions, but were decidedly more pronounced. Only now are we finally beginning to appreciate the magnitude of these changes, and the massive implications for the ecosystem and for human adaptation.

Measuring Palaeoclimate Change in Africa

Quaternary climate change at high latitudes is dominated by the expansion and contraction of large ice sheets, whereas in the tropics the predominant signal is rainfall variation. Changes in effective precipitation are reflected in fluctuating water levels and the chemical composition of closed lake basins. These variables provide accurate records of regional and inter-regional tropical palaeoclimate.

Geotechnological advances over the past 25 years have contributed to the reconstruction of African palaeoclimatic sequences. Information from ^{14}C, uranium/thorium and thermoluminescense dating of lake sediments and high altitude peat bogs has been complemented by studies of pollen, molluscs and diatoms contained in these sediments. This work has been supported by a range of geomorphological studies of rivers, dunes, lakes and mountains, and geochemical and isotope analyses of organic and inorganic carbonates, palaeosols, groundwater and deep ocean cores. Continual refinement of existing techniques, combined with development of new methodologies has ensured the progressive resolution of palaeoclimatic patterns.

Palaeoclimatic data are locally derived, and they are therefore limited by the geographical distribution of suitable sampling sites. In Africa, sites are concentrated particularly in lake sediments, a high proportion of which are located in the East African Rift Valley. However there is a good agreement between spatially diverse lacustrine sites in Northern Africa and offshore marine sediments from the Atlantic, the Mediterranean and the Red Sea, which suggests that Late Quaternary climate change was homogenous across Northern Africa. Table 4.1 below documents the major sites for palaeoclimatic and palaeoenvironmental sampling in Northern Africa and surrounding area over the past 30 years.

4.1.2. Long-Term Holocene Climate Changes in Northern Africa

Pluvial Conditions and Environmental Change in the Early Holocene

The End of the Ice Age

The structure and chronology of major African climate changes correspond to global climatic oscillations. These are well established from numerous oceanic records and continental sequences (Berger and Labeyrie 1987). The Last Glacial Maximum (LGM), dating between c.21 000-15 000 BP, was a period of extreme aridity, with lowered lake levels and little available surface water. Sea levels were universally lowered due to the immense body of water locked into glaciers and ice caps. Levels of Northern African, lakes were significantly lower than at present, and the Nile was reduced to a trickle (Grove 1993). The 100 mm rainfall isohyet which defines the southern limit of the Sahara Desert lay 500 km closer to the equator, at 13°N and the Sahelian zone had a southern limit of 10°N, compared to its present position of 14-15°N (Talbot 1980). Aeolian evidence suggests that high velocity winds prevailed over much of Northern Africa during this period. These arid conditions and high winds, combined with lowered temperatures (3-7°C below present annual means) restricted plant growth, and confined tropical vegetation to isolated refugia (Hamilton 1982). The last glaciation terminated globally c.15 000-12 000 years BP as a result of astronomical forcing (Berger and Loutre 1991). Deglaciation occurred in two steps at c.13 000-12 000 and 9000 years BP in temperate latitudes (Petit Maire et al 1991, Gasse et al 1980). This correlated to a massive increase in the prevailing rainfall in the tropics. The ameliorating process in the tropics was interrupted by a return to aridity c.12 000-10 500 years BP, that corresponded to the Younger Dryas in northern latitudes (Roberts 1990, Roberts et al 1993, Hughen et al 1996). By c.9000 years BP, full interglacial pluvial conditions were established across Northern Africa. In areas of extreme aridity between 22-24°N, the onset of pluvial conditions was delayed until the mid 9th millennium BP (Fabre and Petit Maire 1988, Gasse et al 1990, Petit Maire et al 1991). Figure 4.1 shows the general pattern of Holocene climate change reconstructed for Northern Africa.

Table 4.1. Major Palaeoclimatic Sampling Sites in Northern Africa

Site	Location	Type of Evidence
Lake Tigalmamine (Lamb et al 1995)	Morocco	lake sediment
Sebka Mellala (Gasse et al 1990)	Algeria	lake sediment
Taoudenni (Fabre and Petit Maire 1988)	Northern Mali	palaeolake sediment
Bougdouma (Gasse et al 1990)	Niger	sediment core from palaeolake
Lake Chad (Servant and Servant-Vildary 1980)	Chad, North Central Africa	lake sediment
Lake Bosumtwi (Talbot et al 1984)	Ghana, West Africa	lake sediment cores
Moeris Lake (Wendorf and Schild 1976a)	North Egypt	lake sediment
The Nile and Nile Delta (Adamson et al 1980)	North and East Africa, Mediterranean Sea	geomorphology, sedimentology oxygen-isotope analysis
Lake Victoria (Butzer et al 1972), (Stager 1984)	Central Rift Valley, East Africa	lake sediment
Nakuru-Naivasha Basin (Butzer et al 1972) (Richardson and Dussinger 1986)	East African Rift, Kenya	lake sediment
Lake Borgoria (Tiercelin et al 1981)	East African Rift, Kenya	lake sediment
Lake Baringo (Renaut and Owen 1980)	East African Rift, Kenya	lake sediment
Mount Elgon (Hamilton and Taylor 1986)	Kenya	mire sediment
Kaisungor, Aberdares (Hamilton and Taylor 1986)	Kenya	mire sediment
Sacred Lake (Coetzee 1967)	Mount Kenya, Kenya	lake sediment core [palynology]
Lake Turkana (Butzer et al 1972), (Owen et al 1982) (Mohammed et al 1995)	North Kenya/South Ethiopia	lake sediment
Lake Abiyata (Lezine and Bonnefille 1982)	Ethiopian Rift, Ethiopia	lake sediment
Lake Abhe (Gasse 1980)	Ethiopian Rift, Ethiopia	lake sediment
Ziway-Shala (Grove and Goudie 1971) (Gillespie et al 1983), (Street 1975-1977)	Ethiopian Rift, Ethiopia	lacustrine and terrestrial sediments
Lake Besaka and region (Williams et al 1977)	Ethiopian Rift, Ethiopia	lacustrine and terrestrial sediments
Lake Langano (Mohammed and Bonnefille 1991)	Ethiopian Rift, Ethiopia	lake sediment
Lake Awassa (Lamb *pers.comm.*)	Ethiopian Rift, Ethiopia	lake sediment
Guinea Gulf; Dakar (Pastouret et al 1978) (Rossignol-Strick and Duzer 1979)	West Africa coast, Atlantic	off-shore marine cores [oxygen isotope/pollen analysis]
Eastern Mediterranean (Rossignol-Strick 1983)	Mediterranean Sea	off-shore marine cores [sedimentology]
Arabian Sea (Van Campo et al 1982)	Northern Arabian Sea	off-shore marine cores [oxygen isotope-pollen record]

Early Holocene Pluvial Conditions

The African environment is highly sensitive to relatively small changes in precipitation compared to higher latitudes. The onset of pluvial conditions in the Holocene had a profound impact on the distribution and extent of surface water in Africa. Lake levels rose across Africa by up to 180 m above present levels, with many overflowing their basins. Figure 4.2.a shows the distribution of raised lake levels across Africa. Palaeolake Chad expanded across North East Nigeria to occupy an area of 350 000 km^2 comparable in size to the Caspian Sea, compared to the present area of 30 000 km^2 (figure 4.2.b). The Ugandan lakes and Lake Victoria overflowed into the White Nile headwater region (Williams and Adamson 1980), the Ziway-Shala lakes in the Ethiopian Rift Valley merged and overflowed into the river Awash, and Lake Turkana flooded into the White Nile via the Sobat drainage system and the Lotigip mudflats (Butzer 1971, Butzer et al 1969). Riverine and terrestrial sediments in the Nile Valley (Butzer et al 1972, Williams and Adamson 1980) and marine cores from the Nile Delta (Petit Maire et al 1991) indicate a phase of erratic Nile floods in both the Blue and White Niles during the 12th millennium BP, causing downcutting of the river and sustained swamps in the Gezeira region of Central Sudan and the 'Sudd'.

Precipitation levels have been reconstructed by computer modelling of estimated temperature, runoff, evaporation and seasonality coefficients (eg. Street Perrott et al 1990). Rainfall increases of 40-165% of present values have been calculated as necessary to maintain maximum lake levels in East Africa, and of up to 400% in arid regions between 22 and 24°N (Butzer et al 1972, Street 1979, Fabre and Petit Maire 1988, Street and Grove 1976, Yan and Petit Maire 1994). Seasonal rainfall was more pronounced, causing more intense summer monsoons in the tropics, and possibly increasing winter rainfall in Mediterranean North Africa (Lamb et al 1995).

This substantial rainfall increase affected not only lake levels and river volume, but also the entire tropical African ecosystem. Street and Grove (1976) note that

> *'... biological productivity and biomass greatly increased in tropical Africa and the surface albedo was substantially diminshed by the expansion of open water, marshes and forest. Elephant, giraffe and antelope ranged comparatively unchecked across the Sahara, which must at times have existed only as relict desert areas isolated by corridors of gallery forest and swamp along major wadis. Overflowing lakes established links between previously separate drainage systems... Traces of hippopotamus and crocodiles are widespread even in the central Sahara...'*

The Onset of Drier Conditions

Pluvial conditions persisted across intertropical Africa until the mid 6th/early 5th millennium BP (or c.6800 years BP in the Sahara region (Fabre and Petit Maire 1988)). The climate subsequently became increasingly arid (figure 4.1). Ephemeral and permanent water supplies dried up. Tropical flora and fauna retreated from arid regions, contracted vertically up mountain slopes and formed isolated refugia.

Long-term climatic oscillations became increasingly stabilised after the 5th millennium BP, and modern conditions of rainfall and seasonality were established after c.3000 years BP. Figure 4.3. illustrates the profundity of these changes with reference to the areal extent of East African lake basins in the early Holocene and present day.

4.1.3. Intermediate-Term Climate Change in Tropical Africa

This long-term sequence of Late Quaternary climate change in Africa is similar throughout the tropics (Roberts et al 1993, Zahn 1994). In Africa a complex sequence of intermediate-term climatic oscillations can also be observed, superimposed over the long-term, millennia-scale arid-humid transitions. These phases are intense, abrupt and last between 150 and 400 years (Lamb and van der Kaars 1995). They occur throughout the Northern tropics, in Africa and Asia, as well as in Northern sub-tropical Africa (Yan and Petit Maire 1994, Lamb et al 1995), but they are **not** present in this intensity in palaeoclimatic records from higher latitudes, although evidence for abrupt, short-lived climatic events of lower amplitude is now beginning to emerge from high resolution analyses of polar ice cores (Stager and Mayewski 1997, Alley et al 1997). The forcing mechanism of these events is separate from the Milankovitch/astronomical cyclicity of long-term fluctuations. A number of factors may be involved that have a limited impact on temperate climates, such as volcanism, solar variability, sea surface temperature changes, salinity or cryospheric instability (Grove 1993, O'Brien et al 1996).

Palaeolimnological research (i.e. the study of lake sediments for palaeoclimatic and palaeoenvironmental reconstruction) in the 1960s and 70s identified a major lacustrine regression between c.8000 and 6000 years BP in lakes across Africa such as Bostumwi, Chad, Naivasha-Nakuru and Ziway-Shala (Butzer et al 1969, Grove and Goudie 1971, Butzer et al 1972, Street and Grove 1976, Williams et al 1977). Lake levels fell by up to 180 m below the Holocene maximum, indicating a corresponding maximum fall in precipitation of

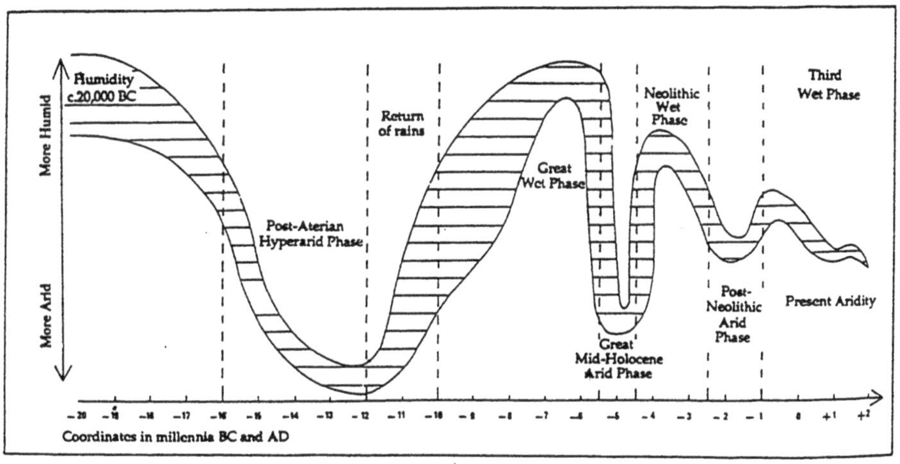

Figure 4.1.a. Summary pattern of Holocene palaeoclimate in North Africa (after Muzzolini 1993).

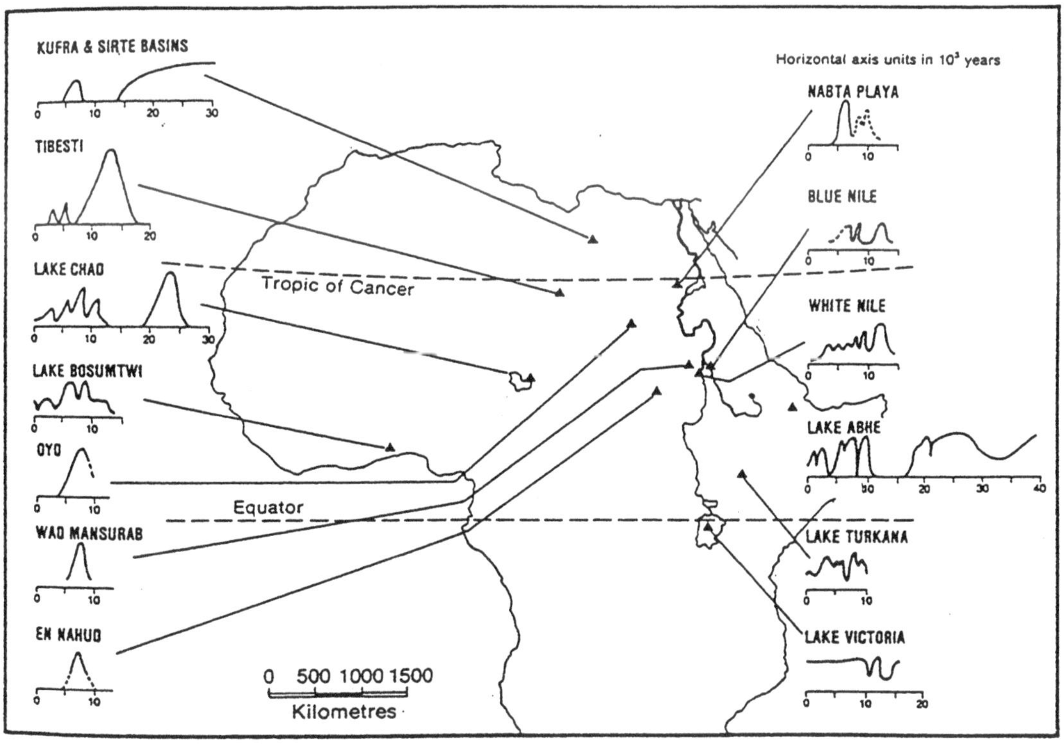

b. Late Quaternary lake level changes across Northern Africa. Peaks denote high lake levels, high flood levels, greater runoff or increased groundwater recharge (after Williams 1988).

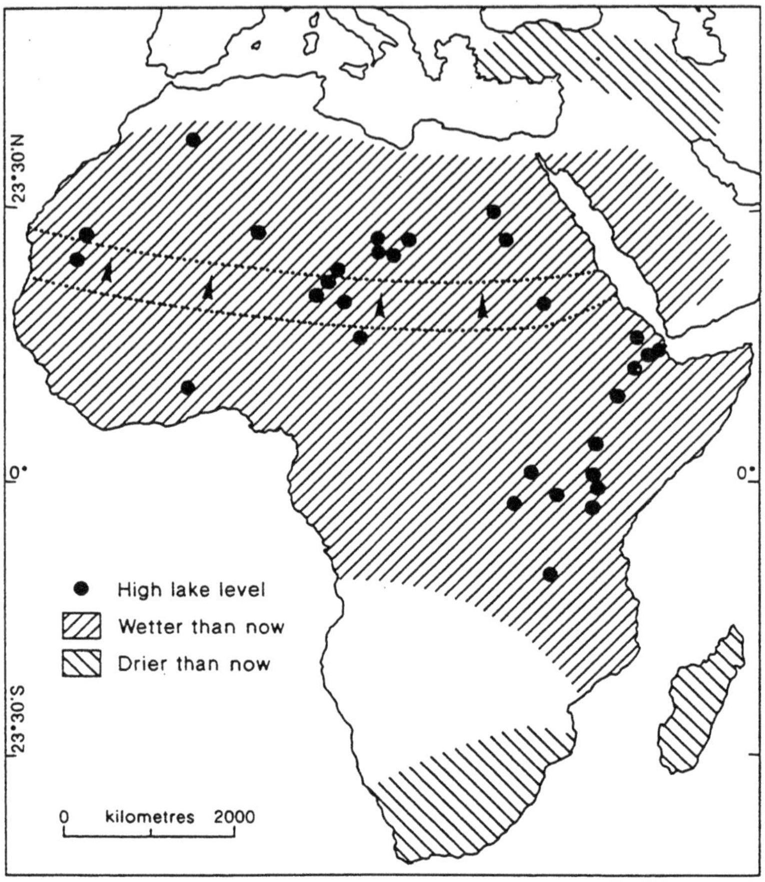

Figure 4.2.a. Location of high lake levels and extent of increased humidity in Africa between c.9500 and 8500 BP (after Grove 1993)

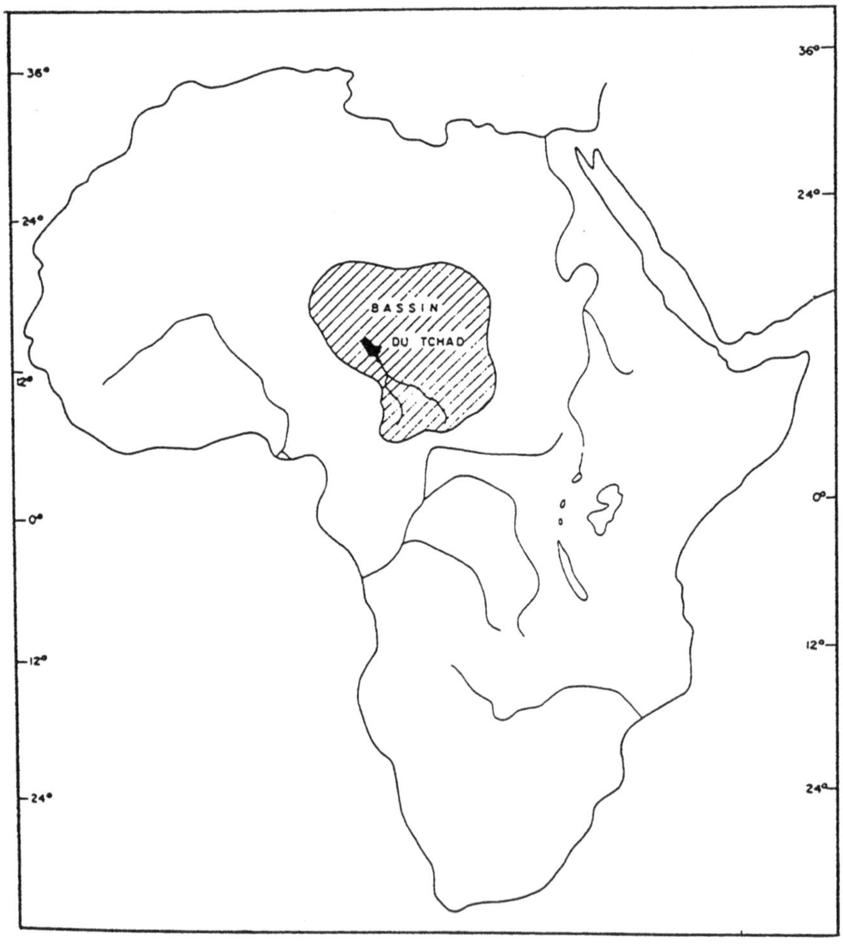

b. Extent of palaeolake Chad in the early Holocene (shaded area) in comparison to the present day lake (black) (after Servant and Servant-Vildary 1980).

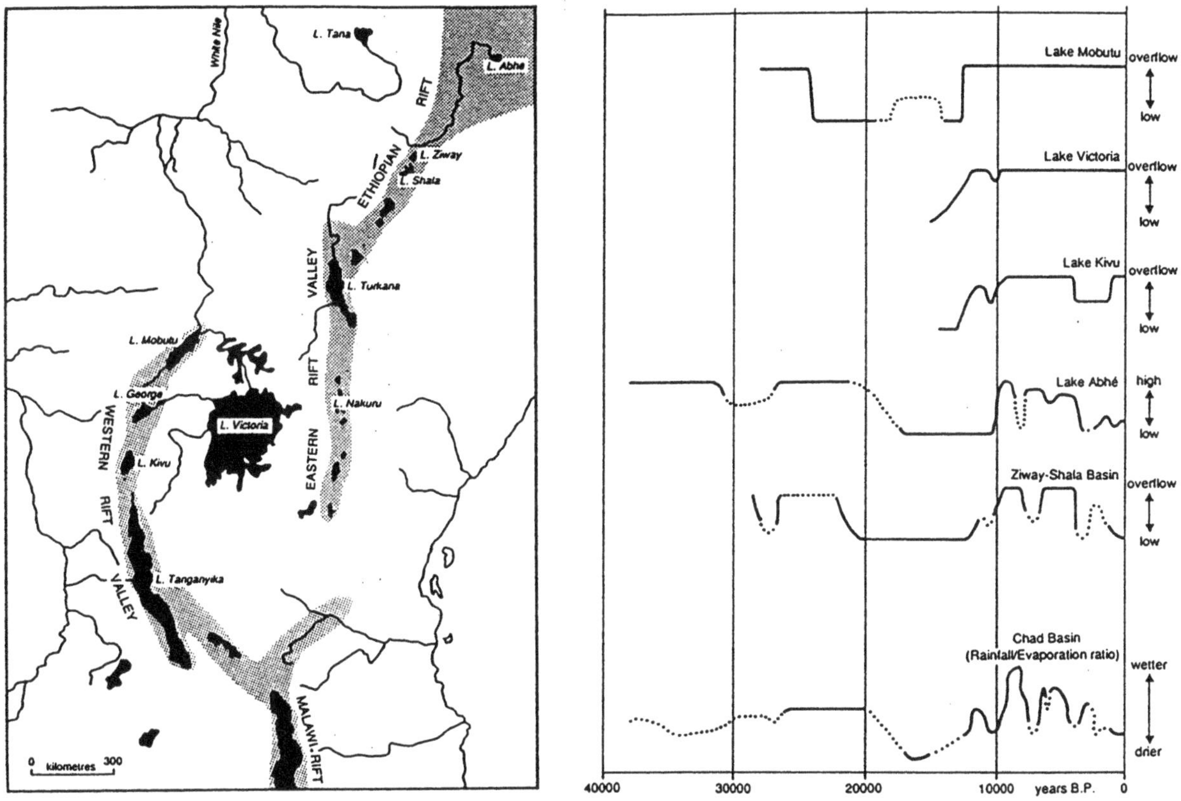

Figure 4.2.c. Lake level changes in selected basins in East Africa (after Hamilton 1982, Dawson 1992).

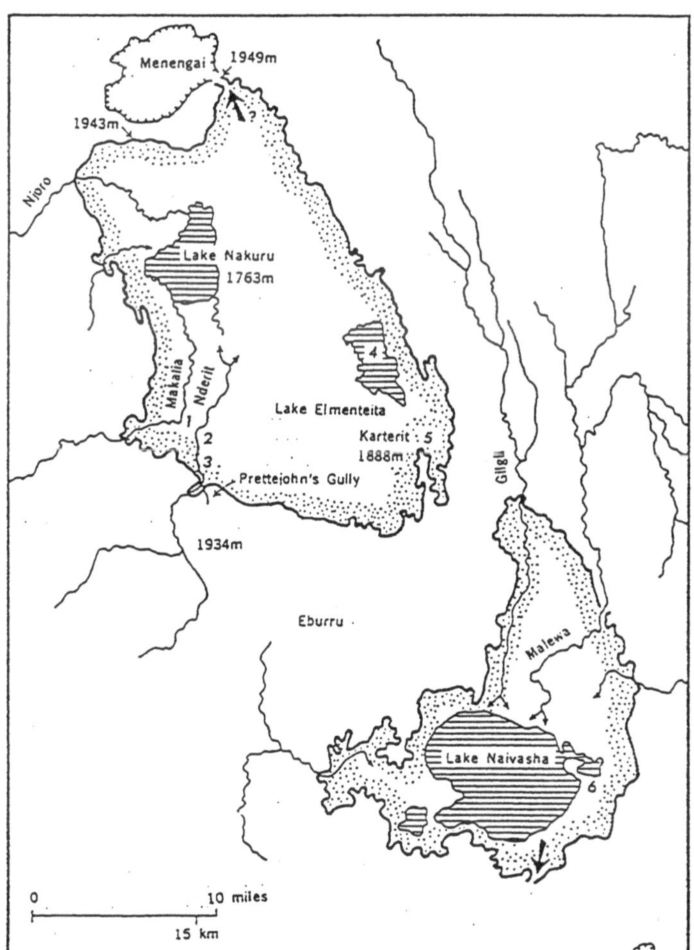

Figure 4.3. Changes in surface area of East Africa lakes between the early Holocene and present day, illustrating the impact of palaeoclimate change (after Butzer et al 1972).

50% below present levels[1] (Tiercelin et al 1981, Talbot et al 1984, Richardson and Drussinger 1986). The chronology of this event varied regionally within this period, but lake levels were generally low between 8000 and 7200 years BP, with a maximum regression centred around 7400 years BP (Roberts 1990, *pers.comm.*).

To summarise, better resolution in recent years has shown that this negative event was widespread across Northern Africa. It has also shown that a series of intermediate-term events with a duration of 150-400 years have repeatedly punctuated the long-term climatic trend during the past 8000 years (Gillespie et al 1983, Fabre and Petit Maire 1988, Lamb et al 1995, Lamb and van der Kaars 1995). The sequence of intermediate-term fluctuations from Tigalmamine in Morocco is illustrated in Table 4.2 below (Lamb et al 1995).

Evidence of Intermediate-Term Climate Change in Ethiopia

Further fluctuations are also suggested from the Ethiopian data, between 6200 and 5800 years BP (Gillespie et al 1983) and 560 years BP (Bonnefille and Mohammed 1994). Studies of relative diatom concentrations from Mount Badda bog in the Ethiopian Highlands indicate repetitive arid-humid cycles of 400-450 years duration during the past 3500 years (Gasse 1978).

Palaeoclimatic and palynological data indicate that these intermediate-scale events were severe. For example there was a reduction of 2-2.5°C around 560 years BP in Ethiopia (Bonnefille and Mohammed 1994), and between 3840 and 3770 years BP the Nile floods averaged 9.0 m higher than the 20th century mean, with three times the discharge volume (Butzer 1976). Additionally, these events had an extremely abrupt onset, and the climate may have swung from humid to arid conditions within a few decades (Gasse et al 1990).

It is almost impossible to contemplate the significance of these changes to the human population that occupied Africa during the Holocene, whose environment and resources changed entirely within a generation.

Summary

Both long-term and intermediate-term climate change can be detected in the Holocene record from Ethiopia, as described in Section 4.2 below. In addition to these two scales of change, a short-term rhythm persisted in Ethiopia during the Holocene period. The evidence for this is presented in Section 4.2.2, and its implications are discussed with reference to the ethnographic data in chapter 2.

Table 4.2. Chronology and Duration of Arid Phases from Tigalmamine, Morocco (after Lamb et al 1995).

Date (^{14}C yr BP)	Duration of arid phase (years)	Date (cal year BP)	Duration of arid phase (years)
10 550-9200	1350	12 480-10 230	2250
7050-6650	400	7860-7500	360
4450-4300	150	5010-4860	150
2900-2750	150	2990-2830	160
1950-1750	200	1880-1660	220

4.2. ETHIOPIAN PALAEOCLIMATE

4.2.1. Holocene Climatic Sequence in Ethiopia

Late Pleistocene - Mid Holocene Climate

The past 25 years have seen extensive palaeolimnological research on lake sediments and shore lines in the Rift Valley in Ethiopia, described in Table 4.1. The distribution of the major sampling sites is shown in figures 4.4.a. and b. These records have received additional support from a range of analytical studies, including variations in Blue Nile discharge from alluvial sediments in the Gezeira region of central Sudan (Livingstone 1980, Williams and Adamson 1980), pollen sequences from mire sediments in the Highlands (Gasse 1978, Street and Grove 1976, Bonnefille and Hamilton 1986, Bonnefille and Mohammed 1994), and geomorphological studies of buried soils in the Northern Highlands (Brancaccio et al 1994, Berakhi et al 1994, Butzer 1982).

The data from these independent research projects corroborate well to provide a detailed reconstruction of Ethiopian palaeoclimatic events during the past 12 000 years. Humid conditions were established in Ethiopia after 12 000 years BP, with a major pluvial episode from c.10 000-8500

[1] A change in lake levels may result from the variation of several factors such as surface run-off, evapotranspiration and temperature. Temperature oscillations can have profound effects on the biogeographic spectrum and are thus particularly important in palaeoenvironmental reconstruction. Temperature changes alter lacustrine diatomic floral populations and the organic and mineral composition of lake waters. These alterations can be detected from lake sediments. For example a severe cooling phase can be detected in Late Pleistocene sediments from Lake Abhe, dating to c.30 000 years BP (Gasse 1980). This is indicated by a reduction in organic matter and a change in the diatomic flora from tropical to non-tropical taxa. No such changes are detected in Holocene lake sediments, and from the early Holocene onwards the temperature in Ethiopia appears to have experienced a maximum fluctuation of 2-3°C around the current annual mean.

years BP, centred around 9200-9500 years BP. Mean annual precipitation increased up to 47% above present levels, with an average increase of 25% during this period. Temperatures fell by 3-4°C and stabilised at 2°C below present (Street 1975). This may have been due to thicker cloud cover during the summer months, especially at higher altitudes (Roberts *pers.comm.*). The levels of the Rift Valley lakes rose by a maximum of 180 m above present, and many previously separate lake basins became interconnected. For example, as figure 4.5 shows, the Ziway-Shala lakes in the Rift Valley formed a single massive body of water which overflowed into the Awash river (Grove and Goudie 1971).

Today 70-80% of the total annual precipitation in Ethiopia falls in the 'heavy rains' of the summer months (June-September) which are influenced by the movements of the Inter-Tropical Convergence Zone (ITCZ). The remaining 20-30% of the rainfall occurs during the 'little rains' in the spring, which arise from the Indian Ocean monsoon system (Grove 1978). This perennial rhythm was exaggerated during the early Holocene. Monsoonal rainfall was more intense, and the seasonal disparity between winter aridity and spring/summer humidity was even more pronounced (Williams and Adamson 1980, Messerli and Winiger 1980, Gasse et al 1980).

Mid - Late Holocene Climate

Pluvial conditions were maintained until the start of the 5th millennium BP, interrupted only by a severe arid period between 8500 and 6500 years BP which centred around 7500-7200 years BP. During this time lake levels fell by 50-160 m of their early Holocene maxima (Grove and Goudie 1971, Gillespie et al 1983). This phase was intense and abrupt. It lasted only a century before high lake levels were restored by 7000 years BP. A second short regression phase, of lower intensity, occurred between c.6200 and 5800 years BP (Gillespie et al 1983).

Lake levels again fell rapidly after c.4800 years BP. They remained low throughout the 4th millennium, and by 3000 years BP they were 20-40 m above present and gradually declining to modern levels (Grove and Goudie 1971, Gillespie et al 1983). This gradual arid trend was interrupted by a further lacustrine period c.2500-1000 years BP (Gillespie et al 1983). Geomorphological studies in Tigray in the Northern Highlands show that soil formation ceased between c.5000 years BP and 4000 years BP (Berakhi et al 1994, Brancaccio et al 1994). This indicates a substantial reduction in vegetation growth and decay at this time due to lower levels of effective precipitation.

The pattern of lacustrine activity in Ethiopia during the Holocene is shown in figure 4.6 by the oscillating levels of the Rift Valley lakes Ziway-Shala and Abijata.

Summary

So far, two separate time-scales of climate change have been identified in Holocene Ethiopia:- *long-term* cycles, with a duration of several millennia, an abrupt onset and a relatively abrupt decline, and *intermediate-term* changes which punctuate the long-term cycles, last for several centuries and also have an abrupt onset. The relative chronologies of these rhythms are illustrated in figure 4.6.

4.2.2. Short-term Climate Change

Evidence for Short-Term Climatic Variation in Ethiopia

In addition there is a third scale of climate change. This ***short-term*** rhythm recurs on a decadal or annual scale. Three sources of evidence are available from Ethiopia which allow us to focus on short-term climate changes over the past 3000 years:

1. Historical records of Egyptian Nile flood levels and Ethiopian droughts which discontinuously span the period from the 5th millennium BP-present.
2. Pollen evidence from Dega Sala swamp in the Ethiopian Highlands 3000 years BP - present.
3. Pollen and isotope data from Lake Turkana in Kenya for 2500-600 years BP.

Historical Records

The Nile Flood Records

In 622 AD (1378 years BP) a 'nilometer' was set up on the island of Rodah, near Cairo, to gauge the perennial seasonal maxima and minima of the Nile floods. Previously, flood levels were recorded in a series of ancient Egyptian texts and graffiti. These documents enable us to trace the annual rhythms of the river back 5000 years. For example, 40 'high Niles' were recorded by inscriptions during the 25th Dynasty. These earlier records are unsystematic and discontinuous, but nevertheless they provide a qualitative framework for comparison of flood-levels.

Summer monsoon rainfall on the Northern Ethiopian Highlands (Blue Nile headwaters) and rainfall in the White Nile headwaters in equatorial East Africa combine to give a strong annual rhythm to Nile flow. Runoff from the Northern Ethiopian Highlands at present contributes a mean annual volume of over 63 km^3 to the Nile system through the Blue Nile and Atbara rivers (Hurst and Phillips 1931). This discharge occurs predominately during the three month summer monsoon, and in September the Highlands contribute over 90% of the Nile flow. This seasonal discharge from the Ethiopian Highlands is responsible for the annual inundation of the Nile (see chapter 2, figure 2.7).

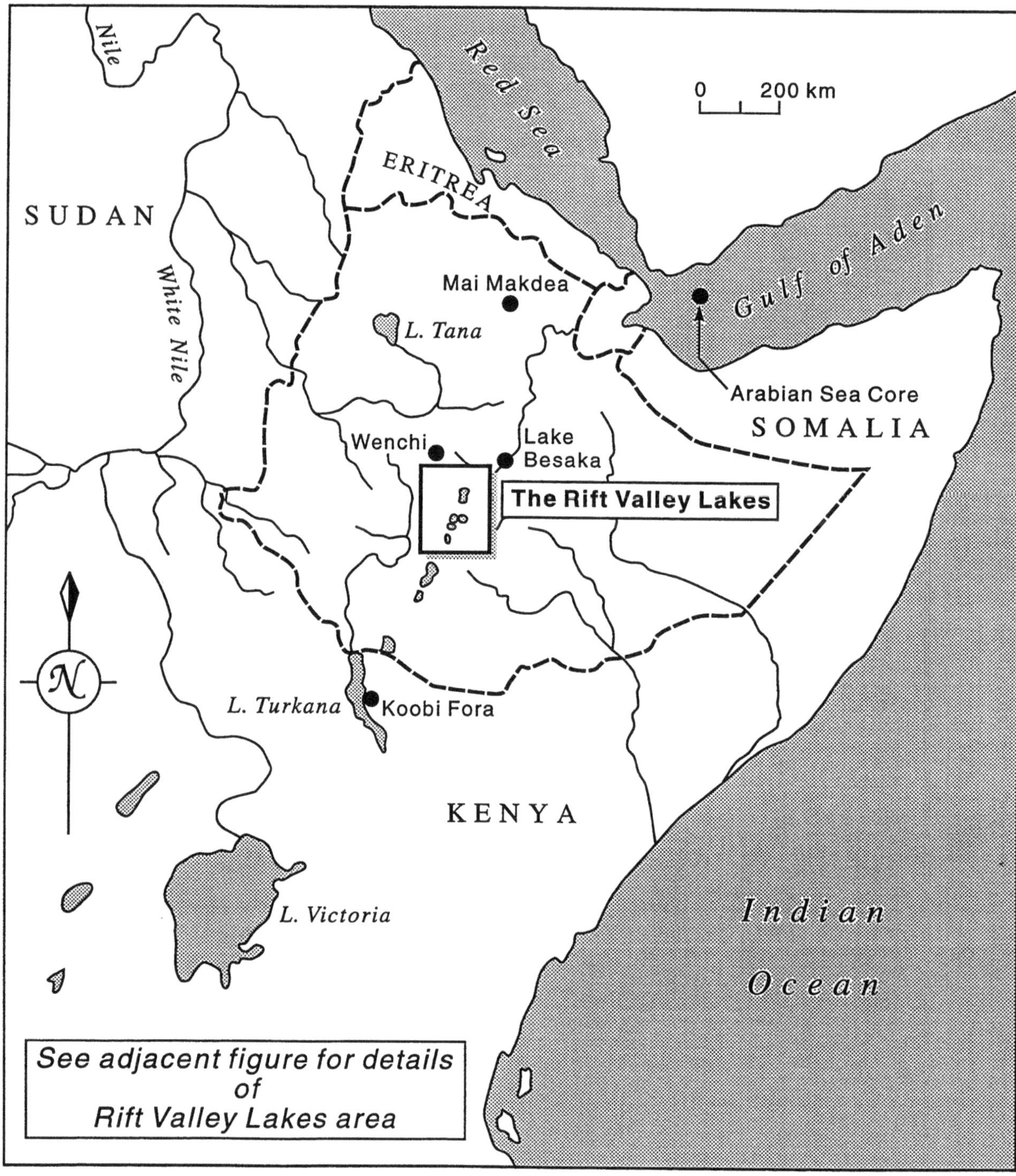

Figure 4.4.a. Palaeoenvironmental sampling sites in Ethiopia.

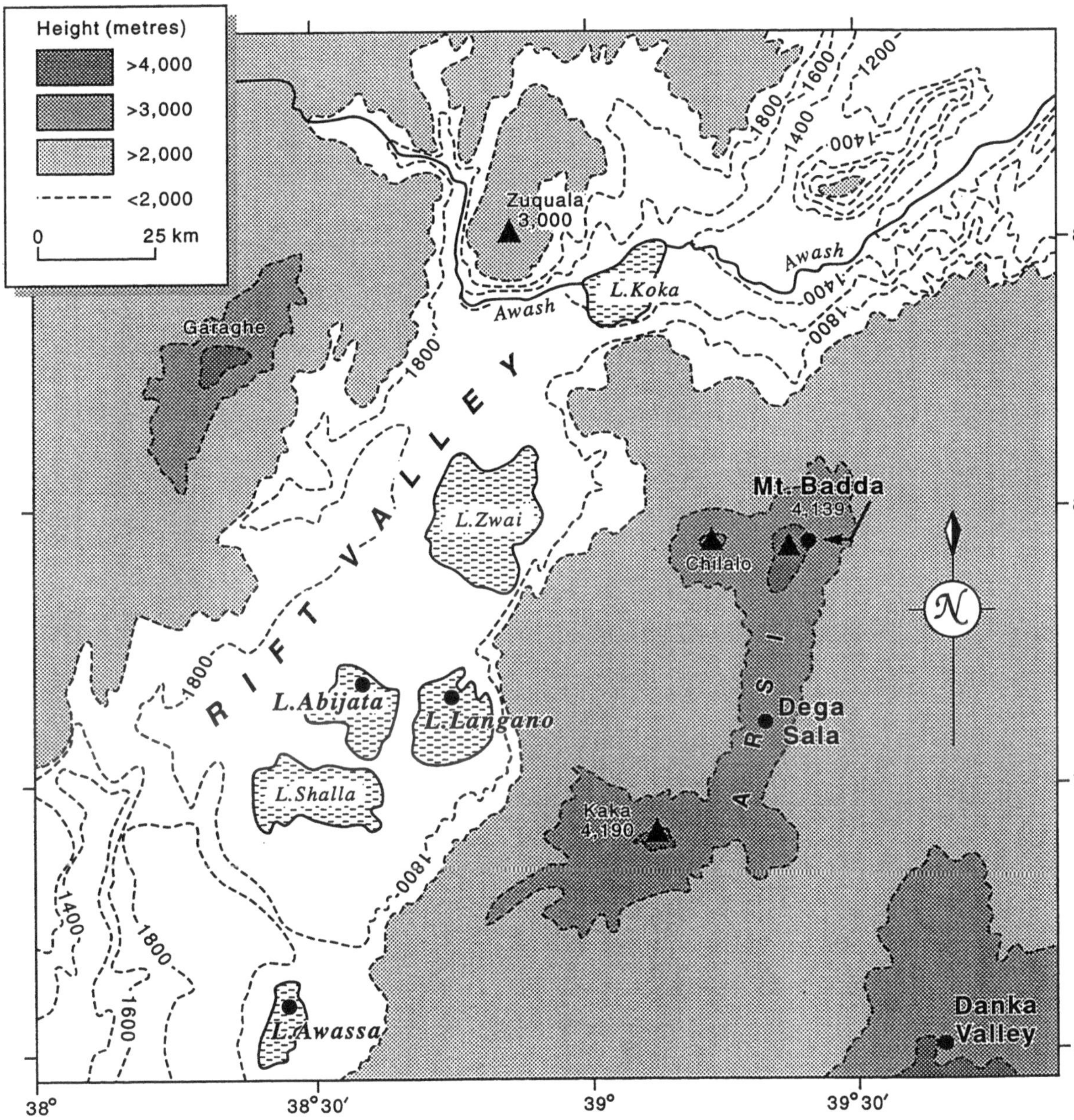

Figure 4.4.b. Detail of palaeoenvironmental sampling sites in the Rift Valley and Southern Ethiopian Highlands.

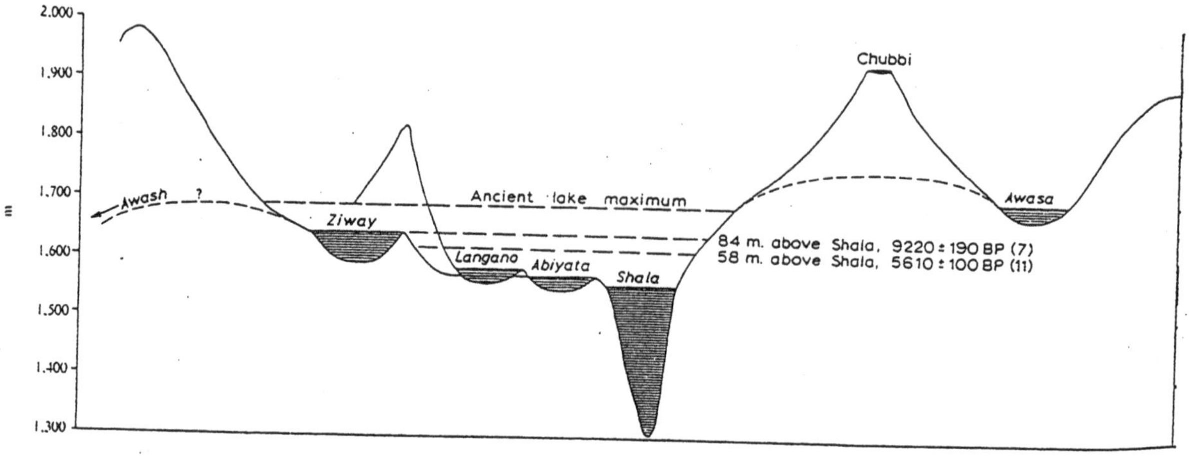

Figure 4.5. Cross section through the Ziway-Shala basin in the Ethiopian Rift Valley showing variations in lake levels during the Holocene (after Grove and Goudie 1971).

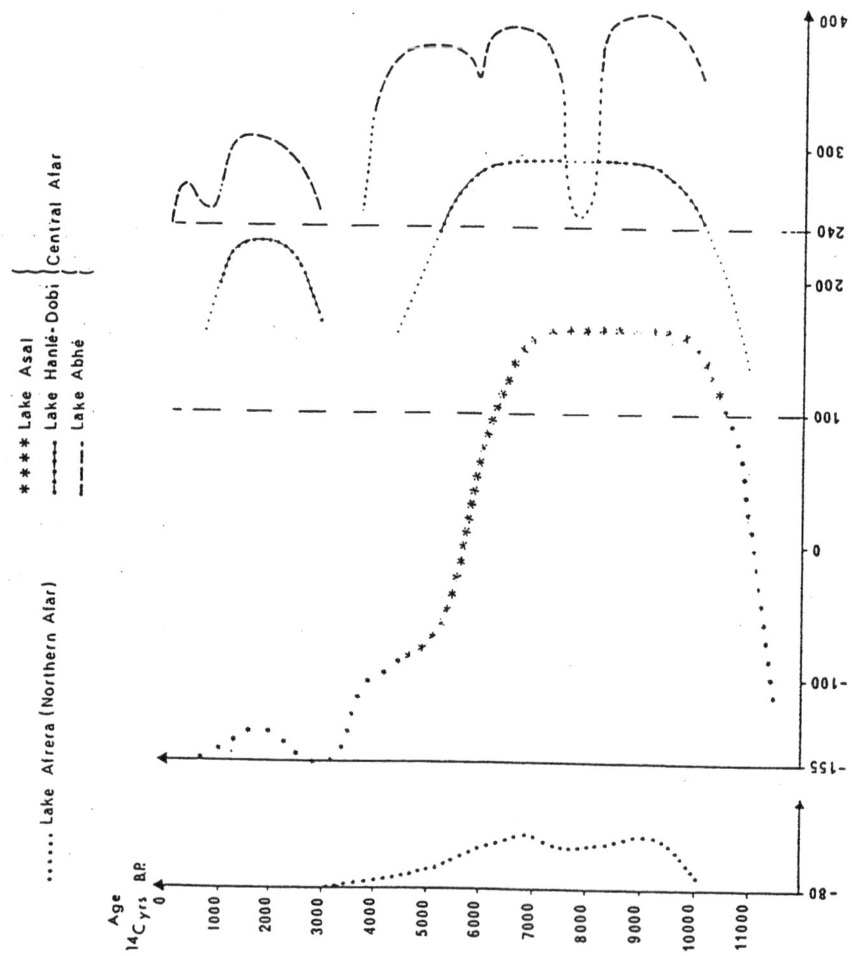

Figure 4.6. Chronological record of lake level fluctuations in Ethiopian Rift Valley lakes during the Holocene (after Gasse et al 1980).

In contrast, the White Nile headwaters have a less seasonally pronounced annual precipitation pattern. The White Nile contributes only about 27 km^3 of the total annual discharge volume. Although over 65% of the Nile flow comes from the interlacustrine region during its lowest months of May-June, Lake Victoria acts as a buffering system to regulate the White Nile flow, and discharge occurs at a relatively constant monthly rate (a ratio of 2½:1 compared to 40:1 for the Blue Nile) (Hurst and Phillips 1931). Thus although the Nile flood pulse is modified to some extent by rainfall patterns in the interlacustrine region, for 75% of the time it provides an accurate reflection of seasonal rainfall patterns on the Ethiopian Highlands (Herring 1979)[2].

Summer monsoon rains in the Blue Nile catchement area on the Northern Ethiopian plateau therefore provide the annual pulse of the Nile flood. Variations in flood maxima in Egypt reflect short-term changes in the dynamic pattern of rainfall on the Northern Ethiopian Highlands. Egyptian documents thus enable us to reconstruct climate changes in Ethiopia over the past 1300-5000 years, as illustrated below:

5000 years BP-AD 622 (1378 years BP) [documented events]:

Date (BP)	*Documented Event*
4913	One of lowest floods ever recorded, set against a backdrop of gradually diminishing floods (1 m decrease between 5000 and 4900 years BP)
4720-4713	Seven years of very low floods
4500-4210	No records?
4210-4185	Series of very low floods
3925-3895	Series of very low floods
3840-3770	Very high floods with erratic behaviour, associated with a return to more humid conditions c.3970 BP
3210-2300	Sequence of very low floods

622 AD-present (1378-0 years BP) [nilometer readings]:
Riehl and Meitin (1979) calculated that Nile flood variations have fluctuated recurrently around the mean discharge value during the 1300 year period from AD 622 to present. Every few decades there have been a series of fluctuations significantly above the long-term average. These 'blips' have each lasted for 50-100 years. They comprise a positive component (increased flood level) and a negative component (reduced flood level) of about thirty years each, with a prolonged period of minor flood level variations between 1050-800 years BP (AD 950-1200). Figure 4.9 shows the timing and relative magnitude of these events.

Ethiopian Historical Records
Ethiopian records are less ancient and less complete than Egyptian documents, but they become increasingly reliable and detailed after the 12th century AD (7th century BP) (Webb et al 1992, Pankhurst 1984). These records also tend to be more ambiguous because they document droughts and famines rather then precise river levels, as shown below:

Date (AD)	*Date (BP)*	*Duration (years)*
1252-1275	748-725	23 Series of drought-related famines
1314-1344	686-656	30 Consecutive drought years
1435-1468	565-532	33 Consecutive drought years
1772-1800	228-200	28 Series of severe droughts
1800-1892	200-108	Consistently poor rains throughout this time, particularly between 1888 and 1892, culminating in extended drought and famine.
20th century		Increasing propensity of drought-related famines due to internal factors. Particularly severe in 1964-1966, 1971-1975 and 1984-1985.

The pattern of drought frequency in Ethiopia correlates well with that described by the Nile flood records. Both show a climatic pulse of 50-100 years duration that is repeated after a gap of a few decades. This pulse comprises two components: an increase in average rainfall lasting approximately 30 years, and a concomitant decrease of the same duration.

[2]Records of droughts and famines are available from oral traditions and historical records of the interlacustrine region, and these have been correleated chronologically to the Nile flood records for the past 1300 years (Webster 1979). This enables us to assess variations in the Ethiopian climate more accurately.

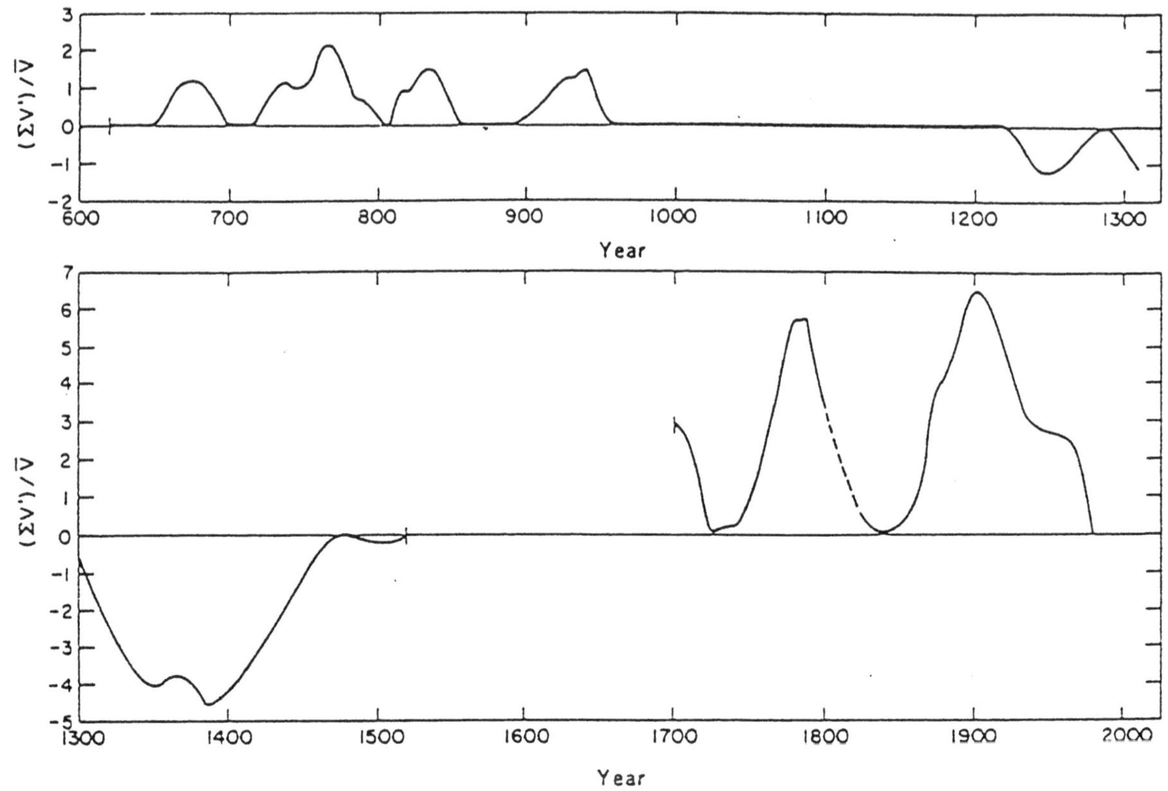

Figure 4.7. Short-term fluctuations in Nile flood levels (vertical axis) relative to the average discharge (ΣV'). Positive slope on the curve indicates that discharge is above the long-term average, negative slope indicates discharge in below the long-term average (after Riehl and Meitin 1979). Measurements after AD 1870 were taken at Aswan.

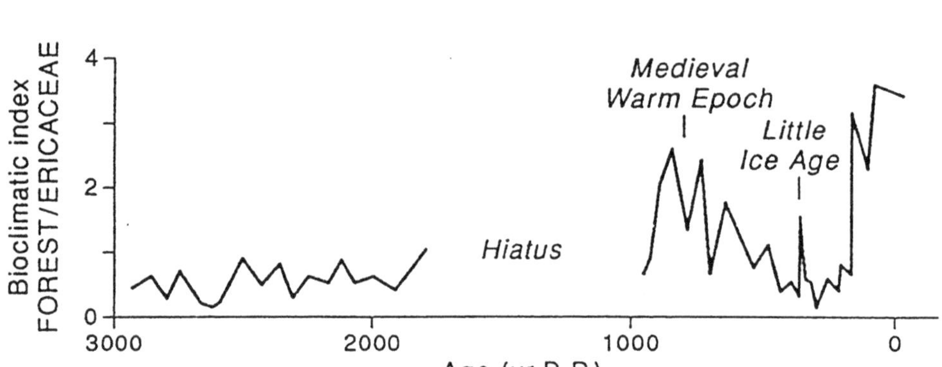

Figure 4.8. Bioclimatic index for the Southern Ethiopian Highlands from Dega Sala swamp (forest/Ericaceae ratio in arbitary units) (after Bonnefille and Mohammed 1994).

Pollen Evidence from Dega Sala Swamp

The swamp is situated at an elevation of 3600 m in the Arsi mountains in the Southern Ethiopian Highlands (see figure 4.4.b.). Seven radiocarbon dates were obtained (Table 4.3) from the organic sediment in the core, from which an age-depth relationship was constructed for the past 3000 years (Bonnefille and Mohammed 1994).

Table 4.3. Radiocarbon Dates from Dega Sala, core D1 (after Bonnefille and Mohammed 1994)

Depth (cm)	Age (yr BP)	Lab no.
1-10	modern	LGQ 343
40-50	modern	LGQ 736
58-60	560±120	LHGI 545
81-90	590±220	LGQ 342
120-130	2050±200	LGQ 737
170-180	1560±284	LGQ 284
190-200	2920±240	LGQ 738

Mountain ecosystems are sensitive to climate change. For instance, for every 1°C change in mean annual temperature, a high altitude vegetation ecozone moves 150 m vertically up- or down-slope (Mahaney 1989). Climatic change can thus be quantified from the relative frequencies of pollen deposited by different types of high altitude vegetation at a given altitude over time.

Fluctuating levels of local Ericaceous scrub pollen and arboreal (montane forest) pollen from the upper timber line in the Dega Sala core enabled Bonnefille and Mohammed (1994) to calculate that temperature has altered significantly in the Southern Ethiopian Highlands over the past 3000 years. Their results, illustrated in figure 4.10, show a general trend towards warmer, possibly drier conditions over the past 3000 years. They also detected a substantial regression to colder, wetter weather around 560±120 years ago, corresponding to the Little Ice Age at higher latitudes (c.AD 1300-1600/700-400 years BP).

Of even greater interest here is the frequency and amplitude of the changes in vegetation and temperature, particularly within the past millennium. These show that substantial shifts in the proportional representation of different plant taxa have occurred on very short-term, decadal time scale. This indicates that there have been significant, recurrent short-term changes in the prevailing temperature.

Pollen and Isotope Data from Lake Turkana

Lake Turkana lies in the Eastern Rift Valley, on the northern-most border between Kenya and Ethiopia, and it is the largest of the Eastern Rift Valley lakes. The lake has a closed-basin system, making it highly sensitive to climate changes which are reflected in massive changes in lake volume, chemistry and biota. Lake levels have varied by 20 m in the past century, and over 140 m in the past 20 000 years. The maximum lake level is determined by an overflow threshold 80 m above the present surface level (Butzer et al 1972, Owen et al 1982). The lake is fed primarily by the Omo river which drains the Southern Ethiopian Highlands to the north, so changes in river discharge due to changes in rainfall on the Ethiopian plateau are ultimately expressed in the rising and falling levels of Lake Turkana.

Mohammed et al (1995) used this system to investigate Holocene climate changes in the southern Ethiopian Highlands through analysis of pollen and isotopes in lake sediment cores dating between 2500-600 years BP. Plant pollen derives predominately from drainage of the Highland Plateau, and from the Rift Valley catchment area. Fluctuating levels of local arid-type vegetation provide background 'noise' for changes in the Highland pollen sequence. For example, the taxanomic sequence is dominated by local arid vegetation, Chenopodiaceae and Amaranthaceae (Cheno/Am). A decrease in the Cheno/Am pollen percentage was interpreted as an increase in local rainfall. A concomitant increase in non-local (eg. arboreal) pollen indicates that rainfall also increased on the Highland plateau. Using this methodology, the following sequence was identified.

Date (years BP)	
2400-1900	Largest recorded Cheno/Am percentage variations for the sequence, correlating with consistently low forest pollen washed down from the Highlands indicate abrupt and strong climate change on a decadal scale during this period
1900-1600	Immense increase in Cheno/Am pollen and very low arboreal pollen indicates severely reduced rainfall, lake regression and very low lake levels
1600-900	Reduced Cheno/Am percentage and slight increase in forest pollen shows increased lake levels due to increased rainfall on the plateau
900-600	Increased Cheno/Am and increased steppe component in forest pollen show a return to drier conditions
600	Localised increase in lake shore/aquatic vegetation.

The pollen-inferred fluctuations correlate well with the $\delta^{18}O$ and $\delta^{13}C$ record, as shown in figure 4.9. The pollen record from Lake Turkana is illustrated in *Appendix B*. This shows progressively lower lake levels in the period c.2500-1600 years BP, followed immediately by a significant rise to high lake levels c.1600 years BP. There followed a gradual regression, then a further significant rise c.600 years BP.

Causality in Short-Term Climate Change

These short-term fluctuations are not confined to Northern Africa. Drought indices and rainfall records from the tropics show that decadal-scale and sub-decadal monsoon variations and drought intervals are a common intertropical phenomenon (Kutzbach 1987, Gasse and Derbyshire 1996). The causal mechanisms are not clearly understood, although there does seem to be a positive correlation between monsoon variation, global palaeoflood activity and temperature change (Lamb 1977, Kutzbach 1987, Smith 1992). More recent evidence from North Atlantic cores shows that sub-decade climatic fluctuations are indeed globally synchronous (Hughen et al 1996). A number of interrelated variables have been implicated in this relationship, including solar variability, volcanism, ocean temperature, snow cover and soil moisture. An imternal feedback system with no external causative factor is the most probable explanation for short-term climate change.

Summary

Palaeoclimatic evidence from Ethiopia has been used to illustrate the coexistence of three chrono-scales of climate variation: *long-term, intermediate-term and short-term climate change*

The chronology and approximate amplitude of long-term and intermediate-term climate changes in Ethiopia during the Holocene are illustrated in figure 4.10.a. above. Short-term climate change is illustrated comparatively in figure 4.10.b. These three phenomena are superimposed to produce a constantly changing global climatic pattern. This pattern has prevailed throughout the Holocene period. The magnitude of the intermediate- and short-term events is amplified significantly in the tropics relative to higher latitudes. These different scales of climate change have differential impact on the environment and on human adaptation. The following section investigates the nature of long-term and intermediate-term climate changes on the Holocene environment in Ethiopia.

4.3. PALAEOENVIRONMENT IN THE ETHIOPIAN HIGHLANDS

Human subsistence behaviour is ultimately determined by the environmental context. Climate, environment and humans exist in a state of dynamic equilibrium. Changes in the prevailing climate alter the environment and the available resources. These changes demand a human adaptive response.

CLIMATE ⇔ ENVIRONMENT ⇔ HUMAN RESPONSE

In Africa the climate oscillated rapidly from arid to humid conditions throughout the Holocene. In order to qualify the anthropogenic response to this instability, we need to identify the nature and extent of palaeoenvironmental change.

4.3.1. Palaeoenvironmental Reconstruction

Various accounts of Holocene climate and prehistory in Ethiopia have speculated broadly on the palaeoenvironmental context of human development (eg. Clark 1954, Phillipson 1977a, Brandt 1982, 1986), but there has been little more than a cursory investigation of the palaeoclimatic data. As discussed above, significant research during the past 15-20 years has heightened our understanding of the magnitude and possible implications of climate change in the tropics. This new understanding should enable us to investigate a broader environmental and cultural framework.

The Nature of the Evidence

Palaeoenvironmental data is largely derived from stratified organic material, including pollen preserved in waterlogged sediments and macrofossils in archaeological deposits. A number of limitations apply to these datasets:

- Different plant taxa produce varying quantities of pollen and are therefore disproportionately represented in pollen sequences.
- Different air-borne pollens can be transported different distances prior to deposition.
- The degree to which local, non-local and regional taxa are represented in a pollen sequence varies with the nature of the catchment zone.
- Different taxa have different response times to changing conditions. Some may take centuries to adapt, giving a poor resolution for short-term climate changes.
- Individual species cannot always be identified or distinguished from one another.

If these limitations are taken into account, pollen sequences can provide an extremely rich source of evidence for the vegetation response to external forces, such as climate change and anthropogenic interference. They provide an indication of

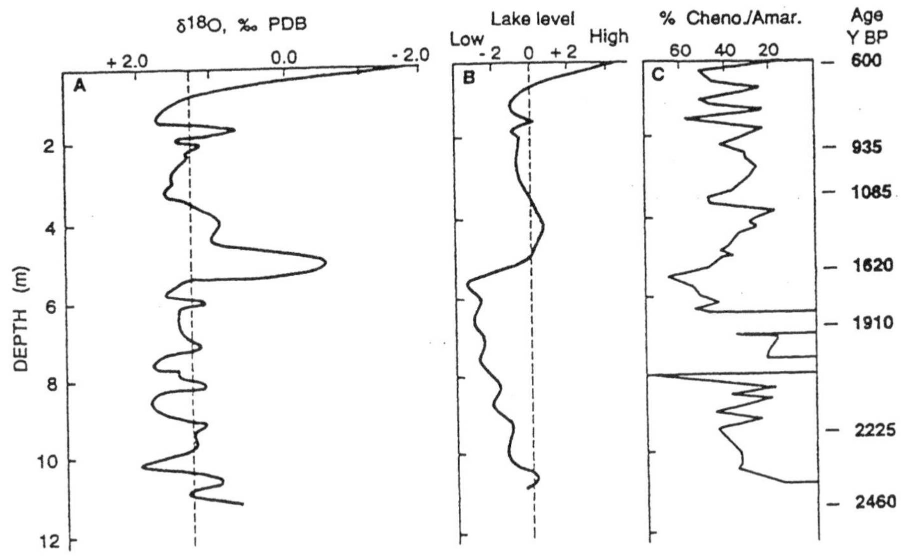

Figure 4.9. Lake level changes inferred from changes in the Cheo/Am pollen percentages at Lake Turkana, relative to changes in $^{\delta18}$ O composition (after Mohammed et al 1995).

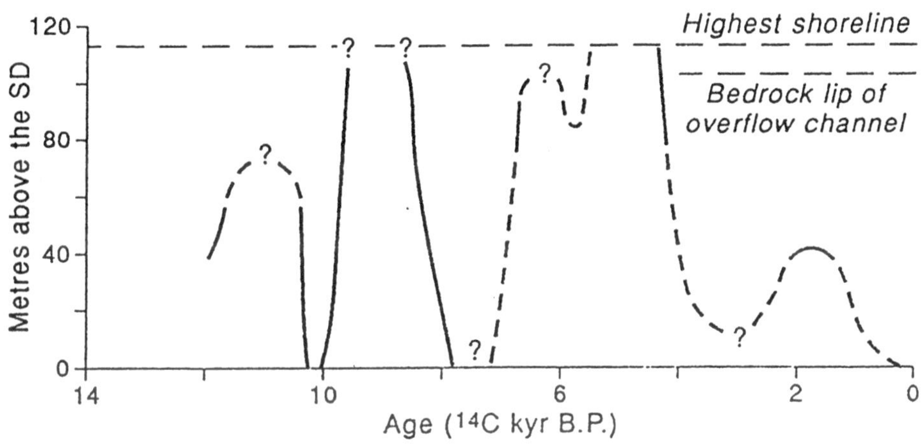

Figure 4.10.a. Long-term and intermediate-term climatic fluctuations expressed by changes in lake levels in the Ziway-Shala basin (after Gillespie et al 1983, Lamb *pers.comm.*).

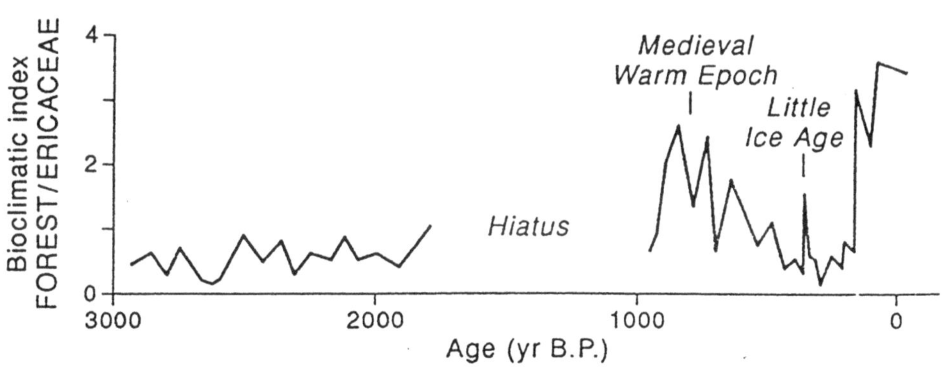

Figure 4.10.b. Short-term fluctuations inferred from changes in pollen deposition at Dega Sala swamp (after Bonnefille and Mohammed 1994, Lamb *pers.comm.*).

the relative changes in vegetation types, and the direction and magnitude of these changes.

Palynology in Tropical Africa

Palynological studies were introduced to African archaeology by van Zinderen Baker (1951, 1976). Although there has since been extensive sampling in Africa, only a minute proportion of the total continent has been studied. Palynological interpretations are therefore extremely broad and often conflicting.

Studies throughout the tropics indicate that arid-humid fluctuations during the Quaternary were responsible for massive changes in the nature and distribution of forest and savannah vegetation types (Hamilton 1976, 1982, Whitmore and Prance 1987, Maley 1989). The general response pattern is an increase in savannah-type grass and scrub woodland during arid phases, and an increase in forest cover, with a concomitant reduction in savannah-type vegetation, during humid periods (as shown in figure 4.11 for example) (van Campo et al 1982).

Palaeoenvironmental Sampling in Ethiopia

Palaeoenvironmental evidence for Ethiopia is derived almost entirely from pollen sequences from peat sediments in the Southern Highlands and stratified lake sediments from Rift Valley lakes that are fed by run-off from the Southern Highlands. These sites are detailed in Table 4.4, and their pollen diagrams are reproduced in *Appendix B*.

Together these cores span the entire Holocene period from 11 500 years BP. Additional pollen sequences are available from mire sediments from the East African highlands, including Mount Elgon, the Cherangani Hills and Mount Kenya and various cores from Rwanda, South West Uganda and Ruwenzoi (Hamilton 1982, Hamilton and Taylor 1986). These corroborate the data from the Ethiopian cores to produce a broad pattern of palaeovegetation changes in Ethiopia during from the Last Glacial Maximum.

Limitations of the Data

There are a number of problems for using these data for a study of the palaeoenvironment of the Northern Highlands. First, the majority of the samples come from the Southern Ethiopian Highlands or the Rift Valley where the vegetation is similar to East African vegetation (figures 4.4.a. and 4.45.b.). These samples are less valid for reconstructions of the different environment of the Northern Highlands. The only sample from the Northern Highlands (Wenchi) was taken from the far southern edge, close to the Rift Valley (see figure 4.4.a). Although the main body of data come from the Southern Highlands, the pattern of change recorded determines our interpretation of environmental change in the Northern Highlands.

Second, the sample sites are not well distributed. This clustering is good for local palaeoenvironmental resolution but is unlikely to be sensitive to more distant, non-homogenous changes in the vegetation spectrum, such as anthropogenic disturbance in the Northern Highlands, 1000 km to the north. Furthermore, the samples from high altitude sites will be relatively more sensitive to changes in high altitude vegetation than sites in the Rift Valley. Rift Valley samples derive from lakes fed by runoff from the Highlands and therefore reflect changes in both Highland and local vegetation.

The resolution for the samples is inconsistent. Slow or fluctuating rates of peat formation and eroded pollen grains at high altitudes, particularly between c.7500 and 3700 years BP, give a poor resolution for this period. There are no comparable data from Rift Valley sites after 6000 years BP.

There are a number of well-dated samples from both the Highlands and the Rift Valley that span the past 3000 years for which the data can be correlated (eg. Lake Turkana, Lake Wenchi). However the data from the early and mid Holocene is poorly dated and can only be used for broad palaeoenvironmental reconstruction. Thick sediment samples had to be analysed from the Mount Badda, Dega Sala and Danka Valley sediment cores in order to obtain a sufficient concentration of dateable material, with over 20 cm depth of sample being used to obtain a radiocarbon date. 20 cm of deposit may span several centuries, and the accuracy of the chrono-stratigraphic calibration for these sequences is reduced. Peat and sediments are not formed at a consistent rate as they are related to environmental variables. Calibration of the core is therefore ambiguous if there are only a few available dates. The Danka Valley core for example has a single date of 7920±80 years BP derived from 20 cm of the sediment, and the pollen sequence can not be used for any accurate diagnosis of vegetation change (Hamilton and Perrott 1980).

The Holocene Palaeoenvironmental Sequence

Although the evidence is far from ideal, it does enable us to make a broad reconstruction of the Holocene environment in the Ethiopian Highlands. The sequence described below Table 4.4. outlines the major palaeobotanical changes in Ethiopia during the Holocene as determined from the palynological evidence.

Table 4.4. Chronology of Sediment Cores for Palaeoenvironmental Sampling from Ethiopia

Sample Source	Location	Core Chronology (years BP)
Mount Badda (Gasse 1978, Bonnefille and Hamilton 1986)	Arussi Mountains (4040 m asl) Southern Highlands	c.11 500 - 0 (3 m peat core from mire)
Lake Abiyata (Lezine and Bonnefille 1982)	Ethiopian Rift Valley (1578 m asl)	c.10 000 - 6000 (162 m lake sediment core)
Danka Valley (Mohammed 1992, Hamilton and Perrot 1980, Hamilton 1982)	Bale Mountains (3830 m asl) Southern Highlands	c.7900 - 0 (2.5 m mud-peat core from mire)
Lake Awassa (Lamb *pers.comm.*)	Ethiopian Rift Valley (1680 m asl)	c.6200 - 2700 (12 m lake sediment core)
Dega Sala (Bonnefille and Mohammed 1994)	Arsi Mountains (3600 m asl) Southern Highlands	c.2900 - 0 (2.2 m peat core from swamp)
Lake Langano (Mohammed and Bonnefille 1991)	Ethiopian Rift Valley (1583 m asl)	c.2500 - 800 (6 m core of lake sediments)
Lake Turkana (Mohammed et al 1995)	Ethiopian/Kenyan Rift Valley (375 m asl)	c.2500 - 600 (12 m core of lake sediments)
Wenchi (Bonnefille and Buchet 1986)	South West, Northern Highlands (2900 m) 150 km west of Addis Ababa	c.1000 - 0 (5.5 m core from crater lake sediments)

The Holocene Palaeoenvironmental Sequence

Late Pleistocene Hyperaridity c.18000-12000 years BP

Upper timber line at 2000/2300 m asl, a lowering of 1000 m from present.

Montane step and scrub extended down to 2000/2300 m.

Montane forest confined in isolated refugia and riparine formations between c.1400 and 2300 m. Xerophytic trees such as acacia more common. Savannah grass and scrub vegetation predominated on thin, patchy soils at this altitude.

Sahelian-steppe encroached up the lower slopes of the Highlands, interspersed with scattered pockets of xerophytic wooded species.

Overview:
A predominately cold, dry, open environment at all altitudes with scattered trees below c.2000 m, particularly around perennial water supplies, and an overall reduction in plant diversity.

Early Holocene Pluvial c.11 500-8000 years BP

Extensive soil formation followed a period of massive erosion at the start of the Holocene, c.10 000 years BP, from sparsely vegetated slopes in the Ethiopian Highlands (Williams and Adamson 1980). Subsequent vegetation upgrowth maintained a prolonged period of pedogenesis over the succeeding millennia which encouraged further vegetation growth. By c.7000 years BP the Highlands were cloaked in deep, fertile altosols, vertisols and inceptisols (Williams and Adamson 1980, Hunri 1989).

Gradual spread of montane forests to higher altitudes[3] and of evergreen forests to lower altitudes.

Retreat of Sahelian steppe and savannah vegetation with forest expansion.

[3] Montane forest species such as *Podocarpus gracilior, Juniperus procera, Olea* and *Hagenia*, and montane savannah species of Gramineae, Cyperaceae, Alchemilla, Tubiflorae and Umbelliferae are consistent during the Holocene in variable proportions, suggesting montane forest-savannah mosaic persisted throughout this period.

Periodic retreat of savannah and forest vegetation corresponding to lacustrine regression phases, particularly c.10 000 years BP.

Upper timber line fluctuated around 400 m below present, at c.3000/3500 m asl.

Increased inclusion of species from moister environments, and greater species diversity.

Overview:
By the end of the 9th millennium BP, the Northern Highlands between c.1500 and 3300 m asl were covered with a mosaic of evergreen thicket and lush montane forest comprising a range of moist and arid adapted species.

Climax vegetation was not fully established until the 7th millennium BP, although botanical diversity increased throughout this period. Prior to the 7th millennium much of Ethiopia was covered with sparse woodland and savannah grassland. This vegetation diminished gradually during this period as the forest canopy became increasingly dense. In addition to this gradual development, vegetation taxa fluctuated rapidly on a century-scale rhythm.

Mid-Holocene Arid Interval c.8000-6500 years BP

Sustained growth and fluctuation of montane species at higher altitudes with a significant increase in *Podocarpus*[4] in the middle of this period. Levels of *Podocarpus* pollen oscillate throughout the sequence as the forest ecozone expanded and contracted with increasing and decreasing aridity.

Moist-adapted species retreated from the Highland forests and were replaced by arid adapted species such as acacia, especially at lower altitudes where these displaced evergreen species.

Increased grass and scrub at lower altitudes.

Slower deposition of high altitude mire sediments.

Overview:
At lower altitudes forest cover was diminished and grass-land expanded to form a more open environment. Above c.1500 m the forest-savannah mosaic expanded and contracted in response to climate changes.

Mid-Late Holocene Pluvial c. 6500-5000 years BP

Climax vegetation established by the 7th millennium BP with a substantial increase in forest area relative to savannah grassland.

Overview:
Dense, lush vegetation cloaked the Highland zone between c.1500 and 3000 m asl and the environment was more enclosed than during the early Holocene. Substantial areas of the Highlands would have been forested, with possibly over 70% tree cover (compared to present 4% cover). Montane forest-savannah predominated at higher altitudes, c.2200-3000 m, evergreen thicket predominated at lower elevations to c.1500 m.

Late Holocene Aridity c.5000-2000 years BP

Fluctuating levels of woodland taxa, with increasing representation of arid species such as *Podocarpus* in the montane component and acacia in the savannah-woodland component.

Expansion and contraction of savannah and savannah scrub type vegetation at lower altitudes.

Upper timber line fluctuated around c.3500 m asl with the development of Ericaceous scrub at higher altitudes.

Rate of soil formation decreases after the 5th millennium BP.

---

[4] This trend also occurs throughout the East African Highlands eg. Mount Elgon (Hamilton 1982), Mount Kenya (Coetzee 1967), Ruwenzori (Livingstone 1967) and Lake Bogoria (Vincens 1989). *Podocarpus gracilior* is a dry montane gymnosperm highly characteristic of the Ethiopian and East African Highlands at altitudes of 1200-2000 m asl.

Overview:
Environment similar to present, although more densely wooded with up to 50% forest cover.

Late Holocene anthropogenic interference (see 5.3.2 below)	Localised anthropogenic clearance of forest and vegetation tentatively inferred from soil erosion in the Northern Highlands between 5000 and 3900 years BP. Charcoal inclusions in the same palaeosols from the 7th millennium BP suggest forest clearance and an even earlier date for human activity. c.2000 years BP, possible human interference detectable in certain high altitude pollen sequences from the southern Ethiopian Highlands where lower altitude wind-blown pollen has been deposited[5]. Charcoal, soil erosion and pollen changes in sediments from Lake Awassa suggest localised vegetation clearance from c.3000-2500 years BP in the Rift Valley (Lamb *pers.comm.*). *Overview:* Possible human interference from the 7th millennium BP in the Northern Highlands, from the early 2nd millennium BP in the Southern Highlands and c.3000 years BP in the Rift Valley.

4.3.2. Evidence for Anthropogenic Disturbance

Geomorphological Evidence

The Northern Highlands

Geomorphological data has recently been collected by a joint American-Italian team working in the Northern Highland province of Tigray (Brancaccio et al 1994, Beraki et al 1994, Bard et al *in press*). Tigray is situated close to the North Eastern edge of the Highland Plateau, with most of the province lying above 2000 m. This study represents one of the few geomorphological investigations in the Northern Highlands (Butzer 1981, 1982, Michels 1988). It is virtually the only study conducted at an altitude conducive to human settlement, and is therefore of utmost importance for identifying prehistoric anthropogenic disturbance.

Geomorphological and stratigraphical analysis of palaeosols in Tigray province identified a number of significant features:

- The onset of soil formation dating from c.8300 BP indicates the development of thick vegetation cover. This phase was maintained until at least the early 7th millennium, although soils continued to be deposited until the start of the 5th millennium BP in certain of the regions investigated.
- A large number of charcoal fragments were associated with peat deposits during this formative stage, particularly between c.7300 and 5000 years BP. Berakhi et al (1994) have suggested that these represent traces of localised anthropogenic clearance activities at a time when the environment was heavily wooded.
- Reduction in soil formation and travertine deposition between c.5000 and 3900 BP years has been tentatively attributed to human factors such as systematic vegetation clearance for early food production (Berakhi et al 1994, Brancaccio et al 1994, Bard et al *in press*). Similar patterns of erosion in the Mediterranean, for instance, are the product of human interference (Vita-Finzi 1969, Delano-Smith 1979).

Although there are difficulties in distinguishing between natural soil erosion factors and those arising from human activities (Goudie 1981), anthropogenic impact may be proposed from at least the start of the 4th millennium BP. This impact progressively increased during the PreAxumite period in the late 3rd millennium BP (Anfray 1967b, Fattovich 1990a, 1990b, Bard et al *in press*). Significantly, this geomorphological record also tentatively identifies non-intensive vegetation clearance during the 6th and 7th millennia BP, suggesting localised human activities.

These proposals must be rigorously tested before any firm deductions can be made. Further analysis of the charcoal, its geological, environmental and topographical contexts is essential for expanding our understanding of this important piece of evidence.

The Rift Valley

The evidence from the Northern Highlands is corroborated by recent palaeolimnological work in the Ethiopian Rift Valley (Lamb *pers.comm.*). Preliminary palynological analysis of

[5] Similar changes are absent from East African Highlands sequences (Hamilton 1982).

sediment cores from Lake Awassa indicate a significant change in the vegetation structure after c.3000 years BP. These changes show an increase in montane flora, especially *Podocarpus, Olea, Hagenia, Juniperus, Celtis* and *Dodonea*. Increased values of *Celtis*-type, *Dodonea* and *Hagenia* species are consistent with forest clearance and secondary forest replacement activities (Hamilton 1982), and may indicate increased human interference from the start of the 3rd millennium BP.

It is unclear at present whether or not these changes are the result of climatic cooling and a consequent lowering of the vegetation bands (Lamb *pers.comm.*). However, a concomitant rise in the charcoal content of the lake sediments was noted, and the rate of soil erosion increased at this time. These patterns suggest anthropogenic forces were operating in the Rift Valley from at least the start of the 3rd millennium BP.

Palynological Evidence

Danka Valley

The Danka Valley core shows possible evidence of human disturbance from the mid-2nd-1st millennium BP in the mountains of the Southern Highlands (Hamilton and Perrott 1980, Hamilton 1982):

- Relatively high values for *Celtis*-type, *Dodonea* and *Hagenia* associated with forest clearance and secondary forest replacement[6].
- Relatively low values for *Macaranga kiilimandscharica* and *Podocarpus* indicative of reduced forest cover.
- Reduced Ericaceae due to high altitude human disturbance.

Mount Badda

The Mount Badda core records massive vegetation changes from c.1850 years BP (Gasse 1978, Hamilton 1982). These changes, which are detailed below, have been attributed to anthropogenic clearance activities associated with food production:

- Increased *Celtis*-type, *Dodonea, Hagenia* and *Myrica* pollen.
- Reduced *Podocarpus* pollen.
- Increased Chenopodiaceae, *Rumex* and *Plantago,* commonly associated with grazed or cultivated land.

The inference that anthropogenic disturbance is detectable only after 2000 years BP in the Southern Highlands must be considered carefully with reference to the pollen profiles from Mount Badda and Danka Valley, shown in *Appendix B*. A number of anomalies in the data suggest that the above conclusion is unacceptable:

- Badly eroded grains and poor stratigraphy between 6000 and 3700 BP prevent an accurate assessment of whether similar changes occurred before 2000 years BP.
- Particularly high concentrations of Chenopodiaceae pollen are present earlier in the sequences until the 8th millennium. This has been identified as *C.procerum* - an upland herb or a weed of cultivation at lower altitudes. The upland and lowland components are not distinguished in this sequence and the changes observed at the start of the 2nd millennium BP do not necessarily result from human disturbance.
- *Dodonea, Celtis* and *Rumex* pollen levels all fluctuated during the Holocene. Increased levels correlate with reduced *Podocarpus* pollen, particularly c.3500-3000 years BP, when they were associated with an increase in Chenopodiaceae.
- There is no apparent increase in Gramineae pollen after 2000 years BP that would indicate cereal crop cultivation and expansion of pasture land.
- If, as Hamilton (1982) suggested, wind-blown pollens from lower altitudes are well represented in the high altitude cores at Mount Badda and Danka Valley, then we should detect similar patterns of change at lower altitude sites. However similar changes are **not** observed in the Rift Valley lake sequences which also span this period (the Dega Sala core is saturated by local Ericaceous pollen and upper timber line arboreal pollen and would not be sensitive to non-localised changes in other pollen types).
- Finally, there is an abrupt humid phase c.2000 years BP (Lezine and Bonnefille 1982, Mohammed et al 1995) which may have been responsible for some of the observed changes.

4.4. SUMMARY: A PALAEOENVIRONMENTAL CONTEXT FOR HUMAN ADAPTATION DURING THE HOLOCENE

The Ethiopian palaeoenvironment has been highly dynamic throughout the Holocene. Recurrent shifts in temperature and precipitation since the Last Glacial Maximum have caused forest refugia and savannah grass-land to expand and contract repeatedly across Ethiopia. Long-term climate trends have influenced slow patterns of plant growth and colonisation, such as the gradual establishment of climax vegetation. In addition, and superimposed over this gradual trend, there is a more rapid, century-scale flicker of vegetation change that correlates chronologically with intermediate-term climate

[6] Similar changes are documented bewteen 2000-1000 BP in pollen profiles from certain other East African mountains, for example South West Uganda and Ruwenzori (Hamilton 1982).

changes in Ethiopia. The general temporal response of vegetation adaptation to changes in effective precipitation (estimated from lake level fluctuations) has been modelled on a long-term time scale by Roberts and Baker (1993).

Two patterns of vegetation change are evident in Ethiopia which correlate to the long- and intermediate-term climatic oscillations that characterise the Holocene period. The Dega Sala pollen profile records rapid, decade-scale changes in the sensitive high altitude vegetation adjacent to the swamp which corroborate chronologically with geomorphological changes. It would appear that the environment responds to all scales of climate change and that it effectively echoes the climatic pattern.

Climate Change and Human-Environment Impact

The impact of short-term climate change was discussed in chapter 2 with reference to the ethnographic record. Small-scale variations in climate can induce a catastrophic anthropogenic response. The extent of this response provides an indication of the implications of more persistent and pronounced climate changes.

In recent years there has been increasing interest in the nature of human impact on the environment, particularly as an archaeological tool for reconstructing systems of land use and subsistence activities. This issue is currently being addressed to an African context (eg. Schultz 1996, Ndiiri 1996, Haynes 1996, Barakat 1996). Palaeoenvironmental research in Ethiopia tentatively indicates anthropogenic activity in the Highlands from the late 8th millennium BP, and in the Rift Valley from the early 3rd millennium BP as shown below. These results provide a viable framework for investigating and interpreting the archaeological evidence from Ethiopia.

c.7000 - 5000 BP	Low intensity forest clearance in the *woina dega* zone of the Northern Highlands, possibly in response to increased vegetation cover in the early Holocene. First indication of human interaction with and impact on the environment.
c.5000 - 3000 BP	Vegetation clearance in the Northern Highlands and possibly also in the Rift Valley.
c.3000 - 1500 BP	Increased intensity of clearance and land use in the Highlands. Possible evidence of agriculture in the Rift Valley and Southern Highlands.

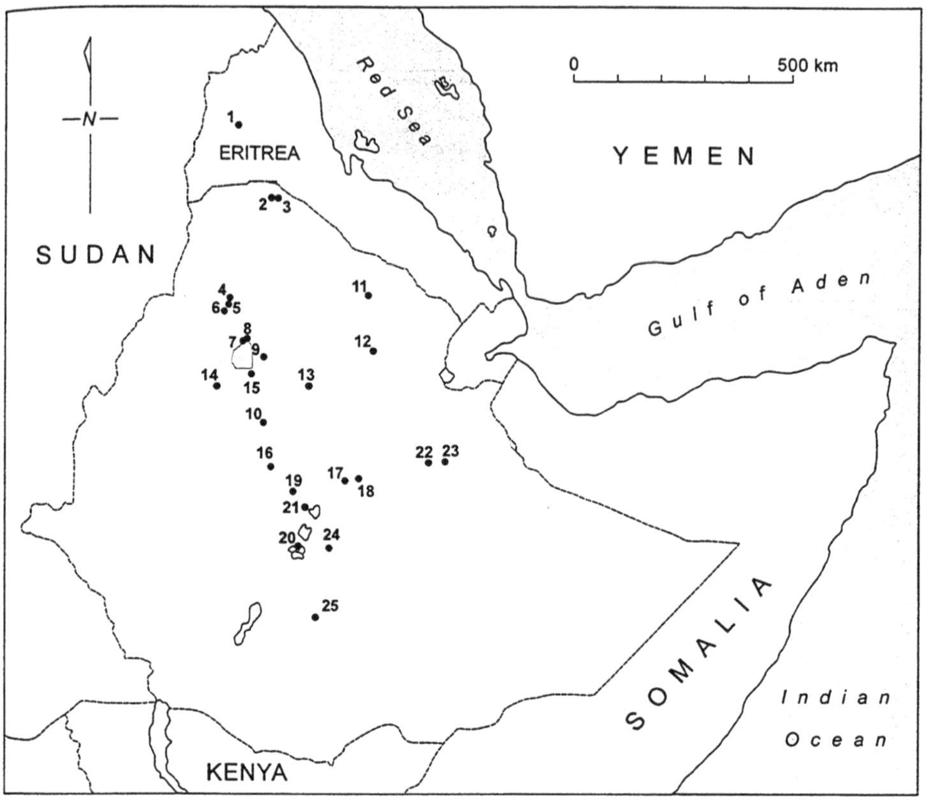

Figure 5.1. Distribution of known archaeological sites in Ethiopia

KEYS TO FIGURES 5.1. AND 5.9.

Archaeological sites shown in figure 5.1:

1. Agordat
2. Gobedra
3. Axum area
4. Cliff of monkeys
5. Wekerdeba Mariam
6. Gondar area
7. Gorgora rock shelter
8. Gorgora peninsula
9. Lalibela and Natchabiet caves
10. Debra Tabor
11. Quiha
12. Zergnat
13. Dessie
14. Dimma
15. Yizana creek
16. Idobi Mariam
17. K'one
18. Lake Besaka
19. Melka Konture
20. Lake Besaka area
21. Aladi Springs
22. Porc Epic
23. Laga Oda
24. Gadeb Plain
25. Yavello

Archaeological sites shown in figure 5.9:

1. Agordat
2. Axum area and Gobedra rock shelter
3. Cliff of Monkeys
4. Gondar area
5. Gorgora peninsula
6. Lalibela and Natchabiet caves
7. Idobi Mariam
8. Quiha rock shelter
9. Dimma
10. Debra Tabor
11. Zergnat
12. Dessie
13. Wekerdeba Mariam
14. Melka Konture
15. Aladi Springs
16. Lake Ziway area
17. Porc Epic cave
18. Laga Oda
19. Gadeb Plain
20. Gorgora rock shelter
21. Yizana creek
22. K'one
23. Lake Besaka
24. Yavello

Chapter 5.

ETHIOPIA IN PREHISTORY: THE ARCHAEOLOGICAL EVIDENCE FOR SUBSISTENCE INTENSIFICATION

5.1. A HISTORY OF ARCHAEOLOGICAL RESEARCH IN ETHIOPIA AND THE HORN

The Foundation of Archaeological Research in Ethiopia

The romantic images that are associated with Ethiopia - the homeland of the Queen of Sheba, the land of Prester John and the resting place of the Ark of the Covenant, for example - have long intrigued and attracted Europeans and have stimulated many historical investigations. Similarly, in 1974, the discovery of early australopithecines in the Afar Rift (Johanson 1975, Johanson and Taieb 1976) promoted extensive palaeontological research in the Ethiopian Rift Valley and established the region's credentials as an evolutionary cradle. These interests have dominated research into Ethiopia's cultural heritage, with the result that Pleistocene and Holocene developments have received alarmingly little attention.

Middle Stone Age (MSA) and Late Stone Age (LSA) artefacts were discovered in the Horn in the 19th century (Revoil 1882), but no Stone Age sites were dug until 1929 when a French team made a small excavation at Porc Epic cave on the Southern Rift Valley escarpment (Teilhard de Chardin 1930).

The Development of a Cultural-Chronological Framework

Much of the research over the next 25 years was conducted by British military personnel stationed in British Somaliland during the Second World War. This was concentrated particularly in Somalia, where excavations at the MSA/LSA site of Gogoshiis Qabe at Buur Heybe established a general cultural development sequence for this region (Graziozi 1940). Work in Ethiopia was more restricted, partly due to geographical constraints, and much of the data were derived from unsystematic surface collections. The archaeological exploration and small-scale excavations carried out by Moysey and Clark in the early 40s underlie the basic cultural-chronological framework for Ethiopia used today. The data were collated by Clark (1954) in *The Prehistoric Cultures of the Horn of Africa.* This remains the only comprehensive review of Stone Age research in the Horn and Ethiopia, although the issue was recently addressed by Brandt (1986). Only four other ceramic-bearing LSA deposits have been excavated in the Ethiopian Highlands since Moysey's work at Gorgora and Quiha rock shelters (Moysey 1943). These include Gobedra, Lalibela and Natchabeit caves in the Northern Highlands, and Laga Oda in the Southern Highlands. The material from Moysey's excavations at Gorgora and Quiha rockshelters has not been published comprehensively, despite its relative importance for our understanding of Ethiopian prehistory. The lithic artefacts from Gorgora were analysed by Leakey (1943), but the pottery was only examined superficially. The material from Quiha was briefly mentioned by Clark (1954), but has not been examined and the excavations have never been published.

1954 also saw the publication of Major Last's findings at Agordat in Eritrea (Arkell 1954). Arkell recorded the artefacts obtained from the surface of four occupation sites at Kokan, Ntanei, Shabeit and Dandaneit. The ground stone artefacts, ceramics and copper objects show distinct affinities with those of the C-Group Nubian and Kerma cultures from the Sudan, leading to Arkell's suggestion that the sites dated to the 4th millennium BP. There has been no subsequent investigation of these sites save for a single sondage dug in 1994 (Brandt *pers. comm.*).

In 1963, field work began at Melka Konture along the Awash River in Central Ethiopia to excavate extensive, stratified Palaeolithic occupation levels in order to define a technological and cultural development sequence. Over the next 20 years this research revealed a dynamic continuity of Stone Age occupation covering the past 1.8 million years (Chavaillon 1976, Chavaillon et al 1979, Hivernel-Guerre 1976). LSA material is known only from surface collections and has not been dated.

Interdisciplinary Research 1970s-1990s

A number of interdisciplinary projects were initiated in Ethiopia in the 1970s in the wake of global interest in the relationship between palaeoenvironmental and anthropogenic factors. Work focused on the Rift Valley where evidence of lacustrine episodes and climatic fluctuation could be derived from lake sediments. The palaeolimnological data provided a context for investigating the relationship between environmental change and human behaviour. In 1970 the 4th International Mission to the Afar identified lacustrine LSA sites associated with early Holocene transgressions (Roubet 1971). Subsequently the Combined Prehistoric Expedition investigated the Lake Tana area of the Northern Highlands. A number of lithic deposits were identified around the lake shore, but the project was relocated in the Rift Valley after the area was found to be extensively disturbed by agriculture (Gallagher 1973). Several MSA and LSA sites were excavated around Lake Ziway during the early 1970s by The

Combined Prehistoric Expedition to the Central Rift Valley of Ethiopia following extensive surveys of the Afar and Ethiopian Rift (Humphreys 1978, Wendorf and Schild 1974).

In 1974 several stratified MSA and LSA sites were located in the Southern Afar Rift. These were test-excavated as part of an interdisciplinary project initiated by J D Clark and M A J Williams, before more extensive excavations at Laga Oda, Porc Epic Cave, K'one, Gadeb and Lake Besaka were undertaken the following year (Clark and Williams 1978) and in subsequent years (Brandt 1980, 1982, Clark and Williamson 1984). Additional research projects identified Stone Age sites in the Ethiopian Rift, in the lower Omo Valley and around Lake Ziway (Brown 1975, Gasse and Street 1978, Humphries 1978).

The Quest for the Origins of Agriculture

Research has been more limited away from the Afar and Central Ethiopian Rift Valley during the past 40 years due to a combination of lower potential for interdisciplinary research, fewer known sites, more rugged terrain and greater agricultural disturbance. In 1970, Jean Dombrowski recorded a number of LSA open sites and cave occupation scatters in the Northern Highlands. Her excavations at Lalibela and Natchabiet caves, east of Lake Tana, were aimed at recovering plant and animal remains and thereby to investigate the origins of agriculture (Dombrowski 1970, 1971). This still remains the only project specifically designed to test the existing theories on food production in Ethiopia. Although these sites dated from the mid-3rd millennium BP and were later than anticipated, Dombrowski's identification of domesticated crop plants at Lalibela cave remains the earliest evidence for agriculture in Ethiopia, and some of the earliest definitive evidence for domesticated plants in sub-Saharan Africa.

In 1974, David Phillipson directed a small-scale excavation of a stratified LSA deposit at Gobedra rock shelter, close to Axum in the Northern Highlands, in order to investigate the economic and cultural precedence for the Axumite civilisation (Phillipson 1977b). Although the excavation was constrained in this respect, it provided the longest stratified LSA occupation sequence in Ethiopia, spanning the period from the late Pleistocene to the 1st millennium BP.

Current Research Programmes and Objectives

Political unrest and upheavals interrupted archaeological research in Ethiopia and Somalia in the 1980s and all investigations were temporarily suspended in 1982 pending new antiquities laws. Current research interest is particularly focused on the North East Highlands following the re-opening of excavations at the important 2nd millennium BP site of Axum in Tigray Province, with a number of localised survey projects and excavations aimed at determining the origins of the Axumite kingdom (Fattovich et al 1988, Bard and Fattovich 1995, Finneran *in prep.*, Bard et al *in press*).

There has been no further work on prehistoric Holocene LSA sites since the 1980s, although renewed interest in palaeoclimatic and palaeoenvironmental change in both the Rift Valley and the Highlands has potential for integrating archaeological investigations into a programme of interdisciplinary research (Lamb *pers.comm.*, Coltorti *pers.comm.*).

Summary

A combination of academic, physiographic and political factors have restricted archaeological research of the later prehistory in Ethiopia. The majority of excavations to date have been small-scale and site-specific, with very few projects addressing specific objectives. As archaeological research elsewhere in Africa is influenced by an increasingly scientific and conceptual approach, the gaps in our knowledge of Ethiopian prehistory are becoming glaringly apparent. Almost nothing is known about major cultural developments during the Holocene period, such as the origins of food production and the rise of commercial trade and state formation. The schematic cultural development framework that was constructed over 40 years ago is now in drastic need of revision. Any contribution to our current state of knowledge is amplified in this context. This chapter presents a review and a reinterpretation of secondary data, which incorporates a detailed analysis of excavated artefactual material from Gorgora rock shelter in the Northern Ethiopian Highlands. This material is currently stored in the National Museum of Kenya, and was examined during a visit to Nairobi in June 1996.

The location of LSA sites mentioned in the text are shown in figure 5.1. Details from all sites noted in the text are summarised in *Appendix C*. Table 5.1 gives a general overview of broad cultural changes during the Late Quaternary period. Figure 5.2 illustrates the chronological span of Holocene archaeological sites.

Table 5.1. Current Overview of Major Cultural and Technological Developments in Ethiopia

Cultural Period	Date (years BP)	Technological Development
MSA	>200 000-30 000	Composite tool technology
LSA	c.27 000-recent	Microlithic technology
Ceramic LSA	c.7000?	Ceramics and food processing equipment
Pastoral LSA†	c.4th millennium	Domesticated animals
PreAxumite*	c.3800/2800	Metal technology Monumental architecture Domesticated plants and agriculture
Axumite*	c.AD 300-800	Written texts
Historical	c.AD 800-present	

* In the Northern Highlands and Eritrea only.

† Only known in Rift Valley at present, possibly earlier in the Highlands.

Figure 5.2. Comparative chronology for occupation of Ethiopian LSA sites

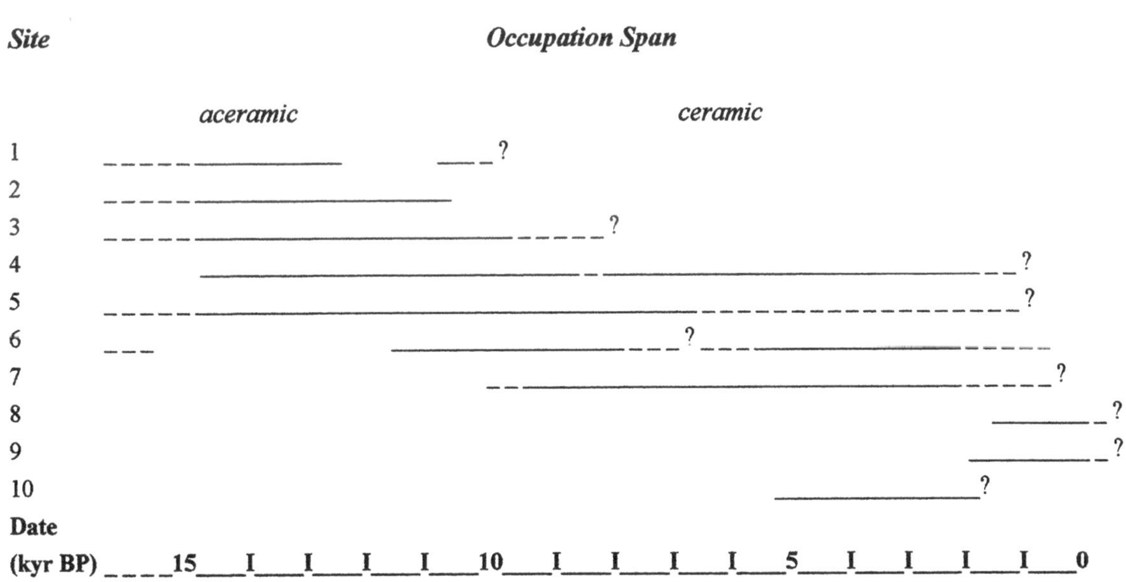

Key to sites:

1. Aladi Springs (Rift Valley)
2. Macho, Lake Ziway (Rift Valley)
3. K'one (Rift Valley)
4. Laga Oda (Southern Highlands)
5. Gorgora (Northern Highlands)
6. Lake Besaka (Rift Valley)
7. Gobedra (Northern Highlands)
8. Lalibela and Natchabiet (Northern Highlands)
9. Agordat (Eritrea)
10. Quiha (Northern Highlands)

5.2. CULTURAL CONTEXT AND CULTURAL DEVELOPMENTS: THE ARCHAEOLOGICAL EVIDENCE

5.2.1. The LSA Lithic Sequence

Conventional Typology and Terminology

From post-Acheulian times (roughly the past 100 000-200 000 years), the African palaeolithic has been conventionally divided into Middle Stone Age (MSA) and Late Stone Age (LSA). These are now recognised as arbitrary labels within an evolutionary technological continuum, with no clear distinction between successive phases. The MSA developed from the Achuelian between 200 000 and 100 000 years BP, or even earlier if the K/Ar date of 400 000 years BP from Kinangrop Plateau in Kenya is accepted (Evernden and Curtis 1965). During the Pleistocene, the size of MSA tools (including blades, points, burins and parallel sided flakes) gradually decreased. An increasing number of tools were blunted or backed to facilitate hafting as composite implements. The MSA lithic industry of Ethiopia shows little change for over 100 000 years (Brandt 1986), comprising a significantly high proportion of blades in addition to other MSA tools.

By the late Pleistocene (c.50 000-25 000 years BP), microlithic, backed tools which characterise the LSA industry were being produced. These tools were either hand-held or were hafted in wood using natural mastic, traces of which are often preserved on the microliths (Clark and Prince 1978). The earliest microlithic industry currently known from Ethiopia comes from Bulbula River, near Lake Ziway. A radiocarbon date of 27 050±1540 years BP for this industry was obtained on associated geological levels (Gasse and Street 1978, Gasse et al 1980).

Ambiguities in the Ethiopian Data

Transitional Industries

However, a rigid evolutionary trajectory may not be appropriate to describe the sequence of technological development in Ethiopia. MSA tools frequently occur in association with late Pleistocene LSA material, for example at Porc Epic cave, Aladi Springs, K'one and Gorgora (Clark and Williamson 1984, Clark and Williams 1978, Leakey 1943). Clark (1954) proposed that a transitional 'Magosian' industry existed in Ethiopia which comprised MSA points, cores and flakes, with LSA microliths and blades. This typology has subsequently been discarded. The current suggestion in that there is no 'transitional' phase and the Magosian industry is the product of extensive stratigraphical disturbance (Clark and Williams 1978, Brandt 1986). At Gorgora, the ceramics should have been affected by the same post-depositional disturbance as the lithics. Instead, however, the ceramics show a clear developmental sequence (see *Appendix C* and Section 5.2.3), suggesting that the stratigraphy at this site was not drastically disturbed.

Typological Problems

In addition, archaeological and ethnographic evidence indicate that there is a high degree of continuity in lithic technology and functionality in Ethiopia. Shaped lithic tools are still manufactured and widely used in Ethiopia today. Obsidian tools are especially common in the Rift Valley and Southern Ethiopia, and have been the subject of several ethnographic surveys (Gallagher 1973, Weedman 1996, Brandt 1996). These artefacts are often closely comparable to those from archaeological contexts, making precise typological identification problematic. Surface scatters of obsidian tools, found recently in the Rift Valley, have been tentatively identified by one authority as having a late Pleistocene/early Holocene date. Geomorphological survey of their context indicates extensive localised volcanic activity during the 3rd millennium BP that would have buried surfacial archaeological material. This suggests that the artefacts in question were actually deposited substantially later than predicted by typological estimates. A similar problem is encountered at Axum, in the Northern Highlands, where Palaeolithic tool types known as Gudit scrapers have been found in Axumite contexts (Phillipson 1977b).

Additional anomalies exist. At Porc Epic cave in the Rift Valley, a MSA deposit was radiocarbon dated on charcoal to 5700±100 years BP (Clark and Williamson 1984). The deposit was sealed by dripstone which gave ^{14}C and Th/U series dates of 4590±60 and 6270±1020 years BP respectively. The excavators proposed that the dripstone had formed significantly later than the final MSA deposition, and that the charcoal sample was too small or contaminated (Clark and Williamson 1984). Neither of these arguments can be substantiated. These dates should not be rejected until further research has been attempted. Subsequent obsidian hydration dates obtained on three MSA-type artefacts from Porc Epic cave gave values of 77 565, 61 640 and 61 202 years BP (Clark and Williamson 1984, p.68). However these artefacts were from an unknown stratigraphy, from the 1933 excavation, and were *assumed* to belong to early MSA levels. This interpretation is conjectural and the possibility remains that MSA lithics may have been utilised or reused at Porc Epic throughout the late Pleistocene and Holocene.

Typological Continuity

MSA artefact types possibly persisted into the late Pleistocene and Holocene periods in Ethiopia. Evidence from K'one, Porc Epic, Gorgora, Aladi Springs and surface occurrences in the Highlands and Rift Valley suggest that after c.30 000 years BP MSA tool types may have been utilised in addition to a microlithic technology. There are no precise dates on these sequences, and this suggestion remains contentious. Nevertheless it does illustrate that a conventional typology sequence for Ethiopia must be accurately dated and revised before it can be accepted. The examples given above indicate the dangers of corroborating cultural development solely with changes in lithic technology.

LSA Terminology in Ethiopia

The LSA has been identified from numerous surface scatters and excavated sites across the Highlands and the Rift Valley (figure 5.1). All known LSA sites date to the Holocene period, with the exception of Bulbula River, and late Pleistocene phases at Lake Besaka, K'one and Laga Oda. The term **LSA** will be used here to refer to Holocene sites.

The LSA microlithic industry of Ethiopia is termed the Ethiopian Blade Tool Tradition (EBTT). This is a complex of obsidian tools and some chert which is characterised by a high frequency of large blades, either backed or with inverse retouch, also burins, scrapers and geometric microliths (Brandt 1986).

The EBTT is known particularly from LSA Rift Valley sites, especially Lake Besaka where it dates from c.20 000-4500 years BP (Brandt 1982). The EBTT is poorly defined in the Highlands due to the limited archaeological work in this region, but it can be said to form a broadly homogenous unit across the Ethiopian Highlands and Rift Valley (Clark 1954). The industry developed from the later MSA industries, with continuity shown in the blades and in the manufacturing technology. Similarities have been proposed between the EBTT and the Eburran and Elmenteitan industries from the late 5th and 4th millennia BP in the Kenyan Rift Valley and with Saharan industries, although these statements remains to be tested (Clark 1954, Kurashina 1978, Ambrose 1984).

Spatial and Chronological Diversity

No clear spatial or diachronic divergence from the EBTT can be identified at present for the Holocene period. However, Clark (1954) recognised a distinct, localised LSA lithic assemblage in the Northern Highlands. This is known only from one excavation at Quiha rock shelter and two surface collections at Zergnat and Amba Sel (Clark 1954). The ceramic-bearing assemblage is dominated by obsidian blades, scrapers and whole or broken microlithic lunates. Many of the tools from Zergnat show edge-wear and edge-gloss similar to that identified on tools from Laga Oda dating between 15 000 and c.500 years BP (Clark 1954, Clark and Prince 1978). Clark suggested that the Quiha artefacts resembled those of 'the Elmenteitan and derivative cultures of Kenya' (1954 p.324). At present it is unclear whether this group of assemblages represent an introduction to the Highlands or a spatial or temporal development from the EBTT.

A distinct lithic industry is also known from a single small excavation at the site of FeJx3 at Lake Besaka (Brandt 1982). This industry appears c.3500 years BP and is characterised by a large number of scrapers with fewer blades, burins and microliths than are found in the previous EBTT assemblages at Lake Besaka. Differences in cores and flakes and in the length-width ratios of the tools indicate changes in manufacturing technology from the EBTT. Domesticated cattle teeth have been tentatively identified from this context (Carter in Brandt 1982), leading Brandt to suggest that this industry was introduced to the Rift Valley by pastoralists during the 4th millennium BP (1980, 1982).

Tool Types and Function

Gallagher's ethnographic studies at Dalacha, Lake Ziway, showed that many tool types found in archaeological contexts are used in this area today (Gallagher 1973). The local inhabitants distinguished only three classes of lithics, in which the size determined the function: small flakes and bladelets (2.0-4.0 cm) for facial shaving, nail trimming and hair cutting; medium flakes (4.0-8.0 cm) for cutting meat and for weapons when needed; large flakes (8.0-12.0 cm) for scraping hides. Only raw flakes and blades were used, rather than shaped tools. The retouch common to archaeological tools was not deliberate, but was caused by use-wear. It is probable that metal tools have replaced more complex, shaped lithic tools in recent years. More recent ethnographic work indicates that a more sophisticated manufacturing technology still persists in Southern Ethiopia (Brandt 1996).

Stone Age types of scrapers are the most common stone tool type used today in Ethiopia and parts of Africa for processing and preparing hides for clothing and bedding (Weedman 1996, Brandt 1996). For example Fuga tanners among the Gurage around Lake Besaka and in the Afar Rift use scrapers similar to those from LSA deposits at Lake Besaka for separating the fat from the skins and for thinning the hides to make them supple (Clark and Williams 1974).

Use-wear and edge-gloss was found on microliths from Laga Oda and Zergnat (Clark and Prince 1978, Clark 1954). Edge-damage on a number of microliths suggested a two-way sawing motion had been used. Comparable results have been obtained experimentally from light whittling and cutting materials such as fish or animal flesh, or soft-stemmed plants (Tringham et al 1974). These activities produce moderate edge-wear with microscopic striations or crushings. Tropical plants are thicker and more fibrous than temperate plants and require efficient tools for cutting and processing. Reconstruction experiments have shown that a range of fibrous and silicaceous tropical plants, including *Agave sisalana* (agave), *Phormium tenax* (type of flax), *Sanservieria* sp (African sisal) and *Equisetum* sp are capable of producing edge-wear and polish that is microscopically comparable to that on microliths from Laga Oda (Clark and Prince 1978). While there is considerable debate about the diagnostic reliability of use-wear and edge-polish, particularly with respect to the type of material worked (eg. Keeley 1980, Newcomer et al 1986), developed edge-polish, or gloss is **only** formed when processing plants with a high silica content (Keeley 1980, Unger-Hamilton 1988). Edge-gloss has been found on LSA tools across Northern Africa dating from the end of the Pleistocene (eg. Laga Oda and Zergnat in Ethiopia, Clark and Prince 1978; Qadan sites in Tushka, Nubia, Wendorf 1968; Iwo Eleru in South West Nigeria, Shaw and Daniel 1984; and Pre-Dynastic sites at Fayum, Egypt, Caton-Thompson and Gardener 1934). This implies that plants with silicated stems, especially certain grasses, were being exploited from at least 15 000 BP.

Summary

The limited evidence currently available from stratified, dated LSA deposits indicates that a broadly homogenous lithic industry persisted across Ethiopia during the Late Pleistocene and Holocene which was derived in part from the MSA. The LSA lithics had a variety of applications, particularly processing of plant and animal material. There was little apparent technological development during the LSA, although the appearance of distinct industries at Lake Besaka in the mid-4th millennium BP and in the Northern Highlands at Quiha, Zergnat and Amba Sel may reflect regional cultural fragmentation and isolation, as well as possible economic adaptations.

Anomalies in the typological sequence mean that we should not rely on lithics for relative dating of surface deposits or for interpretation of culture change. The database is not detailed enough for a comprehensive quantitative investigation of regional and temporal variation. Within the limitations of the data, Ethiopian lithic technology provides a background against which other aspects of material and non-material culture can be assessed.

5.2.2. Groundstone Artefacts

Technological Precedence and Functional Adaptation

Archaeological Evidence for Grindstones in Ethiopia

Grindstones rarely occur in MSA contexts in Ethiopia. The presence of stained grindstones from MSA deposits at Gorgora rock shelter suggests that pigment processing may have been a primary function at this site (Clark and Williamson 1984). A fragment of a grindstone was also recovered from possible MSA levels at Porc Epic cave where it was contextually associated with heat-treated and abraided pigments (Clark and Williamson 1984).

Grindstones and rubbers appeared with increasing frequency in both Highland and Rift Valley sites during the 7th millennium BP (Phillipson 1977, Brandt 1982, Leakey 1943), often at sites where there was no precedence for their use (see Table 5.2). Grindstones continued to be found in archaeological deposits during the Holocene, but none of these later implements were pigment stained (see figure 5.3). Instead, these had smoothed, pitted surfaces with traces of silica patina. This suggests that by the 7th millennium BP, grinding technology was being applied to plant food processing, particularly of silicated grasses. The frequency with which grinding equipment occurs in LSA deposits at sites such as FeJx2 at Lake Besaka suggests that it was a regular feature of the food processing technology from the 7th millennium BP.

LSA Applications and Cultural Implications

Grindstones can be used for a variety of functions. Pigment stains on grindstones from Lake Besaka and Gorgora suggest that these implements were used for grinding inorganic substances and minerals[1]. Surface pitting, abrasion and polish on tools from Gorgora, Gobedra and Besaka was comparable to use-wear from processing hard, starchy roots and seeds, indicating that similar plant materials were being processed at these sites in the mid Holocene (Phillipson 1977, Brandt 1982).

Starchy roots and tubers of aquatic plants may have been important resources at sites with access to swampy environments, such as Lake Besaka. Gobedra was better situated for exploitation of seeds and grains, which are more prolific in a Highland environment and are still exploited widely in this region today. Grinding stones may also have been used to process a wide range of non-edible vegetable material, for example for use as medicine, cosmetics and poison.

Grinding technology increases the range of plant resources that can be utilised within a particular environment. Many indigestible plants, including certain species of yam, can be rendered edible if processed in this way (Jones and Meehan 1989). Ethnographic and archaeological research show that even highly toxic tubers and grains can form an important component of the diet if they are detoxified (Stahl 1989). Furthermore, grinding is a process through which a useful, versatile resource, such as flour, can be produced from small, almost unusable seeds and grains including grasses and

[1] Linguistic evidence from interlacustrine Africa also indicates that grindstones may have been used for processing salt (Bernal *pers.comm.*).

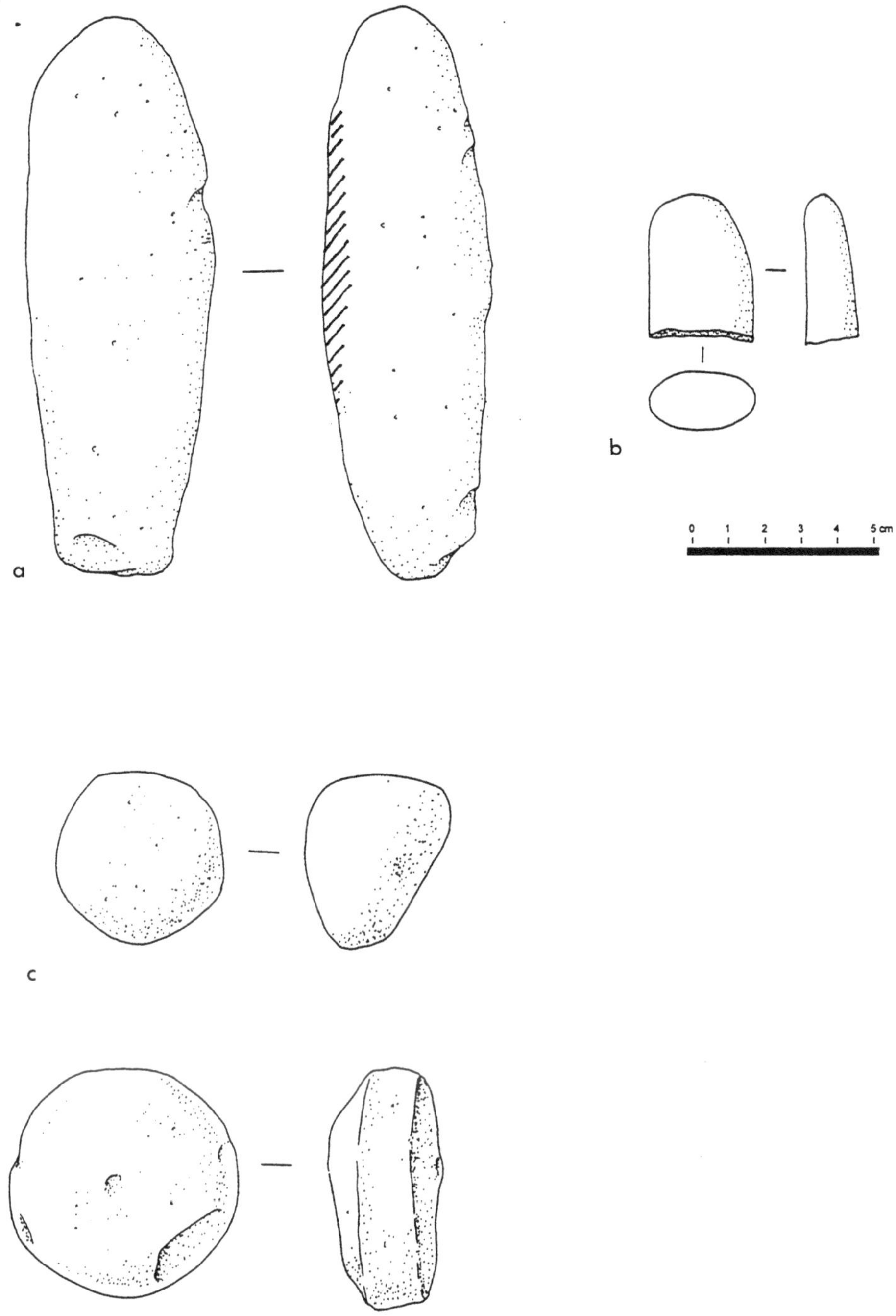

Figure 5.3. Groundstone artefacts from Gorgora rock shelter, Northern Highlands.

(a) large, ovoid grindstone stained with pigment, Level 3 (shaded area shows location of pigment staining)

(b) fragment of a groundstone 'rod', Level 3

(c) small, circular grindstones with smooth, pitted surfaces, Level 2.

legumes (Butler 1995). Ethiopian resources such as tef and noog are unlikely to have been important before grinding technology was introduced.

Grinding technology represents a pre-agricultural adaptation in Ethiopia. Applying this technology to the subsistence context in the Highlands provided a means for intensifying the resource potential of the environment. By increasing the range and density of available resources that could be obtained from within a particular catchment, there was a concomitant reduction in the mobility requirement of the subsistence group. This had profound implications for cultural change, and possibly represents one of the fundamental steps towards food production in Ethiopia.

Table 5.2. The Appearance and Use of Grinding Equipment in Ethiopia

Site	Earliest Occurrence of Grindstones	Contextual Association and Possible Function
Porc Epic cave Rift Valley escarpment	MSA levels (undated - possible early Holocene date)	Associated with pigments and cave paintings. The grindstone itself was smoothed on one surface and reddened due to burning, but not stained with pigment.
Gorgora rock shelter Lake Tana, Northern Highlands	1. MSA/LSA 'transitional' levels (undated - possibly early Holocene).	1. Large ovoid grindstone stained with pigment (figure 5.3.a.). Fragment of ground stone rod or implement (figure 5.3.b). No evidence of cave paintings. Stratigraphical association with pigment.
	2. LSA levels (undated), possibly early-mid Holocene	2. Small, circular grindstones with abraded, pitted surfaces (figure 5.3.c). Appearance of ceramics.
Gobedra rock shelter Northern Highlands	c.7th millennium BP	Four globular grindstones with flattened, abraded surfaces. Fragments of two ground stone rods or implements. Appearance of ceramics.
Lake Besaka Afar Rift	c.7th millennium BP	Upper and lower grindstone fragments, including one from a possible burial context stained with reddish pigment. Appearance of ceramics, on-site burials with 'status' items, including exotic marine shells from the Red Sea coast.

5.2.3. Ceramic Technology and Food Production in Africa

The Functional and Symbolic Role of Ceramics

Pottery manufacture has commonly been designated as one of the characteristic attributes of agricultural societies (Childe 1956). Ethnographic and archaeological research however has shown that many non-agricultural groups, such as Australian aboriginal groups, North American Indians and North African early Holocene groups, use ceramics and possess pottery technology (Berndt and Berndt 1951, Wendorf et al 1984b, Marquardt 1985). Conversely, the earliest known agricultural societies in the Near East did not develop pottery for several millennia *after* the domestication of local resources (Braidwood and Howe 1960).

The Functional Role

Ceramics perform a functional and a symbolic role within a society. Pottery provides heat resistant, durable, water-tight containers in which solids or liquids can be prepared, heated, transported and stored. These properties enable a wider range of food resources to be exploited within a particular habitat. Cooking renders previously inedible foods more digestible (Arnold 1985), or mitigates the impact of many toxins, such as proteinase inhibitors, lectins and lethyrogens (Stahl 1984), while the possibility for storing seasonally available resources insures against seasonal deficits. Ceramic containers may also have been used for fermenting a variety of plant and animal foods, thus increasing the range of resources offered by a particular environment. Fermentation serves a variety of functions, including detoxification, preservation, brewing alcoholic drinks, aiding digestibility of grains and legumes, and enhancing the flavour of certain foods (eg. Sturtevant 1969, Zamara and Fields 1979, Kazanas and Fields 1981). Ethnographic evidence attests the wide range of resources that are fermented. For example, in the Sudan, practically every part of a slaughtered animal is fermented, many dairy products are processed in this way, a range of unconventional, wild resources (including caterpillars, locusts, jumping frogs and cow urine), sorghum is fermented for drinks and porridge, and to give a sour dough which is used especially for the production of the staple flat bread, *kissera* (Dirar 1994).

In addition, Handwerker (1983) and Lee (1972) suggested that the potential for boiling and preparing liquefied foods allowed earlier weaning of children. This could increase female fertility by reducing hormonal constraint on lactating mothers, whilst decreasing the possibility of infant mortality during a critical growth period, leading to greater potential for population growth. Similar changes in diet and nutrition have been proposed as influencing population growth in other regions of the world (eg. Stanley and Rose 1980).

Although pottery is durable, it is also fragile and not suitable for transportation. For these reasons it is often excluded from nomadic hunter-gatherer assemblages. Ceramics therefore reduce seasonal movement in two principal ways - first by encouraging and facilitating more sedentary behaviour, and second by constraining group mobility.

The Symbolic Role

Pottery also provides a creative medium for non-verbal, cultural expression and communication (Hodder 1982). The work of Hodder (1982) and Braithwaite (1982) in the Sudan argues that ceramic decoration and form play a role in the formation and maintenance of inter- and intra-group relationships. These roles are prominent in the expression of cultural identity, which Hodder suggested may be enhanced by increased cultural diversity, reduced group mobility, and a concomitant desire for territorial demarcation.

Symbolic expression of cultural identity is intimately associated with sedentism, territorial awareness, intensification of resource utilisation and cultural diversity - concepts which are commonly found amongst complex, agricultural societies. Although ceramic innovation is not coeval with the emergence of food production, it does derive from similar cultural processes i.e. a shift in subsistence strategy that is related to changing cultural perceptions of resource procurement and processing.

Development of Ceramic Technology in the Old World

Ceramic technology is known to have developed independently in the Old and New Worlds. The earliest known agricultural societies in the Near East only began to manufacture pottery in the mid-9th millennium BP in Anatolia (eg. 8500 years BP at Cayonu. Mellaart 1975) and during the 8th millennium BP in the Levant, over 2000 years *after* the appearance of permanent settlements and domesticated resources.

Conversely, in Africa ceramic technology *preceded* food production by as much as 2000 years in the Nile Valley and 4500 years in East Africa (Table 5.3). So although ceramic innovation may correlate spatially and temporally with the domestication of cattle in the Eastern Sahara during the 10th millennium BP (Wendorf et al 1984b), present evidence suggests that North African pottery may predate the earliest use of ceramics in the Near East by a thousand years. Given this chronological discrepancy, the proposal that ceramic technology may have been introduced from Africa rather than independently 'invented' in the Near East should not be rejected until further data are available.

The Emergence of a Northern African Ceramic Tradition

Between the 10th and 7th millennia BP, pottery formed part of a broadly homogenous material tradition across Northern and Eastern Africa. During this period, ceramics appeared in association with barbed bone harpoons and net or line sinkers (Adamson et al 1974, Sutton 1974, 1977). The pottery from this immense area belongs to a highly distinctive decorative tradition called Early Khartoum, or Khartoum Mesolithic, after its type site at Khartoum in Central Sudan (Arkell 1947, 1949). The predominant decoration types within this tradition were 'dotted-wavy line' and 'wavy line' ware. The surface designs of these wares were incised with catfish spines. Early Khartoum-type pottery developed into the equally distinctive 'Khartoum Neolithic' tradition, which occupied a similar geographical area during the 7th millennium BP (Arkell 1953).

The sites, which were located adjacent to palaeolake shores, playas and palaeochannels, often in regions which are desert today, were occupied during the major pluvial period of the early Holocene. The absence of evidence for food production, and the ubiquitous presence of fishing equipment at these sites indicate a foraging economy based on seasonal or perennial exploitation of aquatic resources. This pattern led Sutton (1977) to propose the existence of an 'Aqualithic' culture in Northern Africa in the Early Holocene. Homogeneity in the ecological and economic context of the sites and the ceramics decoration has led to speculation over whether the pottery tradition was the product of a broadly unified culture group (Sutton 1974), or the result of convergent technological adaptation to comparable environmental conditions (Phillipson 1977).

Table 5.3. Earliest Occurrences of Pottery in Northern Africa

Site	Location	Date of earliest of pottery levels (years BP)	Ceramic Tradition
North Africa			
Tagalagal	Niger	c.9500	Early Khartoum type esp.d.w.l.[2]
Ti-n-Torha	Libyan Sahara	c.9300	Early Khartoum type esp.d.w.l.
Sarurab	Central Nile Valley	c.9300	Early Khartoum type esp.d.w.l.
Amenki	Ahaggar, Sahara?	c.8700-8000	w.l[3].
Launey	Ahaggar	c.8500	w.l.
Abu Darbein	Central Nile Valley	c.8600	Early Khartoum type[4]
Timidouin	Ahaggar	c.8000	w.l.
El Damer	Central Nile Valley	c.8000-7500	Early Khartoum type
Aneibis	Central Nile Valley	c.8000-7000	Early Khartoum type
Uan Tabou/Fozzigiaren	Acacus, Fezzan	c.8000-7000	w.l. and d.w.l.
Uan Muhaggiag	Tassilli, Sahara	c.7500	w.l. and d.w.l.
Delbo	Ennedi, Sahara	c.7500	w.l.
Gabrong/Zouar	Tibesti, Sahara	c.7500	w.l.
Saggai	Central Nile Valley	c.7500	w.l.
Shaqadud	Butana, East Sudan	c.7500	Early Khartoum type
Gash Delta sites	Eastern Sudan	c.7000-6000	Amm Adam type
Shum Laka	Cameroon	c.7000	Stamp decorated
Iwo Eleru	South West Nigeria	c.6500	
East Africa and the Horn			
ZU-4 Turkana	Lake Turkana, Kenya	c.8500	w.l.
ZU-6 Tarkana	Lake Turkana, Kenya	c.8000	w.l.
Lowasera	Lake Turkana, Kenya	c.8000?	w.l.
Lothagam	Lake Turkana, Kenya	c.7500	3 styles inc.w.l.
Eilye Springs (surface)	Lake Turkana, Kenya		w.l.
Salasun Rock shelter	North Tanzania	c.7000	Kaynsore or Nderit ware
Gobedra	Northern Ethiopian Highlands	c.7000-5000	Thin coarse ware (plain and decorated). Thick fine ware
Gorgora	Northern Ethiopian Highlands	c.7000?	Thick coarse ware (plain and decorated) Smoothed coarse ware (plain and decorated)
Lake Besaka	Ethiopian Rift Valley	c.7-6000	Red slipped, decorated ware
Gogoshiis Quabe	Southern Somalia	c.7-6000	
Guli Waabayo	Somalia	c.7-6000	
Guli Garesso	Somalia	c.7-6000	

5.2.4. The Ceramic Evidence from Ethiopia

The Emergence of Ceramic Technology in Ethiopia

At present only a handful of early Holocene ceramic sites have been excavated, and the dates from these are extremely limited. Table 5.4 details the actual or conjectural dates of these sites. In the Highlands, the most reliably dated pottery-bearing deposit comes from Gobedra rock shelter. Further LSA sites have been excavated in the Highlands, at Quiha rock shelter, which is undated, Gorgora rock shelter, also undated, and Lalibela and Natchabeit caves, dating from the mid-3rd millennium BP. Excavations at four sites at Lake Besaka (FeJx1-FeJx4) provide the most complete stratified sequence for the Rift Valley, with undated, possibly later ceramic bearing deposits at Porc Epic cave and at Laga Oda on the Southern Rift Valley escarpment.

[2] dotted wavy line pottery

[3] wavy line pottery.

[4] this includes wavy line pottery, dotted wavy line, comb stamped and vee-decorated ware.

Table 5.4. Location and Possible Date of Ceramic LSA Sites in Ethiopia

Site Name	Location	Techno-complex	Possible date of ceramic development (years BP)
Axum area (Finneran *pers.comm.*[1], Bard *pers.comm.*[1])	Axum, Northern Highlands	LSA ceramic and aceramic	?7000-present (undated)
Gobedra (Philllipson 1977b[2])	Axum, Northern Highlands	LSA aceramic and ceramic	7000-recent
Lake Besaka (Brandt 1980, 1982[2])	40 km SW of Addis	LSA aceramic and ceramic	7/6000-c.2000
Gorgora rock shelter (Moysey 1943[2])	Lake Tana, north shore	LSA ceramic	c.7000-5000* (undated)
Kibish (Butzer et al 1969[2])	50 km N of Lake Turkana, Lower Omo Valley	LSA ceramic	c.5500
Agordat (Arkell 1954[1])	North West Eritrea	chalcolithic? LSA ceramic	4000-2000 (undated)
Quiha rock shelter (Moysey in Clark 1954[2])	Makelle, Northern Highlands	LSA ceramic	c.6th-3rd/2nd millennium* (undated)
Lalibela Cave (Dobrowski 1971[2])	50 km E of Lake Tana, Northern Highlands	LSA ceramic	2500-recent
Natchebiet Cave (Dombrowski 1971[2])	50 km E of Lake Tana	LSA ceramic	2000-recent
Melka Konture (Chavaillon 1976[1])	50 km S of Addis Ababa, edge of Northern Highlands	LSA ceramic	undated
Laga Oda (Clark and Prince 1978[2])	Rift Valley escarpment, Harar 70 km SW Dire Dawa	LSA ceramic	recent (within last 2000 years)
Porc Epic Cave (Clark and Williamson 1984[2])	Central Rift Valley	MSA and LSA aceramic and ceramic	recent (within last 2000 years?)
Yavello (Clark 1954[1])	South Western Highlands	LSA ceramic	undated

[1] surface survey
[2] from excavation
*proposed date based on author's detailed analysis of the material.

Ethiopian Ceramics in a Northern African Context

As Table 5.4 shows, pottery technology emerged in Ethiopia during the 7th millennium BP. Ethiopia can be included in the pattern of ceramic innovation that emerged across North and East Africa during the early Holocene. However the Ethiopian material is very distinct from the early African ceramics. Sherds of wavy line pottery have been found in early Holocene contexts around Lake Turkana in Kenya, for example at Lowesera, Lothagam, ZU-4 and 6 and Eilye Springs (Robbins 1974, 1980, Barthelme 1977, Phillipson 1977c). Barbed bone harpoons have been found around Lake Turkana, and in aceramic early Holocene contexts (c.10 000-8000 years BP) in Ethiopia at Kibish in the Lower Omo Valley (Brown 1975). The Kibish site relates to a palaeolake level of Lake Turkana 80 m above present, at a time when the lake overflowed into the Nile via the Sobat River (Butzer et al 1969, Gasse et al 1980). This waterway provided a direct ecological continuity with the 'aqualithic' cultures of the Nile Valley and North Africa, and these artefacts can be assumed to be part of the broader cultural phenomenon of North Africa.

Two fragments of barbed bone harpoons and wavy line type sherds were recovered from site ES 2 in the Northern Gash Delta, close to the North West Ethiopian border with the Sudan, possibly dating to the 8th millennium BP (Fattovich 1990a).

At present, such 'aqualithic'-type artefacts and ceramics are unknown in Ethiopia beyond the Lake Turkana palaeobasin in the south and the Gash Delta in the north west. With the exception of Lake Besaka, no suitable sites of early Holocene date have been investigated in the Rift Valley, and it is not yet clear whether the Ethiopian Rift Valley was excluded from the 'aqualithic' phenomenon. Unlike the ubiquitous decoration on early Holocene pots from North and East Africa, Ethiopian ceramics were rarely decorated, or modestly adorned with simple incisions or punctations. Possibly continuity exists with the present as today, Ethiopian wares are traditionally plain or simply burnished. Decoration is restricted to valuable artefacts, particularly those involved in social ritual, such as coffee pots and incense burners.

Summary

Ethiopian ceramics differ morphologically and contextually from early Holocene ceramics found across Saharan and sub-Saharan Africa. Since they do not conform to the models of adaptatory convergence under similar ecological conditions and cultural and economic homogeneity, proposed for North and East Africa, a separate explanation should be sought for the appearance of ceramic technology in Ethiopia during the early Holocene.

5.2.5. Detailed Analysis of the Ceramic Evidence from LSA Ethiopia

The data discussed below comes from three roughly contemporaneous LSA ceramic deposits, shown below. The locations of these sites are given in figure 5.1 above. Details of all ceramic sites are presented in *Appendix C*.

(A) **Gorgora rock shelter** close to the north shore of Lake Tana in the Northern Highlands (Moysey 1943)

(B) **Gobedra rock shelter** near Axum in the Northern Highlands (Phillipson 1977b)

(C) **FeJx1-4** at Lake Besaka in the Afar Rift Valley (Brandt 1982).

A. GORGORA

Chronology and Stratigraphy

A total surface area of 8.0-9.0 m^2 was excavated to a maximum depth of 3.30 m. A series of stratigraphical divisions was imposed on the excavation based on depth below surface as illustrated on p.188 below (Moysey 1943, Leakey 1943). The upper four feet (1.20 m) of the deposit (Levels 1-4) comprised a black, humus-rich soil which was homogenous from the surface to a depth of 4 ft. Level 4 downwards to Level 9 (2.40 m) below surface comprised a grey, ashy deposit which was described by the excavator ar 'volcanic' in origin (Moysey 1943, p.198). Concretions appeared in Level 9 and became increasingly common in subsequent levels to a depth of 3.30 m.

No dates are available from this site and re-excavation is not feasible as the deposit has been totally removed (Clark *pers.comm.*). There is, however, a correlation between the depositional processes at this site and palaeoclimate change which is supported by the archaeological evidence. The black soils in the upper 4 ft of the deposit relate to a period of increased rainfall and vegetation growth which stimulated the formation of humus-rich soils. The start of this phase may correspond to the early-mid Holocene c.10th-7th millennium BP when enhanced pluvial conditions in the Highlands inititated a phase of pronounced pedogenesis (Hunri 1989, Brancaccio et al 1994). The occurrence of LSA lithics in this deposit supports the suggestion of an early-mid Holocene date. The uppermost 2 inches (5 cm) of the deposit were sterile, indicating that the rock shelter has not been used in more recent years. The underlying grey, ashy layer in Levels 5-9 was deposited during a period of climatic aridity with sparse vegetation cover and limited soil formation. This phase may correspond to late Pleistocene aridity. A late Pleistocene date for these levels is also suggested by the late MSA ('Upper Stillbay') artefacts in this deposit.

Pottery first appeared in Level 3 at a depth of 3 ft (0.90 m) below the surface, as shown in Table 5.5. One sherd was also recovered from a late MSA context in Level 5. This sherd had affinities with the ceramics from Level 3 and may have been intrusive. The absence of ceramics from Level 4 implies that ceramic technology was unknown during the early Holocene at Gorgora. The subsequent appearance of crude ceramics in Level 3 suggests that pottery technology developed indigenously towards the middle of the Holocene, possibly during the 7th millennium BP. It must be stressed that this deduction is conjectural and in no way reflects the opinion of the excavators. Despite the poverty of the data, it is extremely important as the site represents one of the few excavated LSA sites in Ethiopia.

Table 5.5. Summary of the Stratigraphical Distribution of Artefacts at Gorgora

Approximate Depth	Number of Potsherds	Description	Associated lithics
150-120 cm (Level 5)	1	-	MSA ('Upper Stillbay') 25 shaped tools inc. points, burins, blades and scrapers.
90-60 cm (Level 3)	11	From different vessels, variable thickness. Ring or coil technique obvious. 1 rim sherd, 1 decorated body sherd with parallel incised lines.	LSA with MSA attributes (mixed). 32 shaped tools inc. scrapers, backed blades and microlith crescents 1 large, pigment-stained grindstone
60-30 cm (Level 2)	25	From different vessel types. 5 rim fragments, 1 of 22 mm thickness, 2 thin sherds (4 mm and 6 mm). 5 decorated sherds with incised lines or finger nail impressions.	LSA microlithic industry comprising 27 tools, including backed blades, scrapers and crescents 2 circular grindstones.
30-0 cm (Level 1)	11	From several different vessels including a fine ware. 1 rim sherd. All sherds were undecorated.	LSA microlithic industry, as above. Only 9 tools recovered.

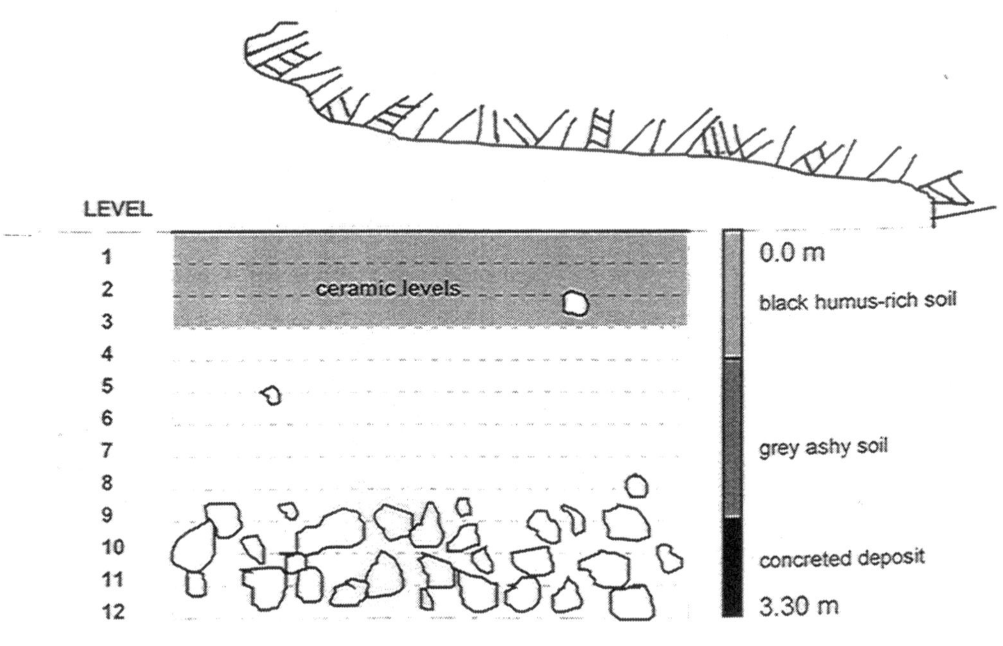

Schematic section of stratigraphy at Gorgora rock shelter

Typological Description and Analysis of Ceramics from Gorgora

The ceramics from Gorgora have only been briefly and incompletely examined (Leakey 1943). The following discussion is based on my own detailed observations of this material in the National Museum of Kenya.

FABRIC TYPES

I Predominant fabric type at Gorgora (53% of all sherds). Fairly hard, containing a very high proportion (c.70-80%) of medium-coarse sand temper, with variations in average size of inclusions found in different vessels. 15% mineral content, especially quartz or muscovite. Colour is dark red-brown to red-brown, generally with smoothed exterior and/or interior surfaces.

II Similar to I, but more friable. 70-89% medium-coarse sand temper. Colour ranges from dark red-brown to brick red.

III Similar to I, but more friable with c.70% medium-fine sand temper. Found only in a single sherd in Level 3. Brick red exterior surfaces and grey core.

IV Coarse, friable fabric, similar to I. 70-80% medium-coarse sand temper with 5-10% muscovite or quartz inclusions. Colour ranges from very dark brown to brick red.

V Similar to I, using different firing technique. Dark brown/red exterior surfaces and black core.

VI Coarse, friable fabric with c.60% coarse sand temper and some larger mineral inclusions up to 3 mm. Known only from a single sherd from Level 2. Dark orange-red exterior surfaces with grey-red core.

VII Friable fabric with low proportion (20%) of fine sand temper and 5% mineral inclusions. Colours range from orange-red to very dark brown.

VIII Soft, fine ware with c.40% fine sand or silt temper and 5% mineral inclusions. Colour ranges from orange-red to buff.

VESSEL FORMS (Figures 5.5 and 5.6 show profiles and reconstructed forms)

1 wide, open bowls

2 shallow bowls or platters

3 narrow necked, possibly globular jars with slightly everted thin or rounded rims

4 bowls or beakers with everted rims

5 straight walled, vertical rimmed bowls or beakers

DECORATIVE STYLES (Figure 5.4)

A simple, horizontal linear incisions

B complex linear incisions

C finger-nail impressions

D surface polished

The vertical distribution frequency of these attributes is given in Tables 5.6.a.-5.6.c. and illustrated in figure 5.7. Table 5.7 proposes a typological framework for the ceramics, and the distribution of these types is given in Table 5.8.

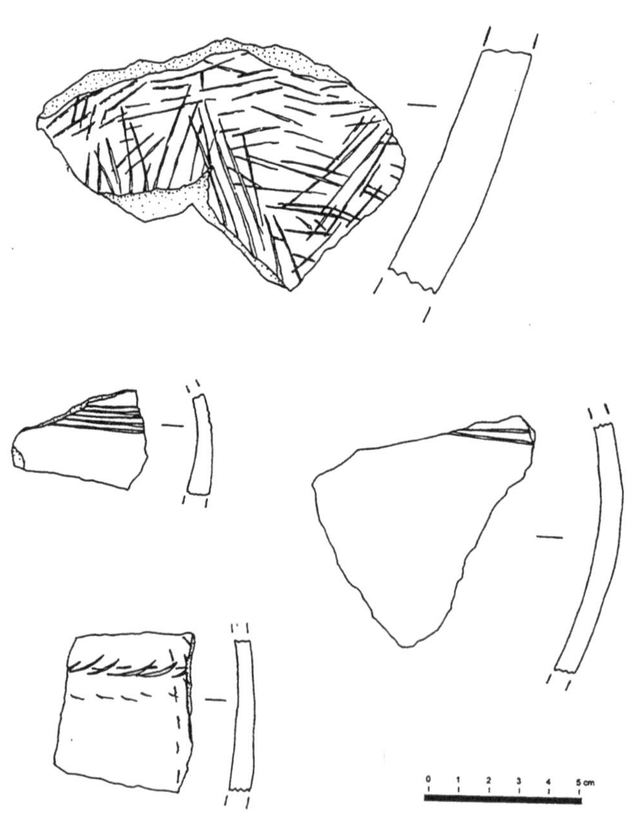

Figure 5.4. Decorated body sherds from Gorgora

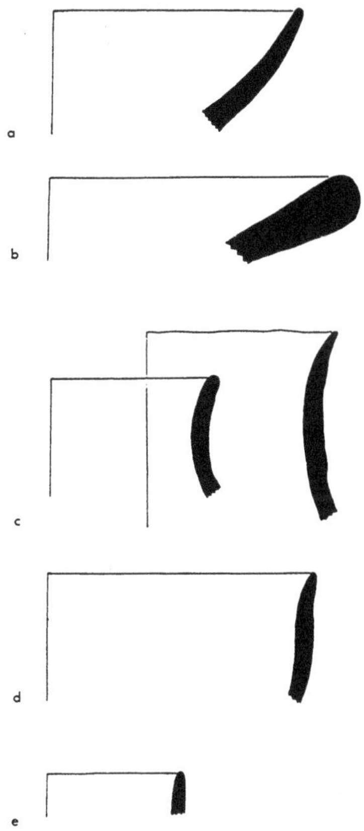

Figure 5.5. Vessel forms from Gorgora.
(a) Form 1 (b) Form 2 (c) Form 3 (d) Form 4 (e) Form 5.

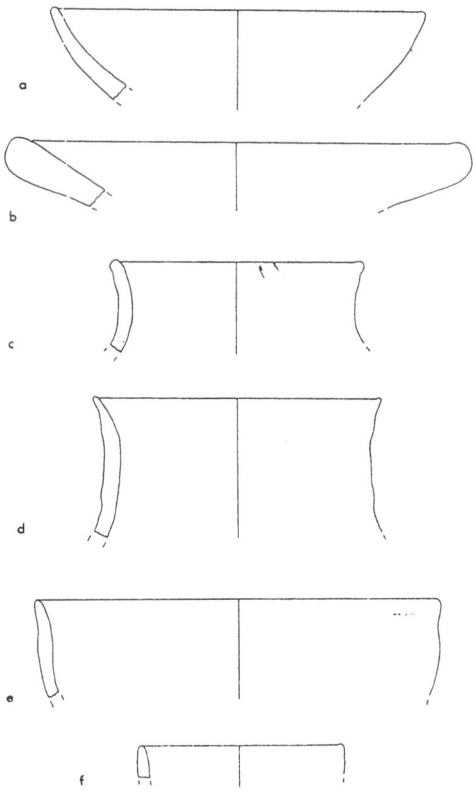

Figure 5.6. Reconstructed vessel forms from Gorgora.
(a) Form 1 (b) Form 2 (c) and (d) Form 3 (e) Form 4 (f) Form 5.

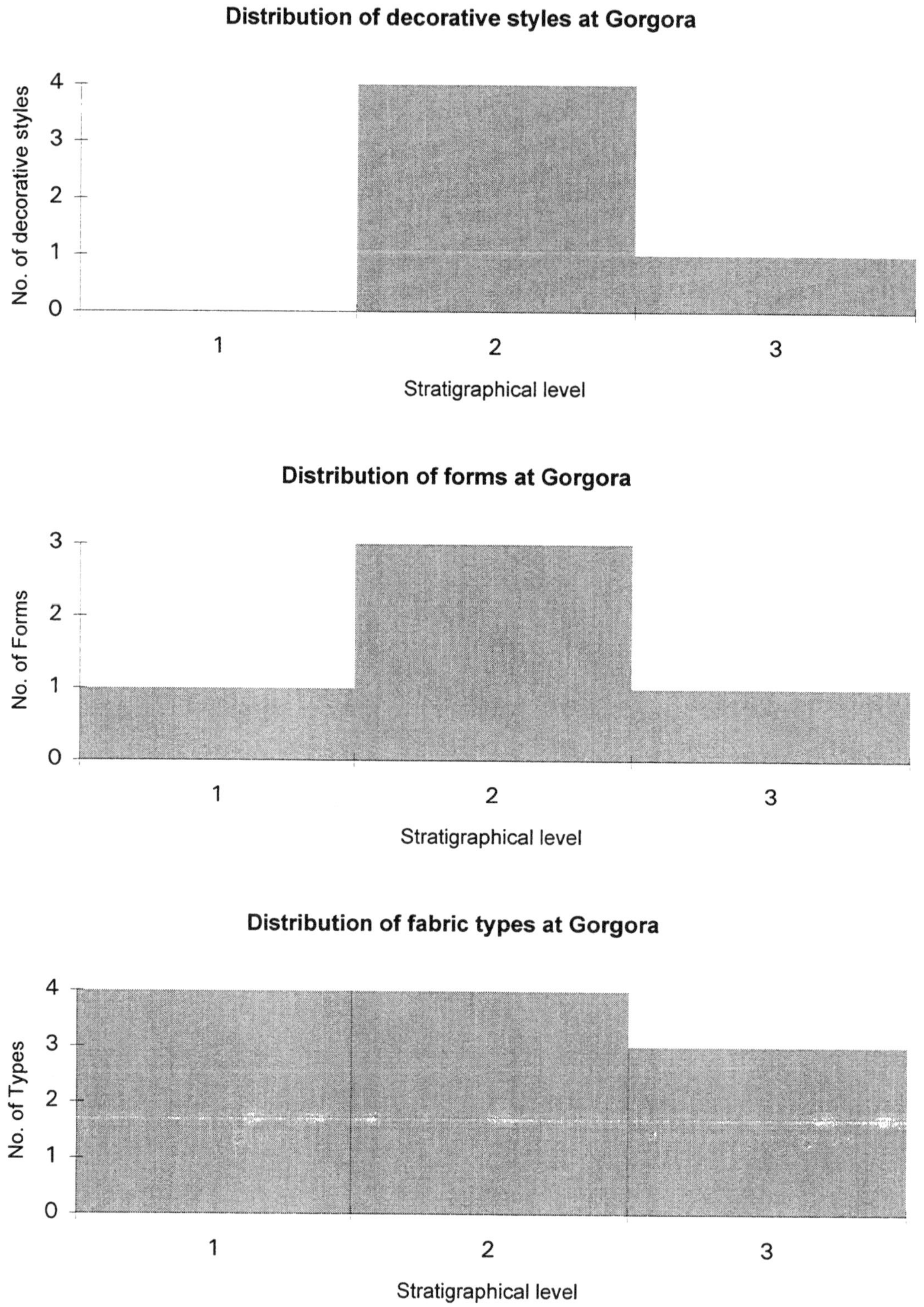

Figure 5.7. Distribution of ceramic attributes from Gorgora rock shelter.

Table 5.6.a. Frequency Distribution of Fabric Types at Gorgora

Fabric Type	I	II	III	IV	V	VI	VII	VIII
Level 1	5			1			2	3
Level 2	13			8	3	1		
Level 3	7	3	1					
[Level 5]		1						
Total	25	4	1	9	3	1	2	3

Table 5.6.b. Frequency Distribution of Vessel Forms at Gorgora

Vessel form	1	2	3	4	5
Level 1					1
Level 2		1	2	1	
Level 3	1				
Total (6)	1	1	2	1	1

Table 5.6.c. Frequency Distribution of Decorative Styles at Gorgora

Decorative style	A	B	C	D
Level 1				
Level 2	1	1	2	1
Level 3	1			
Total	2	1	2	1

Table 5.7. Intersection of Fabric and Decoration at Gorgora to Produce Types

Type	I	II	III	IV	V	VI	VII	VII I	A	B	C	D
1	X								X			
2	X										X	
3					X					X		
4					X							X

Table 5.8. Stratigraphical Distribution of Ceramic Types at Gorgora

Type	1	2	3	4
Level 1				
Level 2	1	2	1	1
Level 3	1			

Discussion of Ceramic Attributes and Affinities at Gorgora

FABRIC TYPES

Fabric types I-VI inclusive are of local origin. Types VII and VIII are much finer and lack the coarse sand temper characteristic of fabrics from this site. These distinctions suggest that fabric types VII and VIII may have been imported. Local fabric types occur throughout the deposit, whereas imported ceramics are exclusive to the upper level. Fabric type I is common to all levels (with the exception of Level 5). Fabric type II is common to Levels 3 and 5, and the possibility that the single sherd in Level 5 is intrusive from Level 3 cannot be ruled out. Fabric type IV is common to Levels 2 and 1. All other fabric types have a discrete stratigraphical distribution.

VESSEL FORMS

Since no complete vessels were found at Gorgora, the forms can only be suggested from the rim sherds. These have a large number of spatial and temporal parallels across Africa and in the Old World, some of which are indicated in Table 5.9 below.

DECORATIVE STYLE

The decorative techniques and styles at Gorgora have few parallels with those from geographically adjacent regions (Table 5.9). Horizontally incised lines occur on ceramics at Gobedra, finger nail impressions are used as decoration in some pottery from the Gash Delta and Lake Besaka.

Summary of the Study of Gorgora Ceramics

It must be stressed that the sample is too small to be statistically significant, and the following deductions are largely conjectural. Temporal variation in vessel forms suggests the following sequence: Ceramics in the earliest level belong to a typically African phenomenon which may date between the 8th and 2nd millennia BP. Non-African traditions, possibly dating between the 4th and early 2nd millennia BP, may have influenced certain shapes in Level 2. Little can be said at present about the single rim from Level 1 which could derive from a variety of forms.

The Gorgora assemblage lacks the characteristic 'red-orange' ware of ceramic assemblages in the Ethiopian Highlands and Sudanese-Ethiopian borders, dating between the 5th and 3rd millennia BP (Fattovich 1978). This ware has been found at widespread sites including Agordat, Jebel Moya, Lalibela cave, Gobedra and Yeha in the Northern Highlands. Its absence at Gorgora suggests that the assemblage may date before the 5th-3rd millennium BP, yet the poverty of the sample size means that no firm conclusions can be drawn. The decorative techniques and modes from Gorgora are unusual within a North East African context. This differentiation suggests that the Ethiopian material may derive from an independent tradition. The combination of attribute affinities and non-affinities in the earlier levels suggests that vessel forms, but not decoration, may have been emulating 'foreign' ceramics, and furthermore that convergent evolution of vessel forms may have taken place in response to similar functional requirements.

The second model is more feasible, given the divergence in decorative techniques and modes shown by ceramics from Gorgora with those from adjacent regions.

Table 5.9. Correspondence Between Ceramic Attributes from Gorgora and Adjacent Regions

	Vessel Form					**Decorative Style**			
Affinity	**1**	**2**	**3**	**4**	**5**	**A**	**B**	**C**	**D**
1	X		X						
2	X								
3	X								
4	X		X	X					
5	X	X	X	X	X				
6	X	X							
7	X								X
8	X		X						X
9	X		X						X
10	X		X		X	X[a]		X[b]	X

Affinities

Location/Type		*Date (years BP)*
1	Khartoum Neolithic	8th-6th millennium
2	Pre-Pastoral Neolithic ware in East Africa	9th-7th millennium
3	Kaynsore ware	c.7000-4th millennium[5]

[5] Kaynsore ware is poorly defined. Current opinion claims that Kaynsore to belong to the Pastoral Neolithic typology (Bower et al 1977, Wandibba 1978), as Kaynsore-type pottery has been identified in 4th and 3rd millennium BP Pastoral Neolithic deposits in Kenya where it apparently post-dates 4th millennium BP Nderit ware (Collett and Robertshaw 1978, 1983a). However Kaynsore-type ceramics have also been discovered in 5th millennium BP deposits around Lake Victoria (4720±150 years BP and 5400±150 years BP) and in 8th/7th millennium BP deposits at Salasun rock shelter in Northern Tanzania (7225±225 years BP and 6596±235 years BP) (Collett and Robertshaw 1978).

4	Elmenteitan (Remnant) ware	3rd millennium
5	Other Pastoral Neolithic wares	late 5th-3rd millennium
6	East African Iron Age	2nd-1st millennium
7	Nile Valley	late 6th-3rd millennium
8	PreAxumite	3rd millennium
9	Axumite	2nd millennium (c.AD 300-800)
10	Other	

[a] Gobedra, Ethiopian Highlands (7th-3rd millennium BP)
[b] Possible affinities with Butana and Gash Group ceramics from Gash Delta (6th-3rd millennium BP)

B. GOBEDRA

Chronology and Stratigraphy

The excavation covered an area of 4 m^2 to a maximum depth of 1.27 m, although large rock fragments were encountered at 0.80 m. Seven natural stratigraphical levels were identified, as illustrated below:

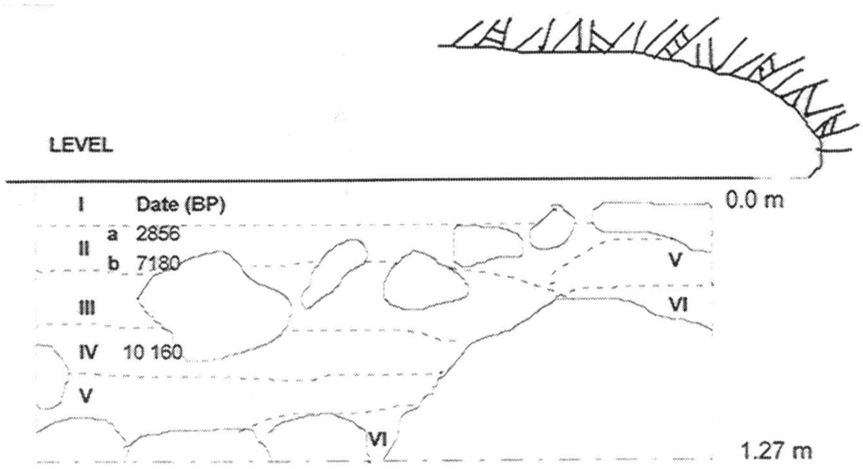

North-south section showing west face of Gobedra excavations (adapted from Phillipson 1977, p.58)

A total of 176 potsherds were recovered from Levels I and II, as shown in Table 5.10. Pottery first appears in Level IIb. Level II is a stony, brownish soil at a depth of 10-12 cm and roughly 30 cm thick. This level comprises two substrata, Levels IIa and IIb, Level IIb being less stony, with greyer, more ashy soil than Level IIa. These levels have been dated to 2856±53 (IIa) and 6875±165 years BP (IIb) from bone collagen and bone apatite respectively. Problems with comparative dating of these materials (see Section 1.2.2) means that these dates must be treated with caution. The date for Level IIb has been ^{13}C corrected to 7180±165 years BP. There is no change in the lithic sequence associated with the appearance of ceramics, although the small scale of the excavation (4 m^2) means that the artefacts are not present in statistically significant quantities.

Table 5.10. Stratigraphical Distribution of Ceramic Types at Gobedra

		Number of sherds	
Type	**Level IIb**	**Level IIa**	**Level I**
Type 1	26 (plain)	87 (28 decorated)	-
Type 2	1 (plain)	23 (plain)	38 (plain)
Type 3	-	1 (exterior burnish)	-

Description of the Ceramic Sequence

Ceramic Type 1 - Thin, coarse ware from Levels IIb and IIa. Sherds average 7-8 mm thickness and include a number of rim fragments and a base fragment. The fabric is grey-brown and tempered with large crushed quartz fragments. The forms are small, open necked bowls or dishes with tapered sides, made using the coil technique.

A fragment of a handle was found in Level IIa. Also in Level IIa, four rim sherds and a significant number of the body sherds (32%) were decorated using comb-stamped, punctated, impressed, grooved or vertically incised designs.

Ceramic Type 2 - Thick, fine ware with sherds averaging 10 mm thickness. The fabric is red-brown and tempered with sand or finely crushed grit. The pottery is undecorated but well finished internally and externally. Forms include large or medium sized (eg. 20 cm diameter) pots or bowls, with open necks and vertical sides.

Ceramic Type 3 - (1 sherd only) Thin fine ware, 4 mm thickness. The fabric is untempered, fine textured and red-brown in colour. Well fired and finely finished with an exterior burnish.

Discussion of the Ceramic Sequence

Pottery first appears in Level IIb at Gobedra. These early ceramics show a level of competence that suggests creative familiarity and manufacturing precedence, which could imply that ceramic technology developed before the 7th millennium BP, assuming the dates are correct. The earliest ceramics have been ascribed to a single typological unit of undecorated, thin coarse ware (Type 1), although more subtle distinctions may have been overlooked. A second ceramic type appears towards the end of the 4th millennium BP, and decorated pottery was common during the 3rd millennium, as shown below. Type 2 persisted into the 2nd millennium.

Type 1	*undecorated (thin coarse ware) from start of 7th millennium BP*
Type 2	*late 4th millennium BP*
Type 1 and Type 2	*decorated 3rd millennium BP*
Type 2	*2nd/1st millennium BP*

Affinities

The pottery at Gobedra is distinct from the wavy line and dotted wavy line ceramics that characterise the early Holocene period in Northern and Eastern Africa, and there are few parallels with East African Pastoral Neolithic pottery. Limited similarities are apparent with the horizontally incised pottery from Gorgora as discussed above, and Fattovich (1982, 1989, 1990) has proposed that fabrics comparable to those of Type 1 ceramics were widespread in North East Africa during the mid-late Holocene, particularly in the Ethiopian-Sudanese border region. A detailed petrographic analysis is required before this observation can be substantiated.

Summary of the Evidence from Gobedra

At present, the evidence implies that the appearance of pottery at Gobedra rock shelter represents an autonomous development that was established by the 7th millennium BP. The pots are distinct from contemporary ceramics in the Nile Valley and East Africa, yet their development is conceptually related to a widespread ceramic phenomenon in Northern Africa during the early Holocene.

The appearance of pottery in Level IIb at Gobedra is stratigraphically associated with a single camel tooth, possibly from a domesticated animal. Domesticated bovid remains were recovered from Level IIa, dating to the 3rd millennium BP. A more detailed discussion of the faunal remains is presented in chapter 3.

C. LAKE BESAKA

Lake Besaka is a small, graben-enclosed lake in the Southern Afar Rift Valley. The lake is currently saline but it is sensitive to small changes in volume, and it supported a rich diversity of aquatic life during the earlier part of the Holocene (Brandt 1980). Four open-air sites (FeJx 1, 2, 3 and 4) were located on the west side of the lake. The excavation of these sites covers an area in excess of 80 m^2, and is the largest existing LSA excavation in Ethiopia. Despite the size of the excavations, the chronological sequence should be treated with caution. Only two precise dates were obtained directly on the archaeological levels. A number of additional dates from the excavation were rejected (see Brandt 1980 for arguments), and the remainder of the sequence was dated relative to the geomorphological stratigraphy.

Three main cultural phases were identified on the basis of changes in the composition of the lithic sequence (see *Appendix C*). These are as follows:

PHASE A (FeJx 2 and 4)	c.22 000 - 11 500 BP	aceramic microlithic
PHASE B (FeJx 1, 2 and 4)	*Metahara Phase* c.11 000 - 7000 BP	aceramic microlithic
	Abadir Phase c.7000 - 4500 BP	ceramic microlithic, grindstones, inhumations, pierced Red Sea shells and ostrich egg-shell beads, pigment
Single date of 4785±120 BP obtained on charcoal near surface		
PHASE C (FeJx 3) Single date of 3460±260 BP obtained on charcoal	c.3500 BP	ceramic microlithic, grindstones, stone bowl fragment, possible domesticated cattle

Description of Ceramics from Lake Besaka

Abadir Phase (c.7th-5th millennium BP)

A total of 193 sherds were recovered, with approximately 14% decorated: There is continu+ity in decoration throughout this phase, although vessel forms vary stratigraphically.

FORM - Sherds have inturned or vertical rims, 5-11 mm thick with an average of 8 mm.

FABRIC - The fabric is fine paste tempered with sandy quartz and muscovite. The pottery was incompletely oxidised during firing to give buff to rusty-red exterior surfaces, and grey to black cores.

DECORATION - Surface decoration includes narrow or wide incisions, comb-stamped impressions on exterior and interior surfaces, linear cord impressions, finger nail impressions and diamond punctates. Some rim lips were decorated with diagonal comb impressions. One sherd had knobbed relief decoration. The majority of sherds were unburnished, with smoothed exterior and interior surfaces. Some sherds were burnished on one or both surfaces, or with a thin red slip or haematite wash on the exterior.

Later ceramic horizon (FeJx 3) c.3500 BP

Two hundred and twenty two sherds were recovered, of which 19% were decorated:

FORM - Sherds generally had inverted rims, 5-12 mm thick. One conical-based vessel was reconstructed from 6 sherds.

FABRIC - The fabric was fine paste tempered with sandy quartz and muscovite. Incomplete oxidation gives buff to red-brown exterior surfaces and black cores.

DECORATION - Sherds were decorated with a range of designs, including criss-crossed or diagonal comb-stamped impressions, finger nail impressions covering much of the body, and linear cord impressions. Sherds from the conical vessel had exterior burnish; all other sherds were unburnished or had a reddish slip on the exterior surface.

Summary of the Ceramics from Lake Besaka

Ceramics first appear at Lake Besaka during the 7th millennium BP. There is apparent continuity in the lithic sequence at this time, indicating that there was no significant cultural or economic change (Brandt 1980). The frequency of sherds found in this early phase was very small relative to the high frequency of lithic artefacts (a total of 193 sherds and c.150 000 lithics i.e. ceramics contributed less than 0.15% of the total volume of artefacts). Despite this relative paucity, all the sherds were well finished and many were elaborately deorated. A diverse range of decorative techniques were displayed within the sample. The absence of crude pot sherds, the small sample size and the diversity of surface decoration represented led the excavator to propose that ceramic technology did not develop indigenously at Lake Besaka, and it is unlikely that ceramics were manufactured at the site. Instead, Brandt proposes that pottery may have been obtained through exchange with the inhabitants of the adjacent Highlands (Brandt 1980, 1986). Such a system operates today, whereby the occupants of the Lake Besaka region obtain pots from the agricultural Highlanders in exchange for secondary cattle products. Interestingly, in LSA interlacustrine Africa, potsherds were gathered from abandoned settlements by aceramic hunter-gatherers, possibly for their decorative and status value (MacLean 1996).

Ceramics from the Abadir Phase at Lake Besaka share few affinities with contemporary wares from Gobedra and Gorgora in the Northern Highlands, as illustrated broadly in the chart below. This is possibly of little surprise considering the distances between the sites, and it suggests that the Lake Besaka ceramics derive from a regionally distinct tradition.

DECORATIVE TECHNIQUE

	cuneiform	knobbed relief	hemispherical incision	cuneiform and linear incision	incised rims
Lake Besaka	+	+	+	+	+
Gorgora	-	-	-	-	-
Gobedra	-	-	-	-	-

Although these techniques are absent from contemporary sites in Ethiopia, similar decoration is known elsewhere in North East Africa. For example, the distinctive knobbed relief technique is common to the Amm Adam Group of the Gash Delta dating to the 8th and 7th millennia BP (Fattovich 1989), and is also known from sites such as GaJi on the shores of Lake Turkana in Northern Kenya, dating to the 5th millennium BP (Barthelme 1977). This raises interesting questions about possible cultural connections between these areas.

The later ceramic horizon, known only from FeJx 3 and dating to c.3500 BP, is significantly different from the Abadir Phase. A change in the associated lithic sequence suggests cultural discontinuity, and this phase is chronologically linked to the appearance of domesticated cattle, as discussed in chapter 3.

Discussion of the Evidence from Ethiopian LSA Ceramics
The Origins of Ceramic Technology in Ethiopia
The evidence from the sites of Gobedra and Gorgora suggests that ceramic technology arose in the Northern Highlands of Ethiopia during the early Holocene around 7000 BP. Pottery became widespread across Northern and Eastern Africa between the 10th and 7th millennia BP where it contributed to the exploitation of aquatic resources within a fishing/intensive foraging culture. However, the early ceramics from Ethiopia are distinct from those of North and East Africa, which belong to a broadly homogenous decorative tradition. Furthermore, there is neither a contextual association between Ethiopian ceramics and features that are characteristic of early ceramic-bearing assemblages across North and East Africa - notably barbed bone harpoons and net or line sinkers - nor do Ethiopian ceramic sites ubiquitously conform to an aquatic location. This discrepancy suggests that ceramic technology was an independent innovation in Ethiopia, distinct from the contemporaneous development of ceramics in North East Africa.

The evidence from Lake Besaka indicates that by the 7th millennium BP pottery was being obtained in the Rift Valley, possibly through inter-group exchange with Highland cultures. Use of pottery as a commodity for exchange implies that it had a perceived value among groups in the Rift Valley during the early-mid Holocene period.

5.3. SETTLEMENT LOCATION AND THE ECOLOGICAL CONTEXT

5.3.1. Past and Present Settlement Concentration
Modern and Historical Settlement Patterns
Altitude variation in Ethiopia creates a range of contiguous ecological niches. Modern and historical settlement in Ethiopia has always been concentrated in the botanically rich *woina dega* zone between c.1800 and 2400 m asl, where the greatest diversity of natural resources thrive. This zone is also the most agriculturally productive region of Ethiopia.

Studies of PreAxumite and Axumite settlements indicate a similar distribution pattern, particularly concentrated between c.2000-2400 m asl (Fattovich 1977). The *woina dega* therefore has been the focus of human occupation and agricultural activity for at least the past 2500 years.

LSA Settlement Location
Evidence for LSA occupation in the Highlands comes from both excavated sites and surface scatters of archaeological material. The spatial distribution of these deposits reflects an adaptive behavioural pattern that is primarily determined by resource availability (Vita-Finzi 1970).

Fattovich (1977) proposed that LSA deposits were confined to the upper *kwolla* and lower elevations of the *woina dega* c.1600-2000 m asl. This deduction was derived from a limited number of sites, particularly in the area which he was surveying in Eritrea and Tigray. As PreAxumite and Axumite sites were principally located in the *woina dega* and lower *dega* (2000-2400 m asl), Fattovich proposed that there was a shift in settlement location between LSA and PreAxumite sites during the 3rd millennium BP. This relocation was seen as an adaptive response to the development of agriculture, which was associated with an increased demand for the more fertile soils and benign climatic conditions of the *woina dega*.

A New Pattern for Holocene Settlement Location
Following a reassessment of the entire database for LSA site location, a coherent pattern emerges for the Holocene period. The results of this reassessment are given in Table 5.11 and figure 5.8. This study contradicts Fattovich's proposition and shows us that:

- The majority of recorded sites are located within a restricted altitude zone. They are concentrated in the *woina dega* between 1800-2400 m asl, and especially around 2200 m.
- There is no observable shift in settlement location between the LSA and the PreAxumite or Axumite sites.

Table 5.11. Location and Altitude of LSA Archaeological Sites in Ethiopia

Site Name	Location	Techno-complex	Altitude (m a.s.l.)	Proximal Water Source
Agordat (Arkell 1954[1])	North West Eritrea	chalcolithic? LSA ceramic	700-1000	o
Axum area (Bard et al *in prep*[2], Finneran *pers.comm.*[1])	Axum, Northern Highlands	LSA aceramic and ceramic	2100-2500	o
Aladi Springs (Clark and Williams 1978 [2])	Central Rift Valley	MSA and LSA aceramic	1600	spring
Cliff of Monkeys (Dombrowski 1971[1])	10 km from Amba Giorgis, Northern Highlands	LSA aceramic	2600	o
Dessie (Clark 1954[1])	Wello, Central Northern Highlands	LSA aceramic	c.2200	o
Debra Tabor (Dombrowski 1971[1])	250 km E of Lake Tana	LSA aceramic	2200-2500	o
Dimma (Dombrowski 1971[1])	50 km NW of Debra Marcos, Gojjam	LSA aceramic, with grindstones	2200-2800	o
Gadeb Plain (Clark 1954[1])	South East Highlands	MSA and LSA aceramic	2300-2400	o
Gobedra (Philllipson 1977b[2])	Axum, Northern Highlands	LSA aceramic and ceramic	2250	springs, streams
Gondar area (Gamst 1965[1], Clark 1954[1])	North West Highlands, north of Lake Tana	LSA aceramic	c.2000-2400	o
Gorgora peninsula (*Personal observations 1994*, Dombrowski 1971[1])	Lake Tana north shore	LSA aceramic, with grindstones	2000	lacustrine
Gorgora rock shelter (Moysey 1943[2])	Lake Tana north shore	MSA aceramic, LSA ceramic	2000	lacustrine
Idobi Mariam (Dombrowski 1971[1])	10 km SE of Debra Marcos, Blue Nile river terrace	LSA aceramic	2200-2500	o
K'one (Clark and Williams 1978[1])	40 km SE of Addis, South Afar Rift	LSA and MSA aceramic	1600	?
Laga Oda (Clark and Prince 1978[2])	Rift Valley escarpment, Harar 70 km SW Dire Dawa	MSA aceramic, LSA ceramic	2000-2300	spring and stream
Lalibela Cave (Dombrowski 1971[1])	50 km E of Lake Tana, Northern Highlands	LSA ceramic	2300	o
Lake Besaka (Brandt 1980, 1982[2])	50 km E of Addis	LSA aceramic and ceramic	950	lacustrine
Lake Ziway area (Clark and Williams 1978[1] Gasse and Street 1978[1,2])	Central Rift Valley	MSA and LSA aceramic	1600	lacustrine, stream
Melka Konture (Kelba) (Chavaillon 1976 [2], Chavaillon et al 1979[2])	50 km SW of Addis	MSA, LSA aceramic, LSA ceramic	2200?	o
Natchebiet Cave (Dombrowski 1971[2])	50 km E of Lake Tana	LSA ceramic	2300	o
Porc Epic Cave (Clark and Williams 1978[2])	Central Rift Valley	MSA and LSA aceramic (with one sherd from LSA)	1800	wadi/river
Quiha rock shelter (Clark 1954[2])	Makelle, Northern Highlands	LSA ceramic	2200	o
Wekerdeba Mariam (Dombrowski 1971[1])	15 km SE of Amba Giorgis, Northern Highlands	LSA aceramic	2600-2700	o
Yavello (Clark 1954[1])	Sidamo, Southern Highlands	LSA ceramic	c.2100	o
Yizana Creek (Dombrowski 1971[1])	20 km SE of Lake Tana	LSA aceramic	2000-2200	stream
Zergnat (Clark 1954[1])	near Magdala, North East Highlands	LSA aceramic	2200	o

KEY

o - numerous streams and springs in the area, [1] - surface material identified by survey, [2] - excavated site

The Ecological Context and Optimisation of Resource Procurement

Figure 5.8 shows that LSA sites were generally located in topographically diverse regions with access to a vertical ecological spectrum which contained a wide range of edible and useful plant resources. Sites were located particularly in the ecotone between two plant resource-rich environments, the evergreen thicket and dry tropical forest of the lower *woina dega* and the montane forest-savannah mosaic of the upper *woina dega* and lower *dega* zones c.2200 m asl. These locations often had access to the open montane environment of higher altitudes. All sites were located in well watered or lacustrine areas with easy access to springs, streams and lakes.

The implications of the observed locational patterning is that it reflects a model of behavioural adaptation specific to Ethiopian biogeography, in which settlements were optimally situated for access to the maximum range of resources. Since settlement distribution relative to the environment is primarily determined by subsistence strategy and resource availability, the patterning of sites in Ethiopia suggests that a broad spectrum of resources was being exploited within a small catchment area, and that there was no change in the resource requirements of human groups throughout the Holocene.

5.3.2. Intensity of Site Use

Section 5.2.1 above observed that there was little change in the lithic technology sequence in Ethiopia during the Holocene. However, the proportional representation and overall frequency of lithic types do apparently vary. Brandt (1982) identified a change in technology and deposition patterns at Lake Besaka (FeJx2 and FeJx1) in the early Holocene c.11 000 BP. This phase was characterised by a significant expansion in the surficial extent of the material deposits relative to the late Pleistocene sites, combined with an increase in the density of surface artefacts and deposited material. The lithic industry showed an increased frequency of microliths, especially geometric crescents, from 30% to 70% of the total number of retouched tools. There was also a decrease in microlith size, and an overall change in the composition of tool types, particularly with an increase in the relative quantity of scrapers.

A change in the nature and intensity of site use is suggested by a reintrepretation of the data from the early Holocene deposit at Gobedra rock shelter. By dividing the total frequency of lithic pieces within a particular stratum by the proposed chronological span of that stratum, the average frequency of lithic material occurring in each millennium can be calculated. Calculating the average percentage frequency of lithic pieces and retouched implements provides a means for comparative analysis. Although this technique is far from adequate, it is a method for determining the gross diachronic variation in lithic deposition and hence in the use of the site. Quantified in this way, the Gobedra data reveal substantial apparent changes in lithic frequency during the period of use of the site, as illustrated in figure 5.9 and Table 5.12. Four different catagroies of lithics were analysed: the total lithic sample (including modified fragments), the retouched lithics, the types of lithic implement, and the geometric microliths. All these catagories of lithics experience an increase in frequency during the 9th and 8th millennia which stabalises during the 7th and 6th millennia. Geometric microliths were absent prior to the 9th millennium BP, but thereafter they follow this general pattern. A further increase is apparent in all catagories in the 5th millennium, which stabalises during the 4th millennium and declines substantially in the 3rd millennium. A second period of reduction can be seen in the 1st millennium BP. The paucity of the data from Gobedra percludes the use of statistical quatification. The observed changes must remain apparent.

Few additional data are available from Ethiopia. Early Holocene occupation horizons generally tend to be represented by isolated surface scatters of material with no diachronic sequence. An increase in tool frequency and diversity during the early Holocene phase at Laga Oda relates to an increase in the size of the excavation, and may not reflect a real change. A similar development in frequency and diversity of tool types at Gorgora rock shelter may date to the early Holocene, but the chronology of this site is conjectural.

This analysis indicates a change in the nature of archaeological deposition of lithic implements from the 8th/9th millennium BP in the Northern Highlands. This change could be interpreted as a shift to more intensive and varied use of the site. More specifically, the appearance of geometric microliths at Gobedra in the 8th/9th millennium BP may represent a shift in food procurement strategies, possibly towards enhanced harvesting of wild grasses. The observed changes in the lithic sequence at Gobedra coincide with the appearance of ceramics and grindstones at sites across Ethiopia from the 7th millennium BP, which, as suggested above, may indicate the emergence of an altered human-environment relationship.

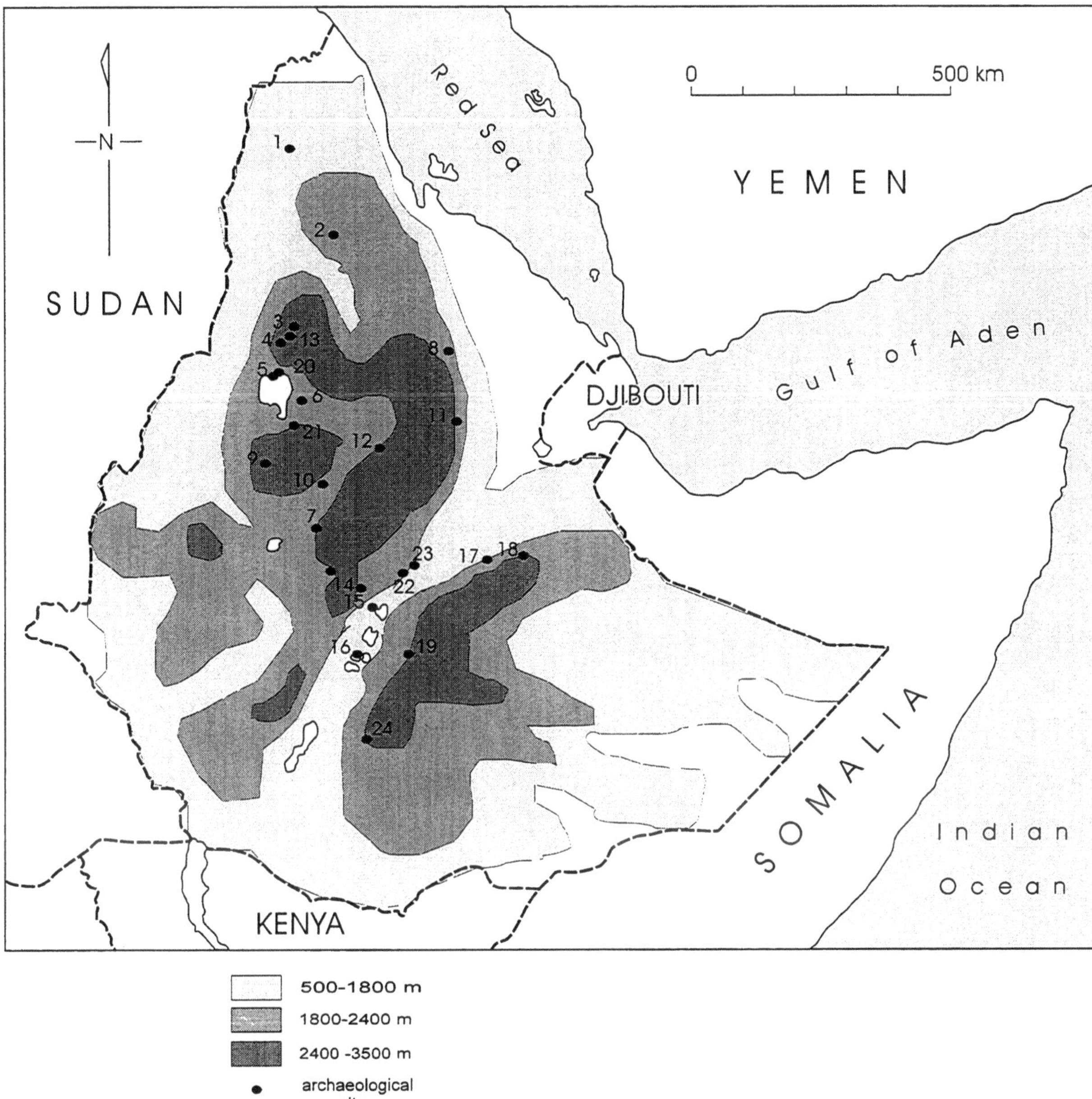

Figure 5.8. Distribution of archaeological sites in Ethiopia in relation to altitude.

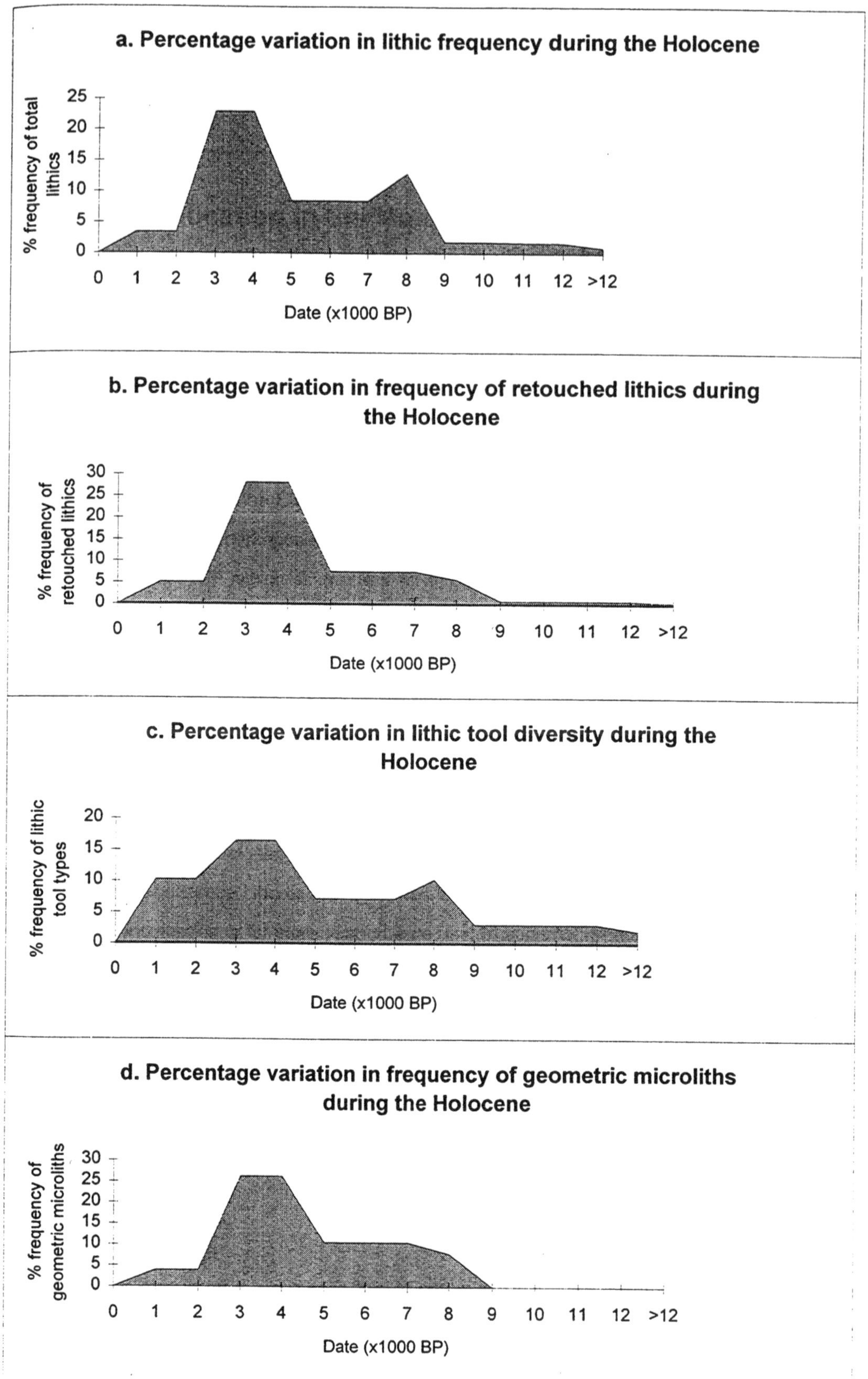

Figure 5.9. Percentage variations in frequency of total lithics (a), retouched lithics (b), geometric microliths (d) and dicversity of implements in the assemblage from Gobedra rock shelter during the Holocene (c).

Table 5.12. Frequency and Diversity of Lithics at Gobedra (after Phillipson 1977b)

Estimated date (x1000 BP)[6]	Total frequency of lithic pieces	% site total	Frequency of retouched lithics	% site total	Frequency of different tool types	% site total	Frequency of geometric microliths	% site total
0	0	**0.00**	0	**0.00**	0.0	**0.00**	0.0	**0.00**
1	917	**3.42**	12	**5.10**	5.0	**10.30**	8.0	**3.96**
2	917	**3.42**	12	**5.10**	5.0	**10.30**	8.0	**3.96**
3	6184	**22.92**	69	**28.16**	8.0	**16.49**	26.5	**26.23**
4	6184	**22.92**	69	**28.16**	8.0	**16.49**	26.5	**26.23**
5	2296	**8.56**	19	**7.62**	3.5	**7.21**	10.6	**10.56**
6	2296	**8.56**	19	**7.62**	3.5	**7.21**	10.6	**10.56**
7	2296	**8.56**	19	**7.62**	3.5	**7.21**	10.6	**10.56**
8	3470	**12.95**	14	**5.71**	5.0	**10.30**	8.0	**7.92**
9	501	**1.91**	2	**0.81**	1.5	**3.09**	0.0	**0.00**
10	501	**1.91**	2	**0.81**	1.5	**3.09**	0.0	**0.00**
11	476	**1.77**	2	**0.81**	1.5	**3.09**	0.0	**0.00**
12	476	**1.77**	2	**0.81**	1.5	**3.09**	0.0	**0.00**
>12	259	**0.96**	1	**0.408**	1.0	**2.06**	0.0	**0.00**
TOTAL	*26794*	**100**	*245*	**100**	*48.5*	**100**	*101*	**100**

Dark shaded areas represent increases in lithic frequency, light areas show stabalisation and reduction in frequency.

5.4. CULTURAL DYNAMICS AND MATERIAL COMPLEXITY: A RE-INTERPRETATION OF THE LSA ARCHAEOLOGICAL RECORD

5.4.1. Intensification in the Material Record

Subsistence Intensification and the Emergence of Sedentism

Indirect evidence from Ethiopia suggests a pattern of cultural and economic development that contrasts with the organic evidence for Holocene subsistence, described in Chapter 3.

Flaked lithic tools display technological and typological continuity during the Holocene period, with the exception of the mid-4th millennium BP at Lake Besaka (Brandt 1982). This indicates continuity in the functional applications of the LSA tool kit. However there is a separate pattern of change in the material assemblage which is superimposed over this background continuity. A major development occurred during the 7th millennium BP which is manifested by the appearance of ceramic technology and grinding equipment within a domestic context. These implements enabled a wider range of foods to be exploited within a particular habitat, allowing a shift to a broader spectrum economy and a less mobile procurement strategy. They also encouraged the more versatile use of certain plants, such as starchy grains and tubers. More intensive exploitation of such resources reduces the necessity for group mobility, and alters the relationship between humans and their environment.

The appearance of grindstones and ceramics during the 7th millennium BP implies a shift in the conceptual approach to subsistence. Increased time was being invested in food processing and preparation for immediate consumption or for storage and future use. In addition, greater time would have been invested in the manufacture of ceramics and grindstones, giving them a perceived value by their owners. Ethnographic and archaeological work shows that these concepts of investment, planning and ownership are alien to mobile, foraging societies (Lee 1972, Lee and DeVore 1968, Smith 1992, Martin 1972, Kent 1989), but are common among sedentary groups whose subsistence is based on food production or intensive foraging. In addition, this ethnographic evidence suggests that highly mobile, hunter-gatherer societies only manufacture non-transportable objects

[6] Dates are extrapolated from the three radiocarbon dates from Gobedra and are therefore only a broad approximation.

with perceived value, such as heavy grindstones and fragile pots, for trade purposes.

Section 5.2.3. discussed the symbolic role of ceramics in cultural dynamics, whereby pottery provides a medium for non-verbal expression of socio-territorial relationships within and between groups. The development of ceramic technology can be correlated with a need for a statement of cultural identity and territorial demarcation, which in turn derives from increased sedentary behaviour and an increased awareness of social boundaries. The appearance of ceramics and grindstones at sites in the Highlands and the Rift Valley suggests that there was a widespread intensification of resource procurement and reduction in group mobility during the 7th millennium BP. These changes can be considered as preadaptations of an incipient food producing society

Settlement Location and Intensity of Site Use

Settlement location throughout the Holocene reflects an adaptive strategy based on optimal exploitation of a broad resource spectrum. An increase in intensity of site use can possibly be seen during the early Holocene in the 8th millennium BP in Ethiopia. This broadly coincides with the emergence of ceramics and grindstones c.7000 BP, and strengthens the hypothesis for a signifcant cultural adaptation at this time. These changes contrast to Northern Africa, where both settlement location and intensity of site use vary throughout the Holocene period (Mohammed-Ali 1982, Marks and Sadr 1988, Haaland 1982). Increased intensity of site use during the early Holocene suggests further that there may have been a general reduction in group mobility during this period. This was possibly a response to more available, localised resources discouraging the need for movement, or denser vegetation cover restricting movement. Reduced group mobility during the early Holocene can be seen as a prerequisite to the subsequent emergence and use of ceramics and grindstones in the 7th millennium BP.

5.4.2. Synthesising the Evidence: a Proposed Cultural and Economic Development Sequence for the Holocene

This examination of the indirect archaeological evidence suggests that the initial, fundamental step towards food production may have been taken indigenously. As dicussed above, a similar pattern of cultural and economic change appeared contemporaneously across Northern Africa, with the appearance of ceramic and grindstone technology. This provokes the question of whether or not contemporaneous cultural rhythms in Northern Africa and Ethiopia were controlled by an objective external factor.

The indirect archaeological evidence from LSA sites in Ethiopia provides a broad picture of cultural and economic developments during the Holocene. These can be summarised as follows:

Early Holocene ***c.11 000-8000 BP***	Increased intensity of site use. Development of a pre-adaptive technology. Possible change in the resource procurement strategy.
Mid-Holocene ***c.7000-6000 BP***	Major cultural shift, with reduced residential mobility and territorial range. Major economic shift with emphasis on food processing activities and exploitation of localised, broad spectrum resources.
Late Holocene ***c.3500-2500 BP***	Ecologically adapted food production with pastoralism in the arid and semi-arid lowlands and mixed agriculture in the well-watered Highlands. Food production was practised throughout Ethiopia by the end of the 3rd millennium BP.

Chapter 6.

PREHISTORIC CONTACT AS A MECHANISM FOR THE INTRODUCTION OF DOMESTICATES

6.1. ETHIOPIA IN AN INTER-REGIONAL CONTEXT

The extent and nature of the interaction between Ethiopia and surrounding regions is fundamental to the emergence of food production. As we have already discussed, the available evidence points towards an intiguing relationship between Africa and Eurasia in prehistory. What part did Ethiopia play in such a relationship? Was it as a centre of innovation and cultural interaction or was it as a cultural backwater?

Western perception of Ethiopia as an isolated, remote country that was 'discovered' by Europeans in the 15th century reflects a colonial view of Africa as a continent subdivided in regimented cultural units. This preception is becoming radically revised in the light of research over the past 20-30 years. There is now substantial written evidence that the whole of Africa was a dynamic entity, criss-crossed by an intricate pattern of exchange routes. The mechanisms and products of these networks changed through time with changing cultural and economic requirements, but the process of dynamic interaction persisted. In this context Ethiopia can be viewed not as a cultural island, but as a cross-roads between Northern and sub-Saharan Africa, and between Africa and the East.

In this chapter ethnographic and archaeological evidence are used to argue that Ethiopia was an integral component of a complex, dynamic communication network throughout the Holocene period. This network provided a mechanism for the movement of objects as well as for the flow of ideas and concepts, and it has significant implications for the emergence and spread of food production in Ethiopia.

Two separate routes and mechanisms are proposed for the introduction of domesticated animals and plants to Ethiopia. The archaeological evidence from Quiha rock shelter in the Northern Highlands is used to examine the appearance of domesticated cattle, while the introduction of domesticated plants is interpreted through early trade contacts between Ethiopia and Egypt.

6.1.1. Ethiopia as Part of a Dynamic Communication Network

The Nature of Contact and Cultural Exchange

Two major mechanisms control the movement of resources and regulate the flow of culturally specific information. The first is the movement of people, either through large-scale migration, such as the Oromo 'invasions' from Southern Ethiopia during the 17th century AD (Hassen 1990), or small-scale movements of groups and individuals (for example the movements of pastoral groups such as the Borana of Southern Ethiopia around the Southern Highlands and Rift Valley during the past 400 years (Wilding 1985b). Second is the movement of objects, such as natural resources and manufactured products, by a process of trade and exchange.

A wide variety of different exchange mechanisms exist (eg. Sahlins 1972), these can be simplified here as directional trade between a specific culture and a specific resource area; ‘down-the-line’ trade, which describes the movement of objects from a core area to a peripheral area, where the value of the object increases in proportion to the distance from its source; and non-directional trade, which describes the non-specific movement of objects between adjacent groups.

Ethiopian Resources and Prehistoric Trade Networks

Ethiopia is rich in natural mineral and organic resources, especially gold, obsidian, ivory, aromatic resins and plants, timber and pasture. These commodities have had been highly valued by different cultures at different times during at least the past 10 000 years.

Obsidian, for example, was an important and valuable commodity for tool manufacture in Eurasia and Africa before the introduction of metal technology, and it formed the basis of a complex web of communication and exchange. Particular sources, such as Melos in the Aegean and the Bukk and Matra Mountains on the Northern Hungarian Plain, became foci for long-range exchange networks (Tringham 1971, Hallam et al 1976). Obsidian from the Panterllaria and Lipari sources off Sicily was widely distributed around the Mediterranean Basin in prehistory, reaching the North African coast before the 10th millennium BP where it was utilised until the 6th millennium BP (McBurney 1967, Tykot 1996). Obsidian sources in Ethiopia were being exploited from surrounding regions from at least the 8th millennium BP (Fattovich 1990a, Zarins and Zaharani 1985). Ethiopian obsidian has been found on the Dahlak and Farasen Islands in the Red Sea, and in coastal Southern Arabian contexts where it dates from the 8th millennium BP. This evidence suggests contact across the Red Sea, possibly via the narrow Strait of Bab el-Mandeb (Table 6.1) (Zarins et al 1981).

Gold became a valuable commodity from the mid-Holocene onwards, when the rise of civilisations in Mesopotamia, Egypt and the Indus created a market for precious goods and luxury items. South West Ethiopia, particularly the Damot region south of the Blue Nile, was especially rich in fine gold. Inferior sources were also known in Gojam province and possibly the Axum region (Beke 1852, Whiteway 1902). These sources were highly valued and have been well documented by Arab and Portuguese traders and travellers during the past millennium. Gold bearing regions became a primary focus for trade to the Red Sea coast and to the Nile Valley via the Northern Highlands during this period, and the recorded routes were of considerable antiquity even at that date (Bruce 1790, Abir 1968, Kurimoto 1995). The economic role of gold in Ethiopian prehistory is unclear, but may be reflected in its historical importance. Ethiopia's gold deposits have been vital to the country's international economy and status during the past 2000 years, and have paid for the majority of the country's imports.

Documentation of Ethiopian exports over the past 2000 years record little change in the major commodities of ivory, gold, aromatic resins such as myrrh and frankincense, slaves, precious woods and tortoiseshell (Pankhurst 1961, 1965, Kurimoto 1995). Many of these resources were obtained within Ethiopia (for example, gold from the South West, precious woods from the North West, aromatic resins from the Northern and Eastern lowlands), whilst others were imported from more distant regions to the south and west, and traded at Ethiopian markets. For example in the 1st century AD, the *Periplus of the Erythraean Sea* recorded that 'all ivory is brought from the country beyond the Nile, through the district called Cyeneum[1] and thence to Adulis'.

Ethiopia's strategic location, allowing access to the African interior and to the Red Sea coast, allied with the natural resource wealth of the country, formed the basis of extensive long-range contacts throughout the historical period. The emergence of a market for precious commodities, such as gold and incense, from the 6th millennium BP in Mesopotamia, Egypt and the Indus meant that Ethiopian sources may have been targeted from an early date.

Historical Documentation of Trade Mechanisms

Historical accounts of travellers and merchants during the past 2000 years show that trade contact within Ethiopia operated through both caravan trade and through intermediary 'down-the-line' trade (Alvares 1881, Bruce 1790, Crawford 1958, Pankhurst 1965). Caravan trade provided the principal artery for the movement of goods between the coast and the interior, while both mechanisms operated across the interior. The *Periplus* records that by the 1st millennium AD, ports in Eritrea and Somalia were internationally renowned and were visited by many ships from different nations every year. By this date Ethiopia was importing products such as cloth, metals and metal objects, glass artefacts, wine and olive oil from Egypt, Syria, India, Arabia and the Mediterranean. These items were transported inland by vast, organised caravans, to markets and fairs across Ethiopia (Alvares 1881, Bent 1893). For example, Marco Polo noted that in the 13th century AD the town of Manadeley in Tigray had an immense market where a great range of foreign produce was exchanged by merchants from Arabia, India, Turkey, Egypt and North Africa (Marco Polo).

Ethnographic and ethnohistorical data from the past two centuries underline the importance of precolonial trade in North East Africa. The Gibe area on the south western edge of the Northern Ethiopian Plateau has been a major economic centre throughout the past thousand years, and probably substantially earlier (Abir 1968). The success of this area lies in its proximity to natural mineral resources, especially gold from the Damot region. Gold from this area was exchanged for iron from Tigray in Northern Ethiopia, for coarse Indian cloth and coloured beads, and for 'clothing, cows, salt and other goods' (Whiteway 1902 p.149). Furthermore Gibe lies at a natural cross-roads between Northern and sub-Saharan Africa, and between Africa and the 'Oriental' East. The area was a focus for East-West and North-South caravan routes during the past millennium. It became a centre for cultural exchange and material redistribution, where typically African products (especially ivory, gold, aromatic plants, slaves and civet) were exchanged for 'exotic' Islamic products such as beads, cloth and spices (Beke 1852, Abir 1968, Hassen 1990).

Long-distance caravan trade also interconnected regions of the African interior. Bruce reported that by the 14th century, caravans operated constantly between Egypt and Ethiopia via Nubia and Sennar (Bruce 1790). Inland trade was also conducted through a second form of exchange. This trade was operated through intermediaries, either over relatively short distances or across vast areas. For example, before the opening of the White Nile trade route in the 19th century, numerous trade items, especially metal and beads, were transported between the Upper Nile Valley and Eastern Equatoria along the western edge of the Ethiopian Highlands. This system operated through intermediary tribal groups rather than by direct caravan trade, and historical linguistic reconstructions suggest that this route was extremely ancient (Abir 1965, 1968, Kurimoto 1995). Ethiopia reputedly had overland contact with the North African coast, Benin and possibly with Angola and the 'Kingdoms of the Atlantic'

[1] Cyeneum lay either to the west of Ethiopia, where it has been identified with Sennar and the Gezeira region of Central Sudan, or to the south of the Blue Nile in the Gibe region.

(Bruce 1790, Crawford 1958). Communication on this scale was conducted indirectly through 'down-the-line' trade or non-directional trade.

6.1.2. Trade in Prehistory

Indirect Documentary Evidence

Although there is substantial documentary evidence that Ethiopia was an integral component of a highly fluid, long-range network of trade and exchange through the past 2000 years, the documentary evidence for prehistoric contact is limited and ambiguous:

The *Periplus* suggested that by the 1st century AD, trade between the Horn and India was already ancient. Phoenician merchants were possibly trading with Ethiopia and the Horn during the early 3rd millennium BP (Conti Rossini 1928), and the Ethiopian *Kebra Negast* (Book of Kings) reports 3rd millennium BP trading ventures from Ethiopia under the Queen of Sheba. Egyptian trade may have extended back to the 5th millennium BP, and Ethiopian products were valued in PreDynastic Egypt (Kitchen 1993). Semi-legendary Akkado-Sumerian documents report ships from the Horn and Ethiopia visiting Mesopotamia in the 5th millennium BP, although the geographical definition of Ethiopia in ancient texts does not necessarily correspond with the modern political boundary (Kobischnov 1966).

Archaeological Evidence

Archaeological evidence from Ethiopia and the surrounding regions provides more substantive support for prehistoric contact. As Table 6.1 indicates, Ethiopia was participating in a wide-ranging communication network that operated across North East Africa and the Red Sea throughout the early-late Holocene.

In addition, there is a strong archaeobotanical argument for extensive prehistoric contact within Africa and between Africa, Arabia and Eurasia. Haaland (*in press*) has proposed that tropical cereals and crops such as cotton reached India and Arabia from the Central Nile Valley, following transportation along the waterways of Northern Africa. Betts et al (1994) have argued instead that there was an extensive, mobile pastoral population in the Northern Arabian steppe that was responsible for the long distance movement of resources between Africa and Eurasia as early as the 8th millennium BP. This 'invisible culture' is poorly defined archaeologically, but has left its imprint in the skeletal web of contact that stretches across the Old World. The long distance network operated further afield in Africa than the Nile Valley. For example, the recent discovery of a copal pendant from Zanzibar or the Mozambique coast in a mid-5th millennium BP context in Mesopotamia indicates that ancient maritime trade routes connected South West Asia and the East African coast (Meyer et al 1991). Copal was being exported from the Somali coast by Arab traders from at least the start of the 2nd millennium BP (*Periplus*).

The recent recovery of plantain remains from an early 3rd millennium BP context in Cameroon is highly significant in the context of early trade. As discussed in Chapter 3, plantains were initially introduced to East Africa from South East Asia. The appearance of plantain in West Africa unequivocally shows that a long-distance communication network was established right across Africa by the start of the 3rd millennium BP (Rossel 1996, Doutrelepont et al 1996).

The Role of Trade in Prehistory

Today, trade is tantamount to survival for groups in marginal areas with specialised and vulnerable subsistence economies - for example the pastoral Afar of the Afar Rift in Ethiopia were able to inhabit one of the most inhospitable environments in Africa purely due to the availability of salt in the Afar Rift, their ability to exploit and transport it, and the sustainability of the salt trade in Ethiopia (Abir 1965). Trade and exchange undoubtedly played a varied and critical economic role in Ethiopian prehistory whuch, among other functions, would have provided a means and motive for colonisation of uninhabitable areas, provided a channel for the flow of materials and cultural information and acted as a buffering mechanism among specialised groups living in marginal environments, under the threat of climatic instability.

Summary

By the early Holocene the inhabitants of the Ethiopian Highlands appear to have been participating in a complex network of inter-regional contact that integrated Ethiopia, the Arabian peninsula, East Africa, the Nile Valley and East Sudan. This network may have involved several different mechanisms which maintained the flow of materials and information between disparate culture groups. The following sections examine how this interactive structure may have been responsible for the emergence of food production in Ethiopia, and when this process may have occurred.

Table 6.1. Evidence for Inter-Regional Contacts between Ethiopia, North East Africa and Southern Arabia before the 3rd millennium BP

Region of contact	Date of contact	Evidence
Arabian Peninsula and Red Sea Coast	11th/10th millennium BP	1. Cowrie shells from the Red Sea coast found in occupation deposits at Lake Besaka in the Afar Rift, Eastern Ethiopia, approximately 500 km from the coast (Brandt 1982).
	8th/7th millennium BP?	2. LSA Obsidian tools from the Dahlak Islands in the Red Sea were possibly made from Ethiopian obsidian (Zarins 1988, Fattovich 1993).
	7th/6th millennium BP	3. Obsidian tools from midden sites such as Sihi and Subr along the Red Sea coast of Southern Arabia are of possible Ethiopian origin (Zarins and al-Badr 1986). Similar obsidian remains have been found in middens on the Farasen islands off the Red Sea coast (Zarins and Zaharani 1985). The majority of obsidian scatters in Southern Arabia are concentrated in the south west corner, where the Strait of Bab el-Mandeb provide the shortest crossing to the Horn (Zairns et al 1981). At present only one source of obsidian is known from this area of Southern Arabia. This local obsidian is distinct from the smoky, grey green obsidian from Ethiopia.
	7th/6th millennium BP	4. Red Sea shells found in a mortuary context at Lake Besaka in the Afar Rift Valley indicate long-range (>500 km) contact between the coast and inland regions (Brandt 1982, 1986). Red Sea shells have also been found in burial contexts in the Ngorongoro Crater in East Africa, dating to the 3rd millennium BP (Leakey 1966) and at Kadero in the Central Sudan dating to the 6th/5th millennium BP (Krzyzaniak 1978).
	7th/5th millennium BP?	5. Camel tooth from Gobedra. Domestic camel possibly brought to Africa from Arabia, where it is known from the late 7th millennium BP at Sihi (Grigson et al 1989). The exact date of the Gobedra remains is not known, and the tooth may be intrusive from a younger level (Phillipson 1977b).
	7th/5th millennium BP?	6. Similarities in rock art style between North Ogaden (east Ethiopia) and Eritrea with sites in Central and Southern Arabia, especially in the Jubbah area (Cervicek 1979).
	5th millennium BP	7. Plant remains from Neolithic sites including wild oats (*Avena abyssinica*) and sorghum from Hili 8 in Abu Dhabi (Potts 1994) and sorghum from Wadi Yanaim in Southern Arabia and Ra's al-Hamra on the Oman coast (Biagi et al 1985, Amirkhanov 1994) suggest possible contact with Ethiopia at this time.
	mid-late 4th millennium BP	8. 'Tihama Cultural Complex' emerges along the Eritrean and Arabian coast with sites at Adulis, Sihi, Wadi Urq' and Subr (Zarins and Zaharani 1985, Zarins and al-Badr 1986).
Gash Delta, Sudan-Ethiopian border	?7th-4th millennium BP	1. No direct evidence until the 4th millennium, although obsidian of probable Ethiopian origin was used for lithic tools from the 7th millennium (Fattovich 1982, 1987).
	4th/5th millennium BP	2. Obsidian in Gash Group assemblages in the Gash Delta probably obtained from the prolific sources in the Ethiopian Highlands. Direct or indirect cultural contact between Highland and lowland regions is inferred (Fattovich 1990).
	4th/5th millennium BP	3. Ceramics from Quiha rock shelter suggest links between the Gash Delta and the Ethiopian Highlands (see Section 6.2).
	4th millennium BP	4. Early PreAxumite ceramics have been found at Jebel Mokram sites in the Shurab el Gash region of the Gash Delta indicating contact between the plateau and the lowlands (Fattovich 1993).

	4th millennium BP?	5. Gash Group-type sherds have been found at Agordat in Eritrea in possible 4th millennium BP contexts. Gash Group sites had a wide network of contacts in the 4th millennium that linked the Upper Nile Valley with Egypt, the Horn and possibly Southern Arabia (Fattovich 1993). Obsidian flakes from Ethiopia have been found at Mahal Teglinos on the Ethiopian-Sudanese border in levels dating to 3800-3500 years BP (Fattovich 1993).
	late 4th millennium BP	6. Similarities between Jebel Mokram ware and Type 2 ceramics at Gobedra (Fattovich 1990, 1993).
North African Sahara	c.5th millennium BP	1. Widespread contact or cultural derivation proposed between the Sahara and the Horn due to apparent stylistic and thematic similarities in the rock art of these areas, particularly in the Uwenat region (Muzzolini 1982, 1995). Possible affinities between LSA lithics of the Horn and Sahara material (Clark 1954, Kurashina 1978).
Lower Nile Valley	5th millennium BP?	1. Pictorial and documentary evidence from Egypt of trade contact between Egypt and Punt. Documentary and palaeoecological evidence suggest that Punt territory included part of the North West Ethiopian Highlands. Remains of domesticated plants indigenous to Ethiopia, including sesame, castor and safflower, found in Egyptian contexts from the 6th millennium BP.
	c.4th millennium BP	2. Surface collections from Agrodat in Eritrea contain an 18th Dynasty Egyptian earring and several polished stone axes that are morphologically comparable to 17th/18th Dynasty copper prototypes (Arkell 1954).
East Africa	5th-3rd millennium BP	1. Affinities between Pastoral Neolithic ceramics and ceramics from Quiha (see Section 6.2).
	pre-mid 4th millennium BP?	2. Fragments of a basalt stone bowl from a pastoralist(?) cultural horizon dating to 3500 BP and a basalt stone bowl from surface collection at Lake Besaka suggest cultural contact with the Savannah Pastoralists of East Africa, where stone bowls and domestic stock are found together from the 5th millennium BP onwards (eg. Barthelme 1977, 1984).
	4th millennium BP?	3. Close affinity between rock art styles from Ethiopia and East Africa, especially the Lake Victoria Basin and Mount Elgon (Chaplin 1974, Wright 1961)
Central Nile Valley	c.4th millennium BP?	1. Some affinities between the rock art of the Horn and depictions of cattle on decorated pots of the Nubian C-Group.
	4th millennium BP?	2. Stone maceheads at Agrodat suggest links with the Nubian Nile Valley cultures (Arkell 1954). Polished stone maceheads are found in the Central Nile Valley between the 6th and 4th millennia BP.

6.2. INTER-REGIONAL CONTACT AND THE INTRODUCTION OF DOMESTICATED CATTLE: THE EVIDENCE FROM QUIHA ROCK SHELTER

6.2.1. Background to Quiha Rock Shelter

Quiha rock shelter is located in Tigray Province in the North Eastern Ethiopian Highlands, close to the eastern edge of the escarpment (see figure 5.1.). A ceramic-bearing lithic assemblage was excavated from the site in the 1940s by Colonel Moysey, but this remains unpublished save for a brief mention in Clark (1954). Clark notes that the LSA lithic material from Quiha is distinct from other Highland lithic assemblages, but similar to that from surface collections from two nearby sites of Amba Sel and Zergnat.

Although the sample size is very small, with a total of only 103 ceramic artefacts, this is one of the few excavated ceramic-bearing LSA deposits from the Ethiopian Highlands with a complex occupation sequence and a detailed examination of the artefacts can make a significant contribution to our current limited knowledge.

Methodological Approach

The material from Quiha is currently stored in the National Museum of Kenya, Nairobi, where it was analysed in June 1996. As there are no original field notes from Moysey's excavation, any attempt at interpretation will be highly speculative and will make certain unsubstantiated assertions. The validity of these must be rigorously tested against future field work in Ethiopia and the surrounding regions.

The excavation at Quiha followed a simple stratigraphical format. The site was excavated to a maximum depth of 1.20 m. Three artefact-bearing levels were identified (Levels 2, 3, 4) which related to the depth in feet below the surface. No artefacts were recorded from the upper level (Level 1) and we must assumed that this was sterile. In addition there were a number of artefacts whose provenance was unrecorded. As it was unclear whether these derived from surface collection or from an excavated context, they have all been treated as one miscellaneous 'level'.

The assemblage comprised ceramics, flaked and ground lithics and faunal remains. The attributes of ceramic and lithic artefacts were examined independently and assessed within an inter-regional perspective. The results of these analyses have important implications for the interpretation of the faunal material, which is discussed subsequently in Section 6.2.5. A detailed quantitative analysis of decorative techniques and modes was not feasible given the broad stratigraphy and small sample size from this site. The classification given here is therefore a simplified version of the taxonomic range at the site. This classification indicates a clear stratigraphical distinction between different ceramic types, based on vessel form and decorative style. This suggests a pattern of site-use based on protracted, sporadic occupation and limited deposition of ceramic material.

6.2.2. The Ceramic Evidence from Quiha

Typological Analysis and Description of the Ceramics

The results of these analyses are illustrated in Tables 6.2-6.4 and figures 6.1-6.2. Table 6.5 presents an comparative, inter-regional assessment of the ceramic attributes.

FABRIC TYPES

I Very hard, well fired fabric. Colour ranges from black to dark red-brown and brick red. 20-35% medium-fine sand temper with larger mineral inclusions, especially muscovite. 5-10% airholes and some internal striations. This is the predominant fabric type, found in majority of vessels at Quiha.

II Coarse, hard fabric ranging in colour from dark brown to red-brown and brick red. 20-35 medium-coarse sand temper with larger mineral inclusions, especially muscovite, upto 4-5 mm. >15% airholes in fabric matrix and on surfaces. Surfaces can be very rough.

III Thin, hard fabric with red or red-brown exterior surfaces and buff core. 20-30% fine sand temper with some larger mineral inclusions, especially muscovite.

IV Hard black-dark brown fabric. c.20% medium-coarse sand temper with 5-10% mica inclusions. Found only in one base from the miscellaneous (surface?) context.

V Soft maroon fabric with easily abraded surface. Colour consistent throughout. 50% fine silt temper with no mineral inclusions. Found only in a 'ritual object'.

VI Hard pink fabric with 10-20% medium-coarse sand temper and mineral inclusions. Internal striations. Found only in a single body sherd from the miscellaneous context. The exterior surface was a highly polished grey-green colour. Dark brown/black slip on the interior surface.

VII Hard, light orange-red fabric with fine sand temper. Found only in a single sherd from the miscellaneous context.

VIII Hard, light pink-orange fabric. 20-30% medium-fine sand temper. c.15% airholes and internal striations. Smooth exterior and interior surfaces. Found only in ridged ware from the miscellaneous context.

IX Hard dark red fabric with c.30% medium sand temper. c.15% larger mineral inclusions especially muscovite or quartz and an unidentified black mineral. Found only in a single base sherd from the miscellaneous context. Possible red slip over the exterior of the base, and the remains of pale buff slip on the interior.

These categories are broad, given the small size of the sample. Over 85% of sherds at Quiha belong to Fabric Types I or II, which may be locally derived. The remaining fabrics, which were exclusive to miscellaneous or surface contexts, may be from imported wares. Fabrics IV and V were especially different from 'local' fabrics. As only isolated examples of Fabrics III-IX were present in this sample, the apparent absence of 'foreign' fabrics from the stratified levels should not necessarily be seen as significant.

VESSEL FORMS (figure 6.1.)

1 closed-mouth globular bowls
2 straight walled 'beakers' with vertical rims
3 closed-mouth, straight walled vessels
4 open-mouth vessels with straight walls
5 carinated bowls
6 shallow bowls with straight walls or with inverted rims
7 straight walled vessels with slightly everted rims
8 footed bases

As Table 6.5 shows, the forms found at Quiha are common to a wide range of diverse ceramic assemblages across Africa. However the range of forms is too small to allow concise inter-regional correlation. Three broad stylistic groups can be postulated:

- Forms 1-4 (and 6?) belong to a North and East African tradition spanning early-late Holocene. These forms were common to early levels at Quiha, but absent from later levels.

- Forms 5 and 6 were common to a mid-late Holocene tradition especially in Nile Valley, also East Africa and the Horn, especially after late 5th/4th-3rd/2nd millennium BP.

- Forms 6 and 8 have non-African affinities, and suggest contact with the Indian Ocean.

Overall, a broad chronological development in vessel forms is apparent from the lower levels to the upper levels and surface material.

Vessel form 5 in Level 2 is predominantly part of a Northern Africa tradition dating from the 4th millennium BP. These forms are common in PreAxumite II assemblages dating to the mid-3rd millennium BP, where they have polished black surfaces (Fattovich 1990). Quiha is located on the southern edge of the PreAxumite area, and this sherd from the upper levels of the excavation can be attributed to a PreAxumite tradition dating to the mid-3rd millennium BP. Several forms from the miscellaneous or surface level also had close affinities with PreAxumite vessels. This is especially true of Forms 6 and 8, which may have derived from later deposits. The footed bases from the miscellaneous or surface level had affinities with PreAxumite, Axumite or non-African (eg. Islamic, Indian) ceramics dating from mid/late 3rd-1st millennium BP.

DECORATIVE STYLES (figures 6.2.a and 6.2.b.)

A woven mat impression
B comb-stamping
(i) rocker-stamped (2 sherds in Level 4, 1 in Level 3, 4 in the miscellaneous context);
(ii) herring-bone, zig-zag or diagonal designs (12 sherds in Level 4, 2 in Level 3, 4 in the miscellaneous context)
C linear incisions - vertical, diagonal or criss-crossed
D linear incisions with cuneiform punctation
E cuneiform punctation
F finger nail incised designs
G channelled or ridged
H surface scraped on interior or exterior, or both surfaces
I polished or slipped

Surface decoration is the most informative diagnostic attribute of the ceramic collection from Quiha. Over 42% of all sherds from Quiha are decorated, using a wide range of decorative styles (*style* here refers principally to technique rather than mode due to the small sample size). Table 6.5 indicates that these styles are typical of African ceramic traditions.

Table 6.2.a. Frequency Distribution of Fabric Types at Quiha

Fabric Type	I	II	III	IV	V	VI	VII	VIII	IX
Surface and Miscellaneous	62	7	5	1	1	1	1	1	1
Level 2	1								
Level 3	21	1							
Level 4	7	14							
Total (134)	91	22	5	1	1	1	1	1	1

Table 6.2.b. Frequency Distribution of Vessel Forms at Quiha

Vessel Form	1	2	3	4	5	6	7	8
Surface and Miscellaneous	1	1	2			2	1	2
Level 2					1			
Level 3	2	1		1				
Level 4	1		1					
Total	4	2	3	1	1	2	1	2

Table 6.2.c. Frequency Distribution of Decorative Styles at Quiha

Decorative style	A	B	C	D	E	F	G	H	I
Surface and Miscellaneous		8	5				2	4	4
Level 2									1
Level 3	3	3	1		1	1		5	
Level 4		14		1				4	
Total (57)	3	25	6	1	1	1	2	13	5

Table 6.3. Intersection of Vessel Form and Decorative Style from Quiha to give Types

	Form								Decoration								
Type	**1**	**2**	**3**	**4**	**5**	**6**	**7**	**8**	**A**	**B**	**C**	**D**	**E**	**F**	**G**	**H**	**I**
1	**X**									**X**							
2	**X**										**X**						
3		**X**												**X**			
4		**X**									**X**						
5			**X**							**X**							
6			**X**									**X**					
7				**X**									**X**				
8					**X**												**X**
9						**X**											**X**
10								**X**									**X**

Table 6.4. Stratigraphical Distribution of Ceramic Types at Quiha

Level	Type: 1	2	3	4	5	6	7	8	9	10
Surface and Miscellaneous				1	1				1	2
Level 2								1		
Level 3		1	1				1			
Level 4	1					1				

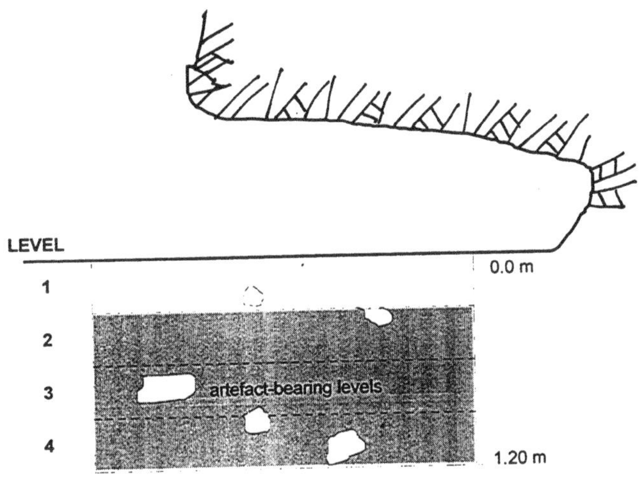

Diagrammatic section of Quiha rock shelter

Figure 6.1. Vessel forms from Quiha rock shelter.
(a) Form 1
(b) Form 2
(c) Form 3
(d) Form 4
(e) Form 5
(f) Form 7
(g) Form 6
(h) Form 8.

Figure 6.2.a. Ceramics from Quiha rock shelter.
Comb-stamped ware
(a), (b) and (d) rocker-stamped ware (Style B(i)).
(c), (e), (f) and (g) zig-zag, herring bone and diagonal designs (Style B(ii)).

Figure 6.2.b. Ceramics from Quiha.
(a) - (e) linear incised designs (Style C).
(f) sherd of unknown provenance, possibly from an oil lamp?
(g) 'ritual' object - possible bottle stopper or lip-plug (Wandibba *pers.comm.*).
(h) linear incisions with cuneiform design (Style D).
(i) cuneiform incisions (Style E).
(j) finger-nail incisions (Style F).
(k)-(l) channelled or ridged ware (Style G).
(m)-(o) surface scraped (Style H).

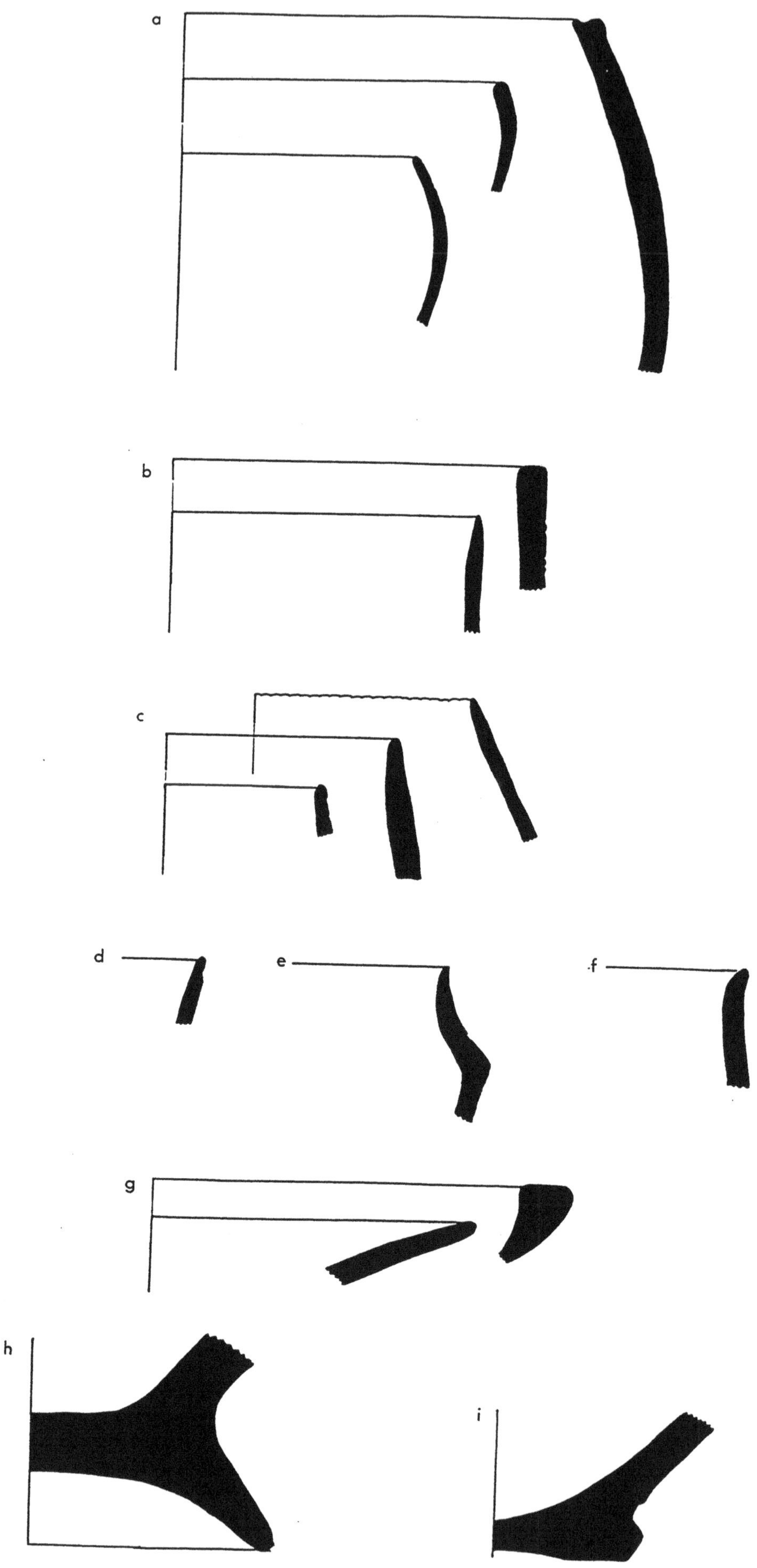

Figure 6.1. Vessel forms from Quiha

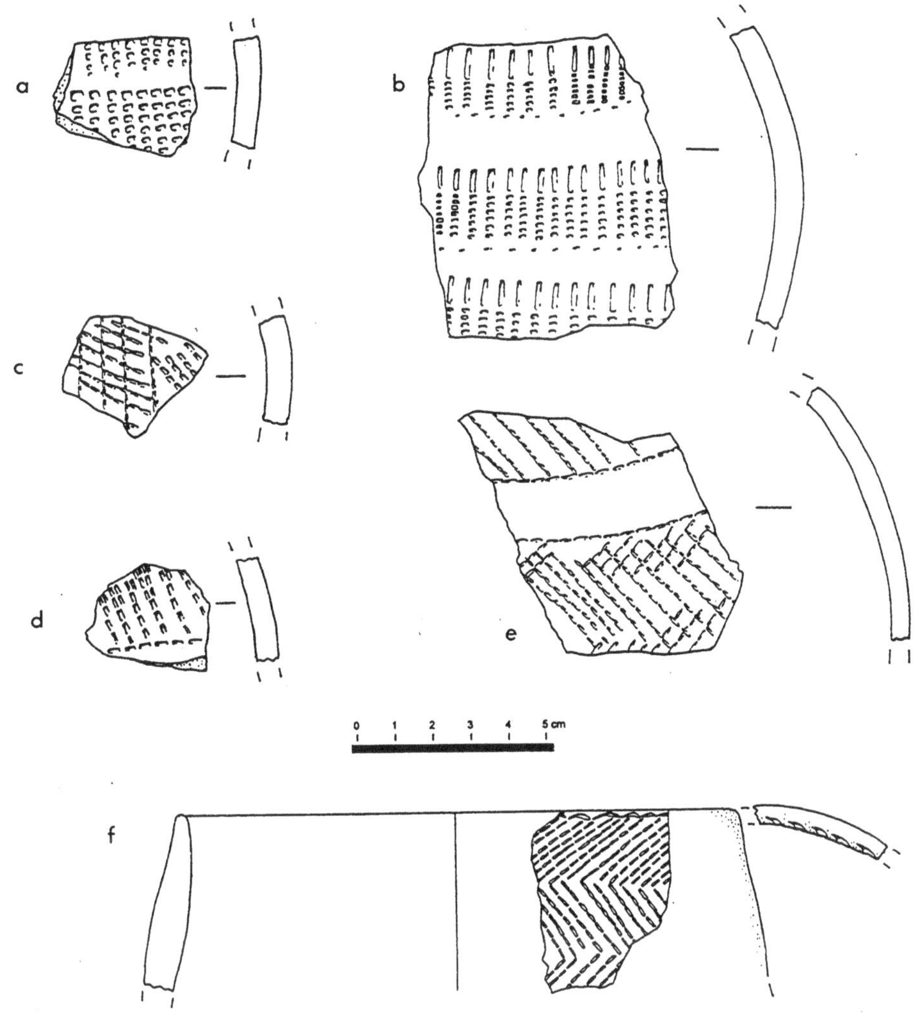

Figure 6.2.a. Decoration on ceramics from Quiha.

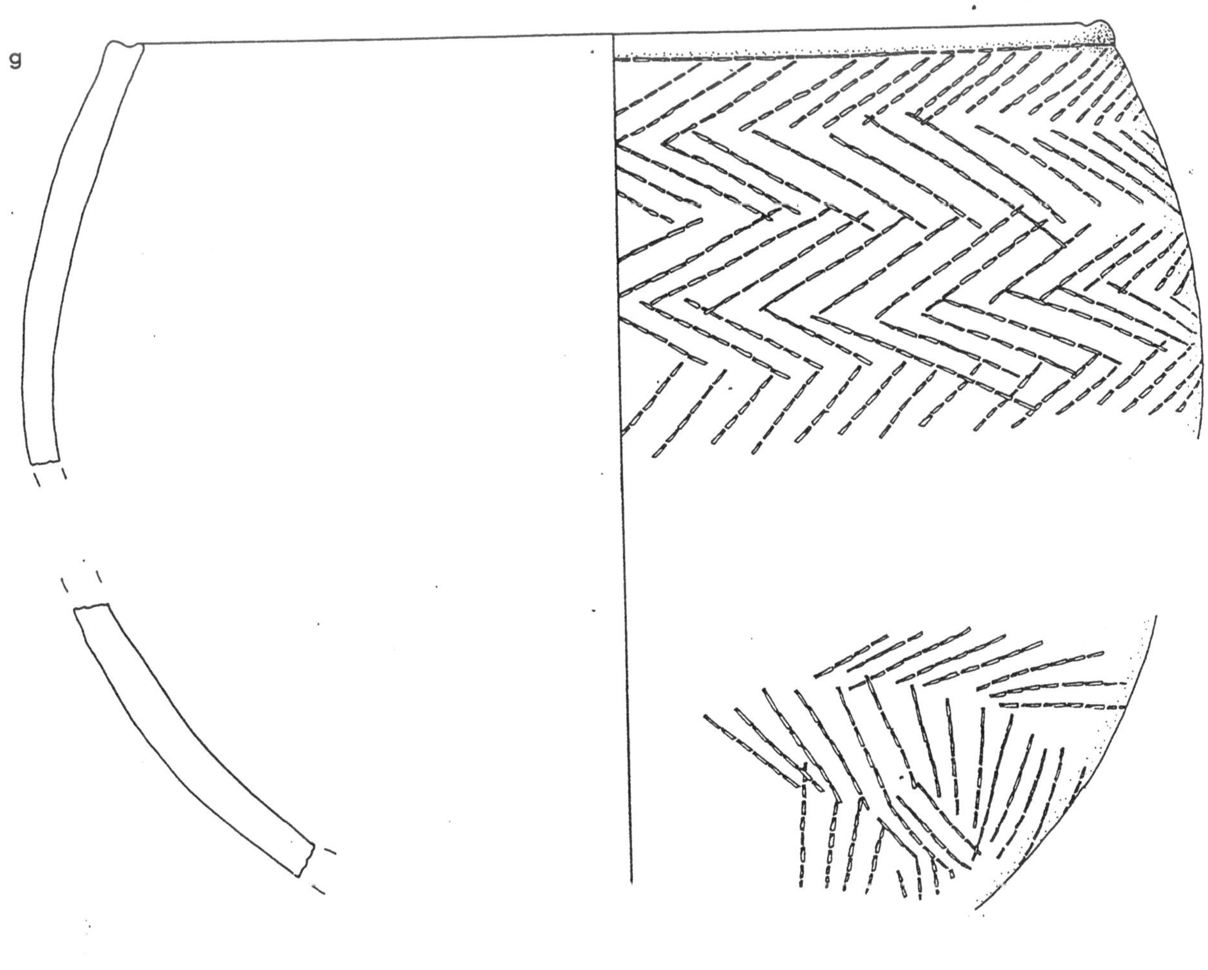

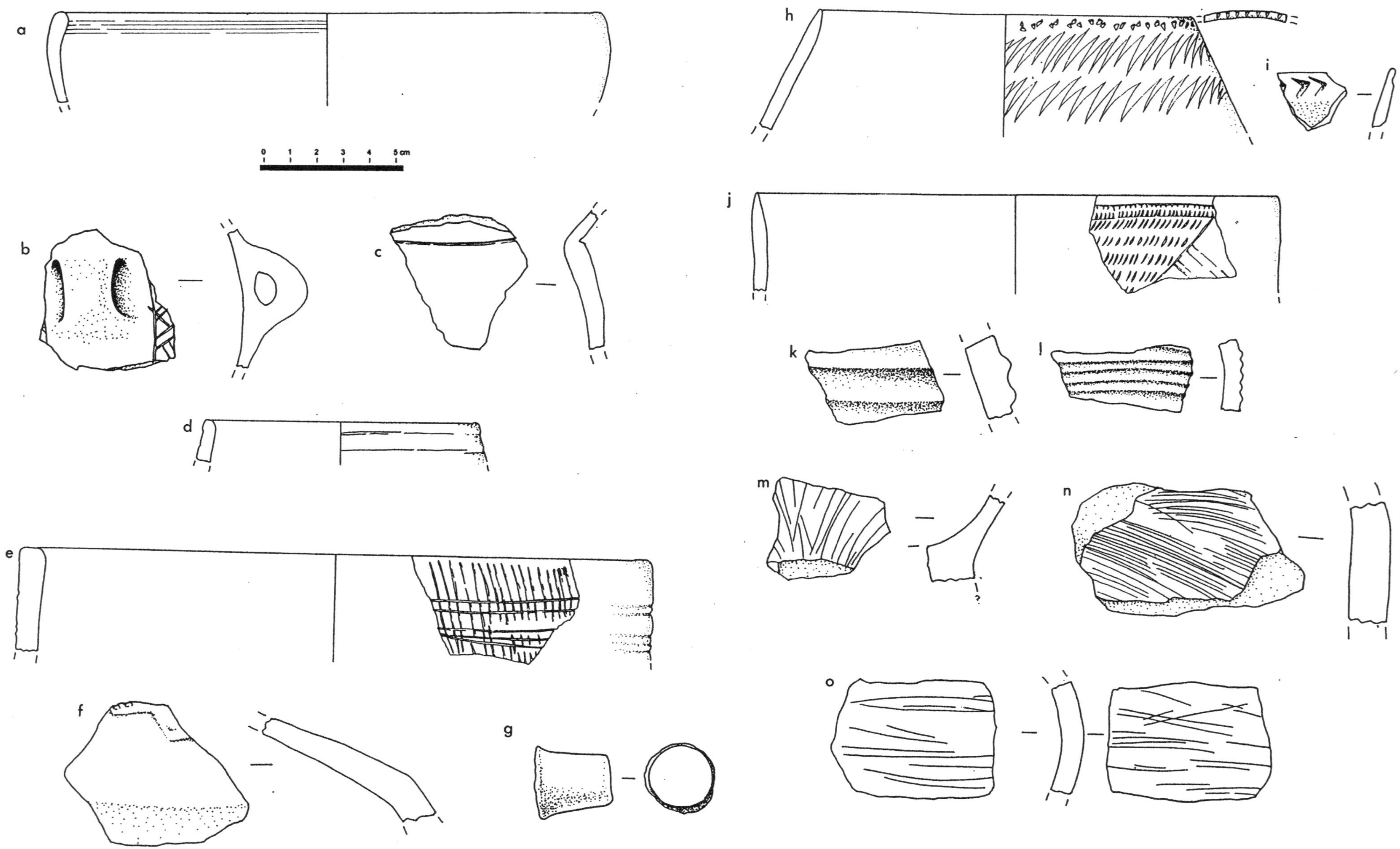

Figure 6.2.b. Decoration on ceramics from Quiha.

Table 6.5. Inter-regional Affinities of Ceramic Attributes

Affinity	Vessel Form 1	2	3	4	5	6	7	8	Decorative Style A	B	C	D	E	F	G	H	I
1									X	X i	X						
2	X	X	X							X i	X						
3										X ii	X	X	X	X		X	X
4	X	X	X	X					X		X 4	X 4		X	X 4	X 1-3	
5	X	X	X	X													
6	X			X			X			X ii	X						
7	X		X	X	X		X				X						
8	X	X	X	X		X	X			X ii	X				X*		
9	X	X	X		X												X
10	X	X	X					X?	X	X ii	X	X			X		
11	X			X	X	X	X	X							X		X
12					X?		X	X							X		X
13						X		X									X
14	X_a	X_a	X_a	X_b	X_{ab}		X_b		X_c	X ii_{fc}					X_{dc}	X_e	X_b

Affinities

1 Khartoum Mesolithic c.10th-7th millennia BP
2 Khartoum Neolithic c.late 7th-early 5th millennia BP. Certain sherds have affinities with Khartoum Neolithic ceramics, such as 6th millennium BP sites of Umm Direiwa, Zakiab and Rabak (Haaland *pers.comm.*).
3 Western Butana - eg. Shaqadud later cave and basin deposits c.6th-4th millennia BP
4 Gash Delta traditions
1 Amm Adam Group c.8th-7th millennia BP
2 Butana Group c.6th-5th millennia BP
3 Gash Group c.5th-4th millennnia BP
4 Jebel Mokram Group c.4th-3rd/2nd millennia BP
5 Pre Pastoral Neolithic of East Africa c.8th-6th/5th millennia BP
6 Kaynsore ware, East Africa c.8th-4th millennia BP (see Chapter 5)
7 Remnant ware, East Africa c.3rd millennium BP
8 Pastoral Neolithic wares c.late 5th-late 3rd millennia BP
9 Later Nile Valley Neolithic and Kerma culture c.5th-4th millennia BP
10 Southern Nile Valley and Southern Sudan c.4th and 3rd millennium BP (eg. Jebel Tukyi, Lokabulo)
11 PreAxumite, 3rd millennium BP
12 Axumite, c.AD 300-800
13 Islamic c.1st millennium BP
14 Other - **(subscripts)**
a Gobedra 7th-5th millennium BP
b eg. PreDynastic Egypt c.6th millennium BP
c Jebel Moya, South Central Sudan c.3rd millennium BP
d Pan-grave type decoration. Found also in surface material at Agordat in Eritrea, possibly dating to 4th millennium BP (Arkell 1954 Pl.XII, no.4). Channelled or ridged sherds of similar type have also been recovered at Lalibela cave, Lake Tana, c.mid-3rd millennium BP (Dombrowski 1971) and from an undated collection from Uahani in the Lake Tana area (National Museums of Nairobi, Ethiopian collection)
e scraped ware also known from Lake Besaka in probable 4th millennium BP deposits (Brandt 1982), possibly traded from the Highlands.
f Eastern Sahara area eg. Laquiha, North West Sudan, close to Bir Kiseiba, c.6th-3rd millennium BP (Cziesla 1978).

Key

symbols
* especially Turkwell ware, although the similarities are not close.

roman numerals
i and ii refer to decorative types Bi and ii as above.

numerals
numerals 1-4 refer to chronological traditions 1-4 in the Gash Delta area as above.

References

Include Addison (1949), Amirkhanov (1994), Anfray (1963, 1967b), Arkell (1949, 1953, 1954), Barthelme (1977), Bower et al (1977), Caneva (1987), Caneva and Marks (1990), Chilodnicki (1984), Collet and Robertshaw (1978, 1981, 1983a), Cziesla (1980), David et al (1981), de Contensen (1963, 1981), Doe (1971), Elamin and Khabir (1987), Fattovich et al (1984), Fattovich (1989, 1990), Haaland (1991), Leclant and Miquel (1959), Munro-Hay (1989), Nelson (1995), Onyango-Abuje (1977), Phillipson (1977b), Schon (1992), Wandibba (1980, 1990).

Discussion of the Implications of Ceramic Decoration at Quiha

The distribution of decorative styles at Quiha shows a similar pattern to the distribution of forms - i.e. Styles A, B and C in particular are common to early-mid Holocene traditions in North Africa and to mid Holocene traditions in East Africa. These styles are found in Levels 4 and 3, with Style B being particularly common in Level 4. Styles G and I, which are absent from the earlier levels at Quiha, are predominant in later assemblages from the Nile Valley, Ethiopia and Indian Ocean.

STYLE A - mat-impressed ware
North African/Nile Valley decorative technique, used throughout the early and mid-Holocene. It occurs exclusively in Level 3 at Quiha and may indicate contact with this region before the 2nd millennium BP.

STYLE B - comb-stamped ware
(i) Rocker-stamped decoration is characteristic of early Neolithic assemblages from the Nile Valley c.6th and 5th millennia BP and suggests that there may have been contact with the Ethiopian Highlands during the mid-Holocene. Sherds from excavated contexts were all highly abraded, possibly indicating prolonged deposition.
(ii) Herring-bone and diagonal comb-stamping is not characteristic of Nile Valley assemblages, although it is found at Shaqadud in the Butana region of the Sudan, and at later Neolithic sites from Southern Sudan. The technique is often used to decorate East African Pastoral Neolithic and LSA ceramics, and an impressed herring-bone design is especially common in earlier assemblages (Leakey 1931). Possible contact with East Africa from the 5th millennium BP may be proposed.

STYLE C - linear incised ware
Linear incised decoration is rare to African assemblages. The single example from Quiha has little potential for comparative analysis.

STYLE D - cuneiform punctation with linear incisions
This type of decoration is only found to the west of the Highlands, especially in the Gash Delta and Butana, suggesting possible contact between this region and the Ethiopian Highlands in the 5th-3rd millennium BP.

STYLE E- cuneiform punctation
Although punctation was widely used to decorate ceramics across North and East Africa, the cuneiform punctation found at Quiha has a restricted distribution in the Western Butana. Broken microlith tools (see below) may have been used to produce this form of decoration at Quiha.

STYLE F - finger-nail incised ware
Again, this technique is restricted to mid-Holocene ceramics in the Butana and Gash Delta region.

STYLE G - channelled or ridged ware
Other than a superficial similarity to late Neolithic/early Iron Age Turkwell pottery, this type of decoration is exclusive to mid-late Holocene ceramics in Northern Africa or to foreign wares. Similar sherds have been found in the Highlands in PreAxumite contexts at Matara (Anfray 1963, de Contensen 1963), in mid-3rd millennium BP levels at Lalibela cave (Dombrowski 1971) and in a LSA surface collection from Uahani, Lake Tana (National Museums of Nairobi, Ethiopian Collection). Affinities between these sherds and ceramics of the Pan Grave culture of the Eastern Sudan and the Jebel Mokram group of the Gash Delta led Fattovich (1990a) to propose contact between these areas and Ethiopia during the 3rd millennium BP. The sherds from the miscellaneous or surface level at Quiha may be part of a 3rd millennium BP PreAxumite assemblage (Fattovich 1990b).

STYLE H - surface scraping
Occurs most commonly on mid-Holocene ceramics from the Butana and Gash Delta.

STYLE I - polished or slipped surface
This style of surface decoration is most common on later ceramics from the Nile Valley - from Egypt, Napata, Kerma and Meroe, or from foreign wares from Arabia. Surface treated wares from Quiha generally occur on later or foreign forms, as outlined above, and the majority undoubtedly belong to a PreAxumite or later assemblage. One unusual body sherd from an unknown context at Quiha had a grey-green polished exterior surface and black slip on the interior

surface. The fabric was imported, and a possible Islamic origin could be suggested.

Summary Interpretation and Chronological Deposition of the Ceramic Assemblage

Nature of the Deposition

The small yet diverse nature of this collection suggests that it was deposited sporadically over a considerable period of time, as a result of intermittent use of the rock shelter by different cultural groups. The rate of deposition is unknown and depends on local conditions of erosion, sedimentation and occupation intensity. For instance, at Gobedra rock shelter approximately 100 km to the north, 10 cm depth of deposit spanned a period of c.2000 years. At Gorgora rock shelter on Lake Tana, the 120 cm (4 ft) level had a probable late Pleistocene/early Holocene date. Hypothetically, the 120 cm depth of deposit at Quiha could represent over 24 000 years of intermittent use. However the nature of the assemblage implies that it was deposited over 6000 years, from the 6th/7th millennium to the 2nd/1st millennium BP

Evidence for Inter-Regional Contacts

The collection from Quiha has diverse affinities with ceramics from North and East Africa. The assemblage aspires to a typically African tradition with rare foreign elements. The similarity between decorative styles from Quiha and from surrounding regions imply that the eastern edge of the Ethiopian Highlands benefited from persistent contact with the Nile Valley cultures from at least the 6th millennium BP, and with East Africa from the 5th millennium BP. The following features are particularly informative:

- *c.6th-4th millennia BP*

  Affinities with comb-stamped wares from sites such as Rabak on the White Nile (Haaland *pers.comm.*). Rare sherds of Early Khartoum and Khartoum Neolithic type have been found in the Ethiopian-Sudanese border region, suggesting that this area also was in contact with the Nile Valley from at least the 7th millennium BP (Fattovich 1990). From the late 5th millennium BP limited quantities of pottery may have been imported from the East and from the Nile Valley (for example black ware from Kerma) (Fattovich 1990a, Zarins and al-Badr 1986, Zarins and Zahrani 1985).
- *c.5th-3rd millennia BP*

  The later occupation deposits at Quiha suggest that contact may have become more focused on the Butana/Gash Delta and the Sennar/Jebel Moya area to the south of the Blue Nile in Sudan during the 4th and 3rd millennia, with a subsequent increase in foreign contacts. There is little evidence of contact with Southern Arabia except possibly in the later stages of the deposit. Ceramics appear relatively late in Southern Arabian contexts, and there is no suggestion of either an independent ceramic innovation or a simultaneous adoption of ceramics and domesticated resources.
- The collection also has a number of features that are distinct from ceramic assemblages elsewhere in Africa which are indicative of an indigenous Ethiopian tradition. Many of the affinities noted in Table 6.5 are superficial and fairly isolated. This may reflect indirect and protracted cultural contact.

Proposed Chronology for the Deposit

The following chronology can be proposed from the above discussion:

Surface - 1st and 2nd millennia BP
Level 2 and top of Level 3? - PreAxumite c.3rd millennium BP
Level 3 - spans late 5th-early 3rd millennia BP
Level 4 - spans 6th and 5th millennia BP.

6.2.3. Lithic Material from Quiha

The Nature of the Assemblage

A total of 211 modified lithic pieces were recovered from Levels 3 and 4, and from the miscellaneous or surface level at Quiha. The industry was dominated by broad blades, circular scrapers and microlithic blades and lunates, many of which were snapped. The majority of the lithics were made from dark grey obsidian, although 13 pieces of chert were also recorded.

Rationale of the Analysis

The lithic assemblage was examined for the following: the stratigraphical distribution of different tool types within the deposit and changes in this pattern over time, and the relative dimensions of different tool types from different levels. These data were used to identify diachronic changes in manufacturing technology, and quantify the diagnostic features for comparison with other lithic assemblages.

Lithic Measurements

Measurements were taken with a pair of calipers on all catageory of tool. Length and width dimensions were recorded along the greatest vertical and horizontal axes, as shown here.

Stratigraphical Distribution Patterns

The frequency and stratigraphical distribution of lithic tools and modified fragments is detailed in Table 6.6 and illustrated graphically in figure 6.4. A selection of the implements is illustrated in figures 6.3.a. and 6.3.b. Due to the small size of the sample, the poor stratigraphical resolution and the lack of dates, all inferences are highly speculative and only a broad comparative analysis can be attempted. In addition, as the provenance of artefacts attributed to the miscellaneous context is unknown, this category cannot be used in any quantitative comparison.

As figure 6.4 shows, there is little change in the relative proportions of different tools types (expressed as percentage frequencies) between Levels 3 and 4 at Quiha (a period spanning c.3000 years between the 6th and 3rd millennia, following the proposals made in Section 6.2.2). This is in direct contrast to the immense apparent diversity of the ceramic assemblage. The gross implications from Quiha are that this technological continuity reflects cultural homogeneity and continuity in functional application during this period. This raises the issue of whether or not there was also continuity in subsistence strategy at Quiha during the earlier phases of use.

Affinities of the Lithic Assemblage

The length-width ratios of three categories of lithic tools from Quiha were measured, for blades, scrapers and microlithics. These measurements are given in Table 6.7 and a comparative metrical analysis is presented in figure 6.6.

Clark (1954) observed that the lithics from Quiha did not conform to the broad LSA lithic tradition of Ethiopia and the Horn. Instead Clark suggested close affinities with the Elmenteitan industry of Kenya. As figure 6.5 demonstrates, there is a correlation between the dimensions of microliths from Quiha and those from Elmenteitan assemblages. Similarities are apparent in the technological attributes of the Elmenteitan industry and the lithics from Quiha, particularly in the use of inverse retouch and small-backed microliths (Nelson 1980).

However Nelson (*n.d.*) maintains that an entire assemblage is more informative as a comparative tool than individual stylistic features. The primary attributes of a lithic assemblage (backed blades, for example) should be used as criteria for assessing the affinities between occupation sites and cultural groups. the characteristic features of the Quiha assemblage - broad blades, circular scrapers and microlithic crescents - are not predominant in Elementeitan assemblages (Ambrose 1980, 1984). Likewise a number of characteristic components of the Elmenteitan assemblage, including long blades, long end-scrapers and large-backed blades, are absent at Quiha.

Summary of the Lithic Analysis

The approach outlined above has shown that there was little change in lithic technology at Quiha, possibly between the 6th and 3rd millennia BP, suggesting continuity in subsistence demands and activities.

In addition, although apparent similarities exist in the manufacturing technology between material from Quiha and from Kenya, significant differences in the range of tool types fundamentally exclude Quiha from being classed as an Elmenteitan-type assemblage. The observed affinities may reflect a parallel technological adaptation to similar ecological and economic conditions, using identical raw materials. The typological framework from Ethiopia is too skeletal at present to determine whether the Quiha assemblage represents a discrete socio-economic unit, possibly introduced to the Highlands, or whether it is a regional facies of the Ethiopian lithic tradition (the EBTT).

6.2.4. Summary and Implications of Lithic and Ceramic Material from Quiha

The nature of the artefact collection implies that it was deposited sporadically as a result of intermittent occupation of the rock shelter, possibly over several millennia between c.6th and 1st millennia BP. The ceramic assemblage is highly diverse, but it falls within a broad North and East African tradition that persisted throughout the early and mid Holocene periods. The apparent homogeneity of the lithic sample may only be a reflection of the poverty of the sample size, and cannot be treated as too significant in the absence of supportive evidence.

When the material from Quiha is examined in an African context, a significant pattern is revealed in which the early sequence at Quiha belongs to a purely African phenomenon and there is no evidence for non-African influences until relatively late in the sequence, possibly the 3rd millennium BP. Whilst there are undoubted similarities between ceramics from Quiha and those from surrounding regions, the distinctiveness of the assemblage is indicative of a localised tradition with widespread contacts. This would suggest that Ethiopia was not isolated from its surroundings, but was actively involved in inter-regional communication during the Holocene.

Table 6.6. Frequency and Distribution of Lithic Tool Types at Quiha

Shaped tools	Level 3	%	Level 4	%	Not known	%
Blades						
long blades[1]	10	*9.00*	6	*8.10*		
retouched blades[2]	5	*4.50*	2	*2.70*		
backed blades[3]	2	*1.80*	2	*2.70*	1	*2.77*
broken bladelets	5	*4.50*	2	*2.70*		
Total blades	*30*	***19.80***	*17*	***16.20***	*1*	***2.77***
Microlith blades[4]	*8*	***7.20***	*5*	***6.75***		-
Scrapers						
circular and disc scrapers	9	*8.10*	6	*8.10*	2	*5.55*
Total scrapers	*9*	***8.10***	*6*	***8.10***	*2*	***5.55***
Geometric microliths						
Lunates	25	*22.52*	13	*17.56*	4	*1.11*
Broken lunates	15	*13.51*	16	*21.62*		
Total geometrics	*40*	***36.03***	*29*	***39.18***	*4*	***11.11***
Outils ecailles	3	*2.70*	2	*2.70*	6	*16.66*
Burins	3	*2.70*	1	*1.35*	3	*8.33*
Cores	1	*0.90*	1	*1.35*	-	-
Inverse retouched flakes	-	-	2	*2.70*	-	-
Total other tools	*7*	***6.30***	*6*	***8.10***	*9*	***25.0***
Total modified waste fragments	*25*	***22.52***	*16*	***21.62***	*20*	***55.55***
TOTAL	**111**	***100***	**74**	***100***	**36**	***100***

DEFINITIONS OF SPECIFIC TOOL TYPES

1 Long blades: Lithic artefacts with roughly parallel edges where the length is at least twice their width, and the length is no less than 2.4 cm. Maximum blade length at Quiha is 5.6 cm.

2 Retouched blades: Blades with secondary working (edge trimming or flaking) along their longitudinal edges.

3 Backed blades: Blades that have been steeply retouched along one long side in order to blunt this edge and produce one sharp, working edge.

4 Microlith blades: blades with length of no more than 2.4 cm.

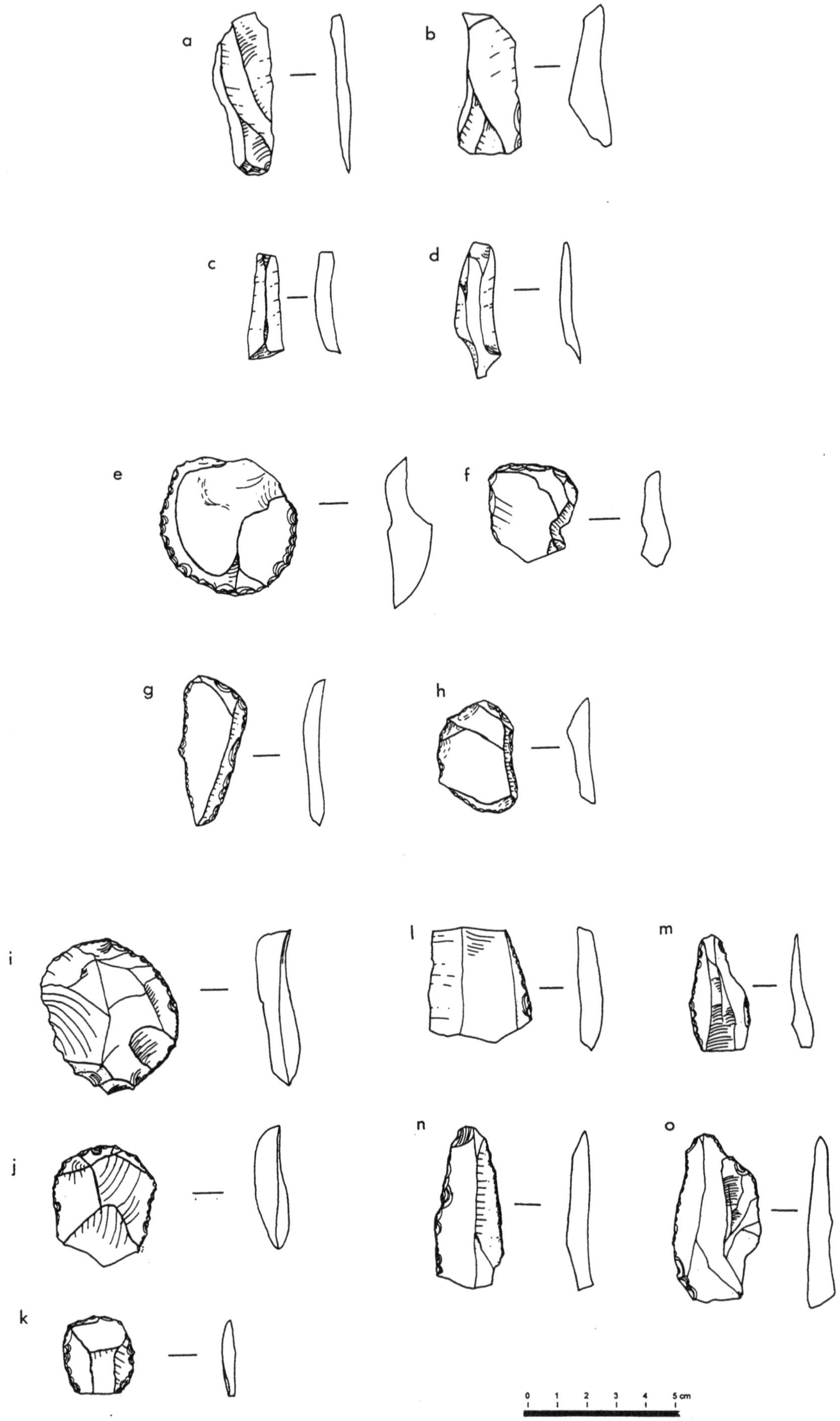

Figure 6.3.a. Obsidian tools from Quiha.
(a)-(d) blades, (e)-(o) scrapers.

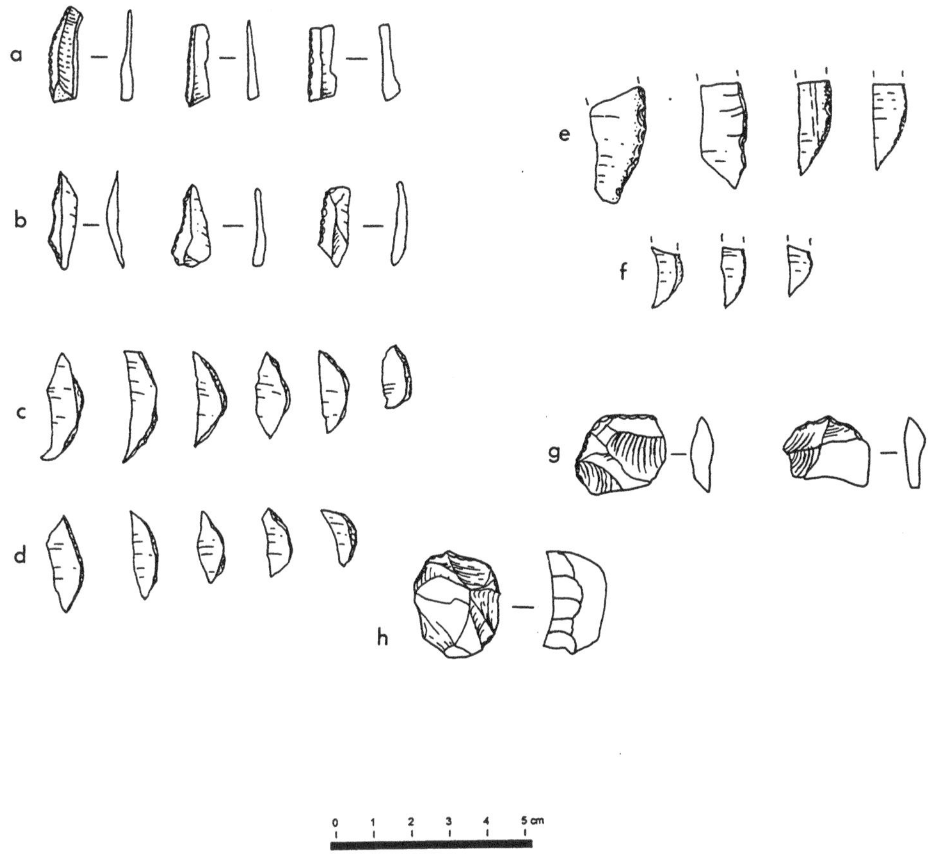

Figure 6.3.b. Obsidian tools from Quiha.
(a)-(b) microlith blades, (c)-(d) microlith crescents, (e)-(f) broken crescents, (g) outils ecailles, (h) core.

Figure 6.4. Relative frequencies of lithic tool types from Quiha expressed as percentages.

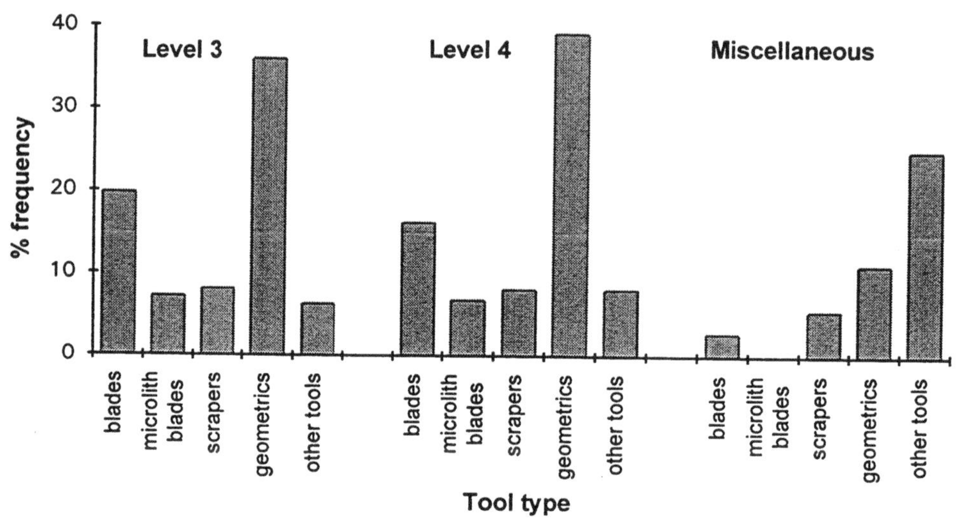

Figure 6.5. Length-width ratios of microliths from Elmenteitan (circles) and Savanna Pastoral Neolithic (squares) sites in Kenya.

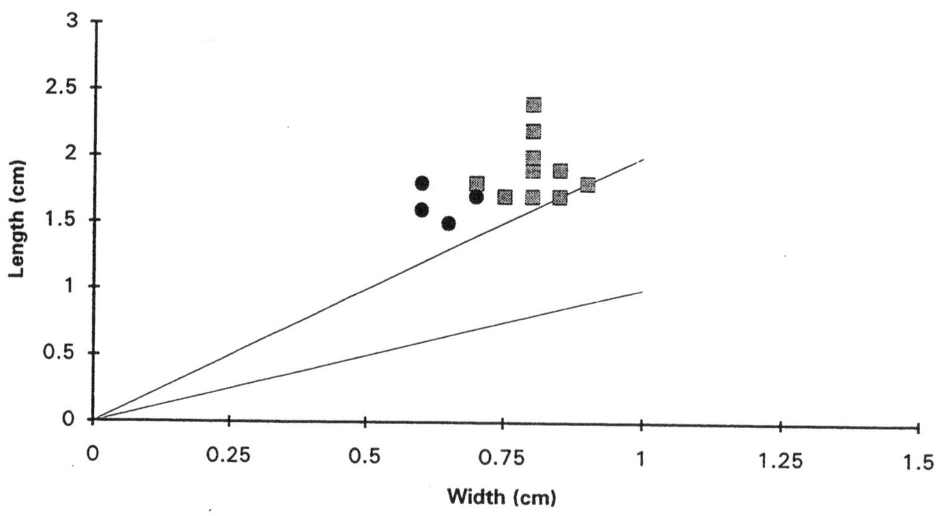

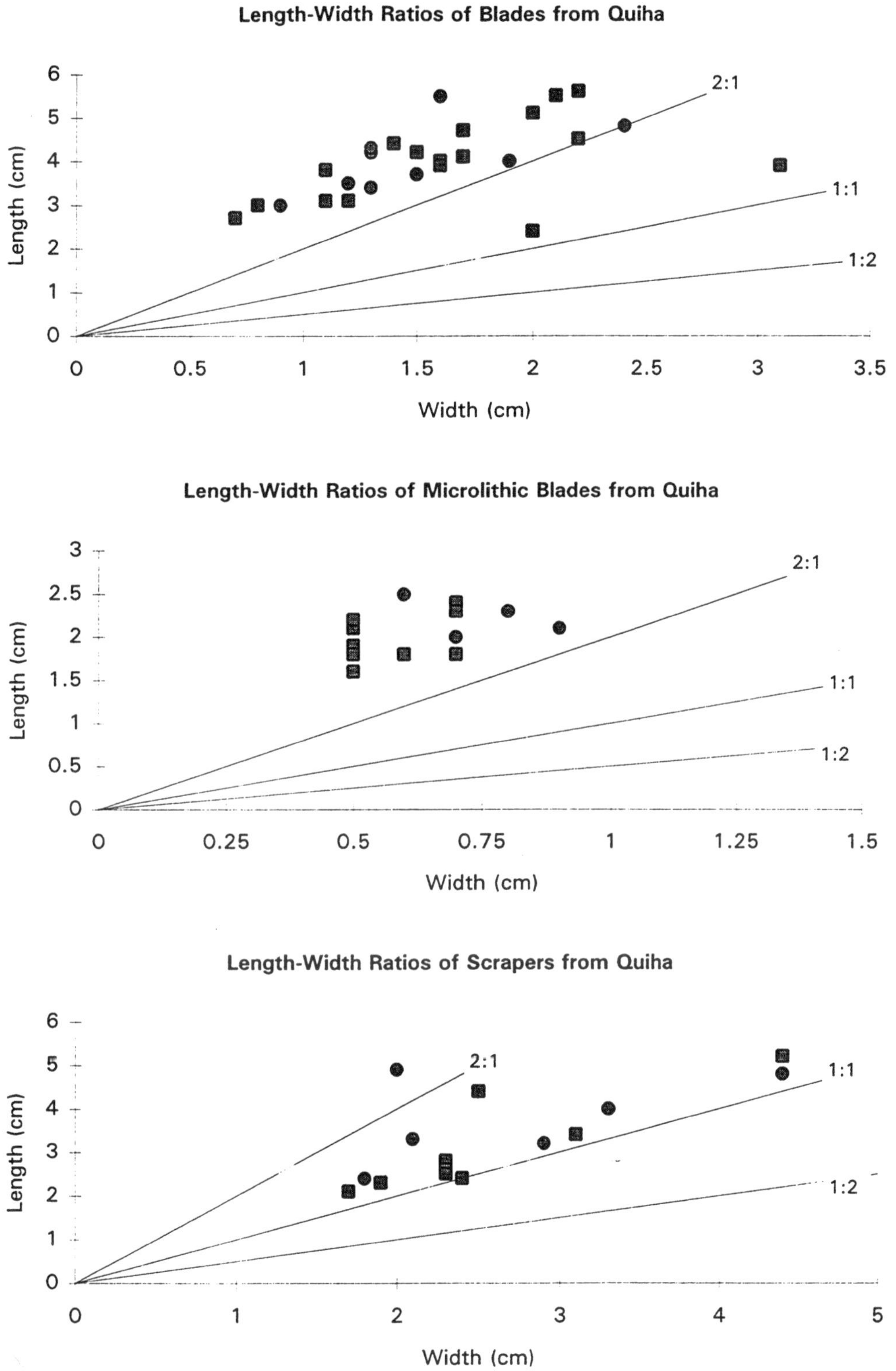

Figure 6.6. Length-width ratios of lithic and microlithic tools from Quiha (circles=Level 4; squares=Level 3).

	Width	Length	Width	Length	Width	Length	Width	Length	Width	Length	Width	Length
Level 3	Blades		Microlithic Blades		Scrapers		*Outil Ecailles*		Retouched blades		Microlith Lunates	
	1.4	4.4	0.7	2.4	4.4	5.2	2.3	2.8	2.1	6.6	0.8	2.9
	0.8	3	0.7	2.3	3.1	3.4	1.9	2.3	1.7	5.7	0.8	3.2
	0.7	2.7	0.6	1.8	2.5	4.4	1.6	2.2	2.1	3.5	0.9	2.4
	2	2.4	0.5	2.1	2.3	2.8			1.4	3.8	0.9	2.3
	2.2	5.6	0.5	2.2	2.3	2.6	Burins		2	3	0.6	2.1
	1.7	4.7	0.5	1.9	2.3	2.5	1.1	3.4	1.4	3.7	0.6	1.6
	2.1	5.5	0.7	1.8	2.4	2.4	0.9	3.5	1.3	3.9		
	3.1	3.9	0.5	1.6	1.7	2.1	0.6	2.6				
	2	5.1	0.5	1.8	1.9	2.3						
	2.2	4.5					Utilized Cores					
	1.5	4.2					2	2.4				
	1.7	4.1										
	1.6	4					Broken Lunates					
	1.6	3.9					0.4	3.1				
	1.1	3.8					0.3	2.2				
	1.2	3.1					0.4	1.7				
	1.1	3.1					0.2	1.3				
Level 4	Blades		Microlithic Blades		Scrapers		*Outils Ecailles*		Retouched blades		Microlithic lunates	
	1.5	3.7	0.8	2.3	4.4	4.8	1.8	2.1	1.2	3	1.7	2.6
	2.4	4.8	0.6	2.5	3.3	4	1	2.4			0.7	2.3
	1.6	5.5	0.9	2.1	2.9	3.2					0.8	1.8
	1.9	4	0.7	2	2.1	3.3	Burins				0.5	1.6
	1.3	4.2			2	4.9	1.5	3.2			0.6	1.4
	1.3	4.3			1.8	2.4						
	1.2	3.5					Utilized Cores					
	1.3	3.4					2.2	2.7				
	0.9	3										
							Broken Lunates					
							1.3	2.8				
							0.9	2.5				
							0.7	1.5				
							0.6	1.2				

Table 6.7. Proportions of Shaped Lithics from Quiha Rock Shelter (in cms)

6.2.5. Faunal Remains from Quiha and the Emergence of Cattle in Ethiopia

A total of 19 mammalian teeth and 1 bone fragment were recovered from Quiha Levels 3 and 4. The final identifications were made by Paul Watene at the National Museum of Kenya, as shown in Table 6.8.

Table 6.8. Faunal Remains from Quiha Rock shelter

Osteological remains	Level 4	Level 3
Bos sp	2 teeth (1 molar, 1 premolar)	9 teeth (6 molars, 3 premolars inc 1 milk tooth from a young animal)
Bovid IV (large bovids including *Bos* sp)	1 incisor upper part of ulna from a mature animal	2 incisors
Equus sp	1 molar	3 molars
Homo sapiens	1 incisor	-

Discussion of Osteological Remains

Cattle

The *Bos* sp teeth all belonged to domesticated cattle. Non-diagnostic remains from large bovids (Bovid IV) may additionally represent domesticated cattle. The rock shelter was therefore occupied by groups with domestic cattle from Level 4 (90-120 cm depth) onwards. The species type has not been identified, although an earlier examination of the faunal material suggested the presence of thoracic humped cattle of the zebu type, based on the analysis of thoracic vertebrae from the collection (Clark *pers.comm.*). There were no vertebrae present when I examined the material, and I am unable to confirm this suggestion.

Equids

The *Equus* sp remains may derive from one of the many wild *Equus* species of the Ethiopian Highlands which were hunted both by foraging and food producing groups. Alternatively, they may belong to domesticated equids. The possibility that domesticated *Equus* were present in prehistoric contexts in the Ethiopia Highlands must be investigated in detail. This has significant implications for our understanding of the domestication and spread of the African wild ass, *Equus africanus*.

The domestic ass was important in Egypt from the 6th millennium BP, and is attested from 4th millennium BP levels at Mahal Teglinos in the Gash Delta (Fattovich 1990). There is no evidence for domestic equids in Ethiopia before historical documentation in the last 2000 years. However according to recent linguistic research, the wild ass was domesticated before the movement of Ethio-Semitic speakers to the Highlands (Blench 1995). If, as discussed in Chapter 1, Ethio-Semitic emerged in the mid-late Holocene, then the domesticated ass may have considerable antiquity in Ethiopia.

Implications of the Faunal Material

The excavations at Quiha recovered domesticated *Bos* teeth from the lowest level, Level 4. Section 6.2.2. tentatively proposed a 6th/5th millennium BP date for this level. The bovid remains were stratigraphically associated with decorated pottery fragments with affinities to early Neolithic ceramics from the Nile Valley. There is no evidence of contact with the Arabian peninsula at this stage. The logical inferences that can be drawn from these observations are, first, that the Eastern Highlands were in contact with groups in the Nile Valley with domesticated cattle during the 6th/5th millennia BP, second that domesticated cattle were present in the Eastern Ethiopian Highlands by the 5th millennium BP, possibly during the 6th millennium BP and consequently that domesticated cattle may have been introduced to the Ethiopian Highlands from the Nile Valley during the 6th or 5th millennium BP.

The ceramic evidence implies that the Ethiopian Highlands had little contact with South Arabia during the mid Holocene period. Lines of communication were instead seem to have been directed towards the Nile Valley region. Domesticated cattle appear to have been introduced from the North West, from the Nile Valley region during the 6th or early 5th millennium BP. Cattle may then have spread through the Ethiopian Highlands during the 6th and 5th millennia. As Clark (1980) suggested, pastoralists then spread through the Rift Valley, reaching the Lake Turkana basin by the 5th millennium BP. Domesticated cattle were subsequently introduced to East Africa from this region during the late 4th and 3rd millennia BP.

The Process of Acquisition

The data from Quiha have provided an insight into the chronology and directionality of the introduction of domesticated cattle to Ethiopia. Unfortunately, these data cannot also describe the mechanisms responsible for this event, and we must move into the realms of speculation.

Ethnohistorical and archaeological analogy provide limited support for a speculative model.

Ambrose (1984) proposed four distinct processes to explain the appearance of domesticated cattle in a given region, which he applied to the Savannah Pastoral Neolithic phenomenon in the East African Highlands. These processes are as follows:
(i) independent domestication of a wild progenitor
(ii) diffusion of pastoral techniques and materials
(iii) small-scale population movements
(iv) large-scale population movements.
Processes (i), (iii) and (iv) are discussed below.

Independent Domestication
As discussed in chapter 3, the absence of a wild cattle progenitor[2] in Ethiopia effectively eliminates the possibility that cattle were domesticated indigenously.

Population Movements
The movement of pastoral groups into the East African Highlands in the 4th and 3rd millennia BP provides a broad anaolgy to the initial pastoral colonisation of the Ethiopia Highlands. Both areas offered vast tracts of unexploited grazing land, with similar climates and available resources. Two opposing models have been proposed to account for the spread of pastoralists into East Africa. Robertshaw (1984, *unpublished report*) suggested that the introduction of domesticates to East Africa was primarily as a result of a moving frontier of pastoralists. These groups progressively expanded their territory through small-scale movements and gradually colonised areas with low existing population densities. Ambrose (1984) argued instead for intermittent, large-scale population expansion. In Ambrose's model, migration was constrained by environmental parameters such as tsetse fly and cattle diseases. Removal of these parameters through changing environmental conditions, or adaptation to them through technological or genetic development, enabled large-scale movement into a new area.

Both these models may be of relevance to Ethiopian prehistory. Ethnohistory and documentary records chronicle the movements of pastoral groups into and within Ethiopia from the Nile Valley and Red Sea Hills during the past 2000 years at least (Hassen 1990, Abir 1968, Wilding 1985). Both large-scale and small-scale movements have recurred during this period, providing a possible broad analogy for the process of colonisation of the Ethiopian Highlands. For example, historical texts document large-scale population movements of Somali pastoralists from Somalia and Beja pastoralists from the Red Sea Hills to the Ethiopian Highlands during the past 2000 years. These movements arose in an environment of political and economic weakness that was precipitated by a period of climatic degradation (Wilding 1985a, Sadr 1990). Conversely, ethnohistorical traditions of pastoral groups such as the Borana of South West Ethiopia record random, small-scale movements during the early 1st millennium BP in response to similar environmental and economic degradation (Wilding 1985b). These small-scale movements are perceived by the Borana as lost tribal wanderings, contrasting to the mass incursions of documented history.

6.3. PREHISTORIC CONTACTS AND THE INTRODUCTION OF DOMESTICATED PLANTS

As discussed in chapter 3, the agricultural system of the Ethiopian Highlands depends on a number of important plant resources that are not indigenous to Ethiopia. Many of these are derived from temperate regions, such as the Near East. The climatic requirements of these crops mean that they cannot be successfully grown south of latitude 25°N. They would not therefore have reached Ethiopia by gradual diffusion along the Nile Valley (von Bothmer 1995). With the exception of a single identification from the Gash Delta, there is no direct evidence that these plants were farmed outside the Lower Nile Valley before the 3rd millennium BP (see Table 3.2, chapter 3). The reports from the Gash Delta are extremely interesting in this context (Fattovich and Piperno 1982). Domesticated barley was identified from single grain fragment from a mid 5th-4th millennium BP context. This observation has yet to be confirmed, but it does support the idea that foreign plants were introduced directly to North East Africa at an early date, avoiding the Nile Valley.

Developed agriculture based on imported crops was well established across the Ethiopian Highlands by the mid-3rd millennium BP. The evidence discussed in chapter 3 posed two main arguments; first that foreign crops were known in Ethiopia at an earlier date than that inferred by the archaeological evidence, possibly as early as the 5th millennium BP; and second that foreign crops did not initially reach Ethiopia from South Arabia.

There are a number of alternative explanations for these issues. For example, domesticated plants may have entered Ethiopia directly from the Near East, or they may have been introduced from India (possibly in exchange for an African crop 'package'?). Although such suggestions are intriguing, they remain purely speculative. The following section outlines

[2] The proposed progenitor of African cattle, *Bos primegenius*, had no known distribution in Ethiopia or the Sahel during the Holocene. The African buffalo, *Syncerus caffer*, whose habitat includes Ethiopia, has never been domesticated (Clutton-Brock 1987).

a more substantial model for the early arrival of imported crop plants to Ethiopia.

6.3.1. Ethiopia, Egypt and the Land of Punt

This section proposes that domesticated food-crops of Near Eastern origin entered the Ethiopian Highlands as a result of trade contacts between Egypt and the Land of Punt during the 5th millennium BP.

The Existence and Historical Geography of Punt

Ancient Egyptian records and hieroglyphs dating between c.2500 and 1170 BC (late 5th-early 3rd millennium BP) document the existence of an East African land which is described through the consonantal framework of hieroglyphic script as *P-w-n-t*, and Punt in modern transcriptions. Punt is the only Africa territory, other than Nubia, to have so ancient a history dating back to the 5th millennium BP.

The historical location of Punt has been the subject of much debate for over a century and it has been discussed at length (eg. Herzog 1968, Kitchen 1971, Phillips 1997). The region lay to the south of Egypt, within the tropics. Punt may have been a generic term for a broad Afro-Arabian circuit of interaction which encompassed Southern Arabia, the Horn and Central and Eastern Sudan. The Egyptians refer to 'chiefs of Punt', and the inhabitants of this land are depicted as being fine featured and brown skinned, similar in appearance to Egyptians, distinct from the black-skinned Negroes from Nilotic regions such as Wawat, Kush and Irem (O'Connor 1982). The products of Punt are typical to the East African savannah and several have a more restricted distribution in East or North East Africa. Certain items, such as giraffe and rhinoceros, pinpoint the location even more specifically within tropical Eastern Africa.

Although the exact geographical position and extent of Punt is still contested, it has been located by toponyms between latitudes 17° and 12°N in a region with a coastal plain and a hilly interior (O'Connor 1982). Documentary, pictorial and archaeological evidence places it securely in Eastern Sudan and Eritrea/Northern Ethiopia between the Nile and the Red Sea (eg. O'Connor 1982, Kitchen 1993). Punt was directly accessible from Egypt both up the Nile and down the Red Sea, although the preferred route was by land and sea, avoiding the Nile cataracts. This way followed the Nile as far as Wadi Hammamt at Koptos, then crossed the desert to the east using pack animals, as far as the Red Sea coast. Recent discoveries of Egyptian remains and stelae dating to the XIIth dynasty at Mersa Gawasis on the coast just north of Quseir suggest that this was an Egyptian port used for trade with Punt (Saied 1977). The route then turned south down the Red Sea and landed on the shores of Punt. Remains of Egyptian-style dressed stone blocks at Aqiq on the Red Sea coast suggest the existence of a possible Egyptian port at this location (Cremaschi et al 1986).

The Deir el Bahari registers depict the Egyptians trading with Puntites on the shores of the Red Sea among dom palms and incense trees. Further texts also allude to expeditions up to 250 km inland (O'Connor 1982). New Kingdom texts recovered from the Sinai indicate that the resources traded from Punt were brought from more distant inland regions (Gardiner et al 1952-1955), suggesting that the Puntites were acting as middlemen, centralising traded goods from disparate regions of tropical Africa.

Egyptian Trade and the Products of Punt

Our knowledge of this land is highly subjective as it is derived entirely from documents on trade. Egypt was engaged in trade with the people of Punt during the 5th and 4th millennia BP, importing many exotic products. The earliest clear record of Punt comes from the Palmero stone and refers retrospectively to a large import of myrrh, electrum and timber from Punt during the 5th Dynasty reign of Sahure c.2450 BC (late 5th millennium BP) (Breasted 1952). The products of Punt were imported to Egypt during the next 1300 years. The fullest account of these commodities comes from Queen Hatshepsut's temple at Deir el Bahari, which was built to commemorate an elaborate expedition to Punt c.1472 BC, sponsored by the Queen (Naville 1894, 1898). This expedition was apparently the first direct trading venture to the 'lands of the south' and its principal aim, other than political propaganda, was to bring back incense trees with which to furnish the steps of the temple.

Trade in incense, myrrh and aromatic herbs was the mainstay of the Egypt-Punt connection, and may have been the reason for which the route was originally established. Aromatic gums, resins and herbs were a vital component of religious and domestic ceremonies in Ancient Egypt, as they still are today, as well as possessing medicinal and cosmetic properties. Other exotic exports from Punt included ivory, ebony and precious woods, several varieties of myrrh (14 varieties were known in Egypt), eye paint (antimony, lead oxide or manganese dioxide), fine gold from the mining regions of Punt, such as the 'green gold' of Amau, baboons, monkeys, dogs, cheetahs, ibex, leopard skins, servants and a dancing pygmy (Kitchen 1993). By the late 4th millennium BP the Puntites were themselves navigating the Red Sea route to Egypt in low boats with triangular rigs that are reminiscent of modern day dhows and feluccas of the Red Sea and Indian Ocean (Kitchen 1971).

The last definitive Egyptian reference to a trade expedition to Punt dates to the end of the 4th millennium BP in the reign of

Rameses III c.1184-1153 BC. This records a 'limitless' shipment of myrrh and other products of 'God's land'. Thereafter trade with Punt apparently ceased, or is not recorded in the narrow range of sources that document the later dynasties. By this date Egypt may well have been importing incense and myrrh from the Levant, taking advantage of the Saba trade route between Southern Arabia and Palestine in the 3rd millennium BP (Artzy 1994).

Archaeological Evidence for the Relationship Between Punt and Ethiopia

Punt in North East Africa

Extensive survey and excavation work by the Italian-American Mission to the Sudan has identified an extensive cultural complex in the Gash Delta region of Eastern Sudan that corroborates with the documentary evidence for Punt (Marks et al 1986, Sadr 1988, Fattovich 1989, 1991, 1993). This region had wide-ranging contacts from the 7th millennium BP that included the Nile Valley and the Ethiopian Highlands (Marks et al 1986, Marks and Sadr 1988). Direct contact with Dynastic Egypt from the 4th millennium BP is implied by the appearance of New Kingdom pottery at the site of Mahal Teglinos on the Ethiopian-Sudanese border (Fattovich 1990), and an 18th Dynasty earring and New Kingdom-style axes at Agordat in Eritrea (Arkell 1954). Furthermore this culture enjoyed a period of growth and prosperity during the 5th and 4th millennia BP, leading Sadr (1988) to propose that it was an economic and political heartland at this time as a result of active trade with the Lower Nile Valley. Egyptian objects occur sporadically elsewhere in North East Africa, including Yubdo in West Ethiopia and Jubaland in Southern Somalia (Fattovich et al 1988, 1989, Fattovich 1993), suggesting that there was an extensive sphere of contact across North East Africa.

Egypt and Ethiopia

Two pieces of evidence in particular locate Punt unequivocally in East Africa and establish the Northern Highlands of Ethiopia as part of the trading region of Punt.

The first is a reference dating from the XXVI Dynasty, c.600 BC (Sethe 1908-9, p.316-21, 524, 695). The text has been transcribed from a damaged stela, but the allusion is unambiguous:

' a great marvel has happened in Your Majesty's time; this has not been seen or heard (before) - rainfall upon the mountains of Punt - .. this month in which its rain was (when) it was not the season (for it) even in the Delta towns... - a Nile flood to sustain your forces.'

The geographical relationship between the mountains of Punt and the source of the Nile inundation was clearly understood by the Egyptians. If this statement is taken literally then it can only be referring to the North Western Ethiopian Highlands, which are the ultimate source of all Nile flood waters. There are frequent documentary references to *Khent Hunnefer*, the mountainous interior of Punt. If the mountains were recognised as the source of the Nile, and of Egyptian prosperity and power, they would be held in awe. Frequent references are made in the Egyptian texts to 'god's lands' in association with Punt, and these may be alluding to the mysterious mountains that controlled Egypt's life-force.

Second, archaeobotanical analyses of specimens of ebony from Egyptian tombs have identified *Dalbergia melanoxylon*, a wood that is endemic to North West Ethiopia, towards the Sudanese border (Beauvisage 1897). This report is unconfirmed, but *D.melanoxylon* has also been identified recently among the cargo of the Late Bronze Age shipwreck of Ulu Burun (Bass 1986, Haldane *pers.comm.*), suggesting that it was being traded around the Mediterranean by the 4th millennium BP.

As Table 6.9. shows, many additional products of Punt are indigenous to Ethiopia. These have been the backbone of Ethiopian trade for at least the past 2000 years (Pankhurst 1961). As discussed above, documentary evidence records that these commodities were being traded from the Northern and Western Highlands since the Axumite period. Gold from the South West was valued especially highly, and may have been exploited from an early date. Ptolemy referred to Ethiopia as 'the Myrrh country', and there are many historical accounts of high quality frankincense and myrrh from different regions of Ethiopia (Pankhurst 1961, Crawford 1958). Timber also was an important resource, particularly for construction, although it was scarce in many semi-arid regions of Africa. The well-forested Highlands of Ethiopia would have been a rich source of this useful commodity.

Table 6.9. Products of Punt that are Indigenous to Ethiopia

Item	Area of Origin
Aromatic gums and resins	
Frankincense	
(*Boswellia* sp including	South East Ethiopia and Eritrea
B.carteri and *B.frecneana*)	
Myrrh	
(eg. *Commiphora abyssinica*	East Ethiopia, South Arabia
Balsamodendron myrrha)	Somalia/Red Sea coast, Arabia
Commiphora pedunculata	East Sudan-Ethiopian border
Aromatic grasses and herbs	
(eg. *Artemisia afra*	Ethiopia
Ocimum suave	Ethiopia
Andropogon nerdus-lemon grass)	West Ethiopia
Precious woods	
Ebony	endemic to North West Ethiopia, close to Sudanese
(*Dalbergia melanoxylon*)	border
Animal products	
Ivory	All found in Ethiopia
Leopard skins	
Baboons/monkeys	
Ibex	
Cheetah	
Ostrich plumes	
Minerals	
Gold	South West Ethiopia, West Ethiopia
Silver	North West Ethiopia
Timber	Ethiopian Highlands

Summary

The Highland region of Ethiopia was significant to the Egyptians. Resources comparable to those that have formed the mainstay of Ethiopian economy over the past 2000 years were imported by the Egyptians from the 5th millennium BP. For example the Pharaohs were already using typical Punt products such as myrrh by that date and Egyptian vessels may even have reached Punt as early as the 1st or 2nd Dynasty (3400-2980 BC - i.e. the late 6th millennium BP) (Pankhurst 1961). As early as the 3rd Dynasty massive shipments of wood were reaching Egypt from forested areas of the Near East and Punt (Breasted 1952).

This does not infer that the Ethiopian Highlands should be definitively equated with the Land of Punt, or that there was necessarily direct contact between the Egyptians and the inhabitants of the Highlands[3]. Ethiopian products may have reached the Red Sea coast and lowland regions along similar trade routes to those that have operated over the past 2000 years. For example, gold from the South West may have been transported to the Northern Highlands for distribution to the lowlands and Red Sea coast, or transported overland to the Red Sea coast in the East.

However, the important assertion made here is that Ethiopia was not isolated in prehistory. On the contrary, Ethiopia appears to have formed a significant part of the elusive Land of Punt. By the mid-late Holocene, the inhabitants of the Highlands were actively participating in a trade network that also involved a culture with a developed system of agriculture. Domesticated plants from this agricultural system possibly appear in Ethiopia at an early date, and they are well established by the dawn of the historical period. Furthermore, unconfirmed reports suggest that similar crops were being grown in the Gash Delta, close to the border with Ethiopia, in what has been identified as the heartland of Punt (Sadr 1988). Surely this North East African trade network was highly significant in the introduction of domesticated plants to Ethiopia?

6.3.2. Trade as a Mechanism for the Introduction of Domesticates

We have considerable evidence for the exports from Punt. But what products did the people of Punt receive in exchange? One Egyptian inscription at Deir el Bahari records

[3] Although the eminent Ethiopianist, Conti Rossini, does equate the region of Tecaru, mentioned in context of Punt in Egyptian texts, with the region of Tigray in the Northern Ethiopian Highlands (Conti Rossini 1928).

that 'all good things are brought by order of Her Majesty to Hathor, Lady of Punt'. But the content of these 'good things' seems to have been limited to what may be described as primitive baubles in comparison to the wealth of Egyptian workmanship. In one document at Deir el Bahari the Puntites receive an axe, a poignard, 2 large bangles, 11 necklaces and 5 large rings (of 'Egyptian blue' glass?). Elsewhere, depictions at Deir el Bahari show the Puntite chiefs bidden to the tents of the Egyptian officials to feast on 'bread, beer, wine, meat, fruits and all the good things of Egypt as the queen has commanded'. 'All kinds of provisions' were taken from Egypt for exchange with the products of Punt (Porter and Moss 1952). We see these being unloaded in large amphorae, similar to those used by the Egyptians for trade within the Mediterranean. The contents of such amphorae have been preserved in exceptional circumstances, as at the 4th millennium BP bronze age shipwreck at Ulu Burun off the Southern coast of Turkey and included oil, figs, ivory, incense and pottery (Bass 1986).

Perishable and desirable objects were taken to Punt from Egypt. This contact may have led to the exchange of edible and useful plants, and so to the presence of Near Eastern-type domesticates in Punt from the 6th/5th millennium BP. If, as argued above, the Ethiopian Highlands formed part of the trading region of Punt, then prehistoric trade networks provided a medium for the transport of foreign domesticates. This model proposes a viable mechanism for the introduction of Near Eastern plant domesticates to the Ethiopian Highlands during the 5th millennium BP. Botanical and archaeobotanical evidence from Egypt and Ethiopia provide support for this model.

The Botanical Argument

There is no archaeological evidence at present to corroborate or refute this model. However botanical and plant genetic research supports the proposal that early contact between Egypt and Ethiopia led to the exchange of Near Eastern plant crops. While this evidence can be used to support a model for the movement of crops from Egypt to Ethiopia, it is open to interpretation. As discussed below, the direction of the movement is unclear. This opens up interesting possibilities about the origins of certain domesticated crops in Ethiopia.

Domesticated Barley in Egypt

Chapter 3 noted that both Ethiopian and Egyptian barleys belong to a primitive variety, which unconfirmed reports suggest was only partially interfertile with Near Eastern barleys (Vavilov 1960). The earliest evidence of domesticated Near Eastern crops in Africa comes from Fayum in Upper Egypt, dating to the 7th millennium BP (Wetterstrom 1993). Four varieties of barley were identified among the earliest archaeobotanical remains from North Africa (Caton-Thompson and Gardner 1934). Included among the remains were two unusual, primitive varieties of barley, *Hordeum vulgare* subsp *deficiens* and subsp *irregulare*. The Near Eastern barley *H.vulgare* subsp *vulgare* only appeared in later Egyptian contexts from the 6th and 5th millennia BP. *Deficiens* and *irregulare* barleys are common in modern day Ethiopia. There is thus a close similarity between the earliest known Egyptian barleys that were grown over 6000 years ago in the Nile Valley, and the modern barley crop from Ethiopia. Unfortunately, the identity of early Ethiopian barleys cannot be substantiated at present.

Barley is deeply entrenched in Egyptian ritual, religion and legend. It was first documented as early as the Ist Dynasty, c.5000 years BP, as a sacred offering from the Pharaohs to the gods and to the people (Darby et al 1977). The Egyptian phonetic symbol for barley, ***it***, also means father. According to ethnobotanical concepts (Alther and Merric 1987), this infers a very ancient knowledge of barley and a deep tradition of cultivation in Egypt that may predate the exploitation of other crops such as wheat from the 6th millennium BP. Ethnobotanical and linguistic research suggest that barley also has great antiquity in Ethiopia.

The relationship between Ethiopian and Egyptian barleys is unclear, and two alternatives can be proposed. Doggett (1970) argued that Egyptian barley moved from ancient Egypt to Ethiopia along the Atbara or Blue Nile river valleys sometime before 5000 years BP. Helbaek (1955,1960) on the other hand suggested that the Egyptian barleys such as *deficiens* and *irregulare* originated in Ethiopia and moved to Egypt before the mid 7th millennium BP, a view that is tentatively supported by the linguistic evidence (Ehret 1979).

These models both imply that Egyptian and Ethiopia barleys derived from a primitive form. This form was either indigenous to North East Africa, or it became separated from the Near Eastern gene pool at an early stage in the domestication process. In the latter instance, the isolated genome could have reached Egypt in the 6th or 7th millennium BP, and was subsequently introduced to Ethiopia in the 6th/5th millennium BP. The Ethiopian barley genome would therefore be comparable to Egyptian barley, but divergent from Near Eastern barley.

Domesticated Wheats and Other Crops

Domesticated emmer wheat may also have followed a similar assimilation process. Helbaek (1960) noted a distinct resemblance between ancient Egyptian emmer wheat and modern Ethiopian types. Emmer wheat is still a popular crop in Highland Ethiopia, and its antiquity is attested from some of the earliest archaeobotanical remains (Hansen *pers.comm.*). Einkorn wheats were unknown in Egypt and Ethiopia prior to the 2nd millennium BP (Harlan 1969), and einkorn is still not grown extensively in Ethiopia. However there is no ethnobotanical or linguistic tradition for wheat in Ethiopia, suggesting that it was less culturally and economically important than barley, and that it may not have been grown until a later date.

The *Periplus* notes that wheat was being imported to the Somali coast from regions such as Egypt, using a route that was already ancient 2000 years ago. It is not inconceivable that this mechanism was responsible for the introduction of Near Eastern crops to the Ethiopian Highlands during the 5th millennium BP.

Chapter 3 and *Appendix A* note that a number of tropical African plants were exploited in Egypt from an early date. Resources such as sesame and castor bean grow wild in Ethiopia. These were important in ancient Egypt, and the wild or domesticated form may have been introduced from Ethiopia as a result of trade contacts. Chapter 3 also mentions representations of the Ethiopian staple, the false banana ensete, have been identified on PreDynastic pottery dating to the 5th millennium BP (Laurent-Tackholm 1951). If these identifications are valid, they provide strong support for a model of early contact between Egypt and Ethiopia.

Chapter 7.

TOWARDS AN INTERPRETATIVE FRAMEWORK: AN ABSENCE OF EVIDENCE OR AN EVIDENCE OF ABSENCE?

7.1. ABSENCE OF EVIDENCE OR EVIDENCE OF ABSENCE?: THE OPTIONS

The evidence presented in this volume can be structured into four separate scenarios to describe how subsistence changed in Ethiopia during the Holocene.

Scenario A

Hunting-Foraging continued until the 3rd millennium BP

The available evidence from Ethiopia supports this scenario. Domesticated cattle are tentatively attested in Ethiopia from the mid-4th millennium BP, and domesticated plants from the mid-3rd millennium. None of these early remains come from indigenous Ethiopian plants or animals. This suggests that wild plants and animals were exploited in Ethiopia throughout the Holocene until the 4th or 3rd millennium BP. Domesticated resources were introduced after this date, stimulating the domestication of local plant resources. Both foreign and local domesticates subsequently became important in Ethiopian agriculture.

Although there is no direct archaeological evidence to refute this model, it is invalidated by considerable indirect evidence from archaeological and non-archaeological disciplines. A combination of ethnographic, linguistic, genetic and archaeological evidence indicates that certain indigenous plant resources were important in Ethiopia by the 7th millennium BP. The available evidence is sufficient to dispense with Scenario A, and it requires no further discussion.

Scenario B

Agriculture developed in Ethiopia following the introduction of foreign domesticated resources from the 5th millennium BP

The evidence presented in this thesis indicates that domesticated plants were introduced to Ethiopia before the mid-3rd millennium BP, and domesticated cattle earlier than the mid-4th millennium. This thesis argued that domesticated cattle were introduced by pastoral groups from the north east as early as the 6th millennium BP, and Near Eastern domesticated plants were introduced as a result of trade relations with Egypt from the 5th millennium BP. Plant domesticates may have been brought from Egypt under the umbrella of trade in exotic African products.

A further, tentative route for the introduction of Near Eastern plants was from India. Domesticated resources may have been introduced as part of an agricultural 'package' which was exchanged for a 'package' of wild African cereals.

Scenario B proposes that indigenous subsistence remained unchanged throughout the late Pleistocene and Holocene. Foreign resources were therefore imposed on or assimilated by an indigenous hunter-gatherer population.

Scenario C

Indigenous resources were domesticated independently at an early date

The indirect evidence suggests that a subsistence system based on local plants was already established in the Ethiopian Highlands prior to the introduction of foreign domesticates in the 5th millennium BP. The evidence additionally indicates that there was a significant cultural development in Ethiopia around the 7th millennium BP. This development involved a shift in food processing activities and food procurement strategies, and a probable reduction in group mobility. The observed changes imply that there was an alteration in the relationship between humans and the environment at this time

Scenario C proposes that the change documented by the indirect evidence can be interpreted as the emergence of food production and the indigenous domestication of local plant resources. This suggestion proposes that local domesticated resources formed part of a principally horticultural subsistence system. This system also relied extensively on wild game and foraged plant resources. The absence of evidence for domesticated plants in Ethiopia undermines the validity of Scenario C.

With the subsequent introduction of domesticated livestock and plants, possibly during the 5th millennium BP, foreign resources were assimilated into the indigenous system of food production to provide a broader resource base. This would have encouraged a more intensive form of mixed agriculture and economic specialisation, which would have enabled specialist adaptation of a range of ecological niches in Ethiopia, including the arid lowlands and the higher altitude *dega* zone.

Scenario D

Cultivation of indigenous wild food plants

Scenario D also proposes that certain indigenous resources were important before the introduction of foreign domesticates. However, this scenario argues that indigenous plants were *not* domesticated before the 5th millennium BP.

Instead, certain wild plants, including the progenitors of modern crops, were cultivated by the 7th millennium BP, using a primitive pre-agricultural technology and land-use system. These resources formed part of a broad spectrum foraging economy which still persists in many regions of Ethiopia today.

This indigenous system was established substantially earlier than the 5th millennium BP, and it was effective enough to persist after the introduction of foreign domesticates and technology. However, the final threshold to intensification and domestication was not overcome in the early-mid Holocene period, and resource exploitation remained below a critical level of intensity until the introduction of foreign resources.

Viability of the Scenarios

The evidence presented and discussed here discounts Scenarios A and B. While considerable evidence exists to suggest that early foreign contact with food producing cultures opened up channels of communication for the introduction of foreign domesticated resources, the available evidence additionally points to an indigenous cultural development around the 7th millennium BP that related to a shift in the subsistence strategy. What is not clear at present is whether this change reflects either the indigenous domestication of local resources, or an alteration in the human-environment relationship that did not involve domestication. Evidence of domesticated plants is needed to establish if food production emerged indigenously at an early date in Ethiopia.

Africa archaeology is notoriously lacking in palaeobotanical evidence, and recovery of plant remains is the exception rather than the rule. Is it possible that there *is* evidence for domesticated plants in early contexts in Ethiopia, but this evidence has not yet been recovered? The following section discusses the problems facing the effective recovery of archaeobotanical evidence from tropical deposits, and uses this discussion to reject Scenario C.

7.2. A PRACTICAL EXPLANATION FOR THE ABSENCE OF DOMESTICATED PLANTS IN THE ARCHAEOLOGICAL RECORD

The problem of the absence of evidence for tropical African plant domesticates before the 3rd millennium BP is fundamental to our understanding of the development of food production in Africa. It is imperative to understand *why* there is no evidence of plant domesticates until such a late date if we are to avoid mis-conceptions, and if we are to establish ways to improve archaeological recovery methods.

Three straightforward explanations can be proposed for this discrepancy in Ethiopia and Africa. First that there were no domesticated plants in Northern Africa before the 3rd millennium BP. Second that domesticated plants existed, but their remains have not been recovered. Finally, that wild plant resources were exploited as if they were domesticated crops, but there was no genetic change to a morphologically domesticated form.

The apparent absence of domesticated plants in Africa may therefore be a function of the archaeological record. The following section presents a discussion of the potential practicalities affecting the recovery of domesticated plant remains in tropical Africa.

7.2.1. Factors Affecting Archaeological Recovery in Tropical Africa

Depositional Factors

The Nature of the Plant Material

Unlike regions such as the Near East, many African plant resources and a number of staples are soft-tissued plants such as yams, tubers, plantains. These plants are rarely preserved and tend to be archaeologically invisible. The majority of plant material preserved in archaeological contexts is carbonised or included in a durable matrix such as mud-brick, daub or pottery. For example, plant impressions in pottery are common in the Central Nile Valley and regions of the Sahara where they have been critical in identifying the prehistoric resource base and the relative importance of wild cereals in the Holocene diet (Jaques-Felix 1971, Stemler 1990). However these conditions for preservation tend to be culturally specific - for example, vegetable material is rarely used to temper ceramics in other regions such as East Africa and Ethiopia. Furthermore, hard seeds and grains are favoured over soft, fleshy plants.

Heating is often required in the processing of many food plants. Temperate cereals such as wheat and barley are heated before threshing and winnowing to facilitate the separation of the grain from the chaff. This process provides excellent conditions for carbonisation and deposition. Although the technique is not normally applied to tropical African cereals, other traditions may contribute to the pre-depositional preservation of plant material. In the Ethiopian Highlands for example, barley or sorghum grains are roasted over an open fire and eaten as a snack (*personal observations 1994*). A similar process may have been used over 8000 years ago at Nabta Playa in the Eastern Sahara where charred sorghum grains and fruits were found concentrated in ashy deposits around hearths within 'dwelling structures' (Wasylikowa et al 1993).

Preservation of plant material burnt in animal dung used as fuel may account for its survival in certain contexts. This is especially true of pastoral groups in arid areas with few trees. Use of animal dung is more likely among groups with domesticated animals. There has been little application of this analytical technique in Africa.

Plant material may also be preserved in the hyper-arid environments of the Sahara. Desiccated remains from sites such as Ti-n-Torha, Adrar Bous and Wadi Kubbaniya have expanded our knowledge of Epi-Palaeolithic and Holocene subsistence in these regions (Barich 1992, Hillman 1989). This factor is spatially inconsistent, and desiccated plant material is rarely found outside the Sahara. Palaeobotanical remains may also be preserved as phytoliths (silicate skeletons formed at high temperatures), but negligible examples are known from Africa.

Site Function

Site use determines the nature of deposited plant material. Carbonised material is more likely to be preserved in domestic contexts where food processing activities are concentrated, as opposed to in temporary encampments. The range of resources represented in domestic base-camps differs substantially from those deposited in temporary seasonal camps, where there is greater emphasis on local, seasonal resources. A significant proportion of excavated sites in Africa are small caves or rock shelters which tend to be temporary, non-domestic contexts, which have less potential for recovering domesticated plant remains.

Non-anthropogenic factors may also be responsible for depositing plant remains in particular contexts, which may affect the archaeological interpretation. For example, caves are popular among animals such as hyenas and porcupines which may excrete or deposit plant remains, or bats which are known to carry banana seeds.

The Nature of Deposition

One of the most noticeable features of prehistoric African habitation sites is the absence of storage facilities. In the Near East, storage structures provide one of the major sources of early palaeobotanical data, whereas in Africa they are limited to isolated examples along the Nile Valley, at sites such as Fayum, Merimde and Jebel et Tomat (Wetterstrom 1993). Structures that have been interpreted as storage facilities at certain early sites, such as Nabta Playa, Bir Kiseiba and Khargha Oasis, have not yielded significant plant remains, and it is possible that they served an alternative function (Wendorf and Schild 1980, Wendorf et al 1984b). Desiccated remains of wild plants such as *Celtis australis* and *Celtis integrifolia* have occasionally been found in jars and pots in early archaeological contexts, for example in the Tenere (Hugot 1968) and in Nilotic burials (Reinolds 1994, *personal observations 1995*). These contexts have not yet yielded domesticated species.

Post-Depositional Preservation and Disturbance

Soil Chemistry and Morphology

The primary factors affecting preservation are soil chemistry and water content/retention. Variations in soil pH, mineral composition and microbial activity provide heterogeneous conditions for preservation of organic material throughout the world. In Africa the soils are very variable, and are difficult to characterise. Broad pedological patterns are discernible, but at a the regional level the picture is highly complex. This has a complex impact on plant preservation (Grove 1978, White 1983). Once the natural vegetation has been removed, the organic matter in African soils tends to be burnt up rapidly by the high temperatures and degraded by the intense bacterial activity (Grove 1978).

Climatic Factors Affecting Preservation and Stratigraphy

Rainfall in Ethiopia varies from less than 1 cm per annum, with the possibility of no rainfall at all for several years, to more than 30 cm per annum, which falls almost entirely during the 2-3 month rainy season (National Atlas of Ethiopia 1988). Monsoon rains affect both to the degradation of organic material in archaeological deposits, and to its stratigraphical location within the deposit. Particles of different sizes will be leached at different rates through the soil. The extent of particular movement is also be determined by regional variations in the soil matrix. For example, the clay mineral montmorillorite found in African vertisols expands and contracts strongly on wetting and drying. This movement upsets the archaeological stratigraphy and may affect the contextual interpretation (Grove 1978).

Bioturbation

The activity of microbes and other organisms within the soil matrix is known to disturb archaeological stratification (Schiffer 1987). Microbial activity is especially intense in the tropics due to the heat and humidity (Grove 1978). Bioturbation may have a profound affect on tropical archaeological deposits. The extent of this disruption is unknown, although post-depositional disturbance is the probable cause of several notable errors in the chrono-stratigraphical identification of domesticated plant remains (Wendorf and Schild 1976b, Wendorf and Schild 1984b, Phillipson 1977). African archaeology would benefit from experimental calibration of organic activity and monsoon-leaching processes.

Sampling and Recovery Strategies

Sampling Methodologies

Lack of systematic archaeological investigation in Africa is a major problem in the recovery of plant material. There have been few intensive surveys. Sampling strategies tend to be inconsistent, and biased towards more accessible environments or geographical features such as river banks, lake shores and roads (eg. Clark 1954, Eggert 1993). Certain regions, including the Nile Valley in Lower Egypt, the forest fringes of parts of West Africa, and the Lake Turkana basin in the Kenya-Ethiopia border region have benefited from extensive archaeological surveys and excavations, whereas vast regions of the Central Equatorial Rainforest, Angola and Namibia remain virtually unexplored.

The Ethiopian Highlands have also been neglected archaeologically, partly as a result of the terrain, and partly due to recent political instability. The majority of archaeological work has concentrated on early hominid remains in the Rift Valley, or Proto-Historical and Historical material in the North Eastern Highlands. Only five excavated sites span the Holocene period during which food production is most likely to have developed, and only one of these was sampled for archaeobotanical remains.

Recovery Procedures

Until recently, little emphasis was placed on the extraction of plant material from archaeological contexts. Only obvious deposits have been recorded, such as the 'silos' at Fayum and Merimde Beni-Salama in Lower Egypt (Caton-Thompson and Gardener 1934, Wendorf and Schild 1976a), and at Jebel et Tomat in Central-Southern Sudan (Clark and Stemler 1975).

Development of efficient strategies, such as screening, water flotation and froth flotation (eg. Pearsall 1989) for extraction of environmental material in Eurasia and America in recent years has led to their increasing application in Africa. These techniques are still in their infancy, particularly in Ethiopia where environmental sampling has still only been applied to a handful of archaeological sites. This lack of systematic use promotes methodological inconsistencies and weaknesses, such as small soil sample size (eg. 2 litre soil samples from Gobedra) or absence of water for flotation when excavations were conducted during the dry season (eg. Natchabiet cave).

Lack of Suitable Methodologies

Many of the methods for recovery of archaeobotanical remains have been developed for temperate conditions, and are not always suitable in the climatic extremes of tropical Africa. This was illustrated most dramatically by Wendorf's team from the Southern Methodist University when trying to extract desiccated plant material from hyper-arid contexts using water flotation. Exposure to water after several millennia of aridity not unsurprisingly caused the plant macrofossils to disintegrate.

Certain African plants have characteristics that would benefit from a more flexible approach. The minute size of tef, for example, makes it extremely difficult to recover. At present only a single, tentative identification of tef exists from Ethiopia (Dombrowski 1971). A more precise method of extraction should therefore be applied on sites where tef is likely to be found.

Identification of Plant Remains

Botanical Analysis

The identification of remains of tropical plants can be highly problematic, particularly if there is a need to distinguish between domesticated and wild plants. For most of the African domesticates, especially the cereals, wild progenitors and primitive domesticates are morphologically very similar. Even today, certain modern cultigens are often difficult to distinguish from their wild relatives. This is true especially for the Ethiopian cereal *Eragrostis tef*, and its putative progenitor *E.pilosa*, which cannot be distinguished between by the naked eye (*unpublished Herbarium report from Addis Ababa 1994*). Consistent phenotypic differences, such as increases in length and breadth in domesticated grain, will only emerge in statistically significant sized samples, and 'positive' identification of domesticates based on small samples are untenable. This has not prevented several authorities from classifying single grains as domesticated species (van Beek 1969, Phillipson 1977, Shaw 1977).

Furthermore, there has been negligible research into how depositional processes can alter tropical plant material. Boardman and Jones (1990) showed that temperate cereal grains such as wheat and barley are morphologically distorted in the temperature range encountered when firing pottery and these distortions are often severe when viewed under a SEM. A similar database for tropical seeds would aid the identification of plant impressions from pottery, and particularly the distinction between morphologically domesticated and wild seeds (Stemler 1990).

Analysis of organic residues in ceramics from archaeological deposits provides a highly informative technique for archaeological investigation of palaeo-diet (Evershed et al 1991, 1992, Charters et al 1993). This approach has delivered some spectacular results. Continued archaeological and ethnographic research will help to define the limitations and applications of this technique, and encourage more widespread use.

Could Domesticated Plant Remains be Recovered from Ethiopian Sites in the Future?

Archaeobotanical Potential in Ethiopia

The possibilities of obtaining evidence of early domesticates in Africa are far lower than in many other regions of the world, due to the pre- and post-depositional conditions for the preservation of plant material, and the problems with extraction and identification procedures.

These problems are especially acute in Ethiopia. Prior to the 2nd millennium BP there is almost no direct evidence for plant exploitation in Ethiopia. This absence of evidence may be a function of the biological processes and archaeological procedures. Few excavations have attempted to systematically recover plant material from deposits in caves and rock shelters from prehistoric Holocene contexts in Ethiopia, and vegetative material has not survived at open air sites such as Lake Besaka and Lake Ziway. Excavations at Lalibela cave, the earliest site from which plant remains have been recovered, date no earlier than the mid-2nd millennium BP. Gobedra rock shelter is the only site spanning the period of interest that has been sampled for archaeobotanical material. As noted above, the results obtained from Gobedra cannot be accepted as a quantitative representation of the available plant remains.

Research elsewhere in Africa indicates that the evidence from Ethiopia does not reflect the true range of resources that were exploited in prehistory. The fact that plant remains *have* been successfully recovered from Ethiopian sites shows that botanical material is preserved in archaeological contexts. If the evidence exists and is waiting to be unearthed, could we expect to find remains of domesticated plants in early contexts in Ethiopia?

Potential Evidence for Domesticated Plants

Despite the numerous problems with obtaining representative archaeobotanical data, considerable quantities of plant material *have* been extracted from prehistoric contexts across Africa. Work in certain regions of Africa, and the Central Nile Valley in particular, has successfully recovered and identified botanical remains from archaeological sites throughout the late Pleistocene and Holocene. These remains have been subjected to extensive quantitative analyses in an attempt to identify evidence of early domesticates. However, despite rigorous testing, this approach has so far failed to identify any tropical domesticated plants earlier than the 3rd millennium BP. There is no direct evidence that indigenous plants were domesticated in Africa before this date.

The appearance of domesticated African cereals in India and Arabia from the 5th millennium BP presents an argument for indigenous domestication. This anomaly, however, can be explained adequately by the export of wild cereals from Africa, and their subsequent domestication in non-African environments. Recent genetic evidence substantiates a model of separate domestication of sorghum in India and Arabia in the 5th millennium BP, and later domestication in Africa in the late 3rd millennium BP.

The absence of evidence for tropical African domesticates implies that they were unknown in Ethiopia until the late Holocene. However, this does not preclude the possibility that *foreign* domesticated plants were present in pre-3rd millennium BP contexts in the Ethiopian Highlands. The proposal that foreign plants were introduced to the Highlands from the 5th millennium BP can be tested against future archaeobotanical evidence.

Implications for Ethiopian Subsistence in Prehistory

Absence of evidence for indigenous domestication excludes Scenario C. The remaining option, Scenario D, proposes that indigenous *wild* plants were being cultivated in the Ethiopian Highlands from the 7th millennium BP. This proposal suggests that there was a shift in the subsistence strategy at this time, and that an initial step was taken towards food production. *Why, therefore, did local resources not become domesticated until substantially later?*

A practical explanation for this question is offered below.

7.2.2. Factors Affecting the Domestication Process

Selection Pressures and the Domestication Process

Genetic Selection and Morphological Change

In order for a plant to be considered as domesticated it has to have undergone a morphological change. If the selection pressures required to produce phenotypic and genotypic change are absent, the plant will not appear as a morphologically domesticated species. Many different mutant characteristics, such as grain size, plant height, disease sensitivity and temperature dependence, exist at low frequencies among wild populations. These features are coded for by recessive genes which are common to all members of a particular species. The genes are only expressed under altered conditions when they disproportionately improve the survival potential of the plant.

Anthropogenic selection pressures are imposed by interfering with the natural process of evolutionary selection (Wilke et al 1972). In order for morphological change to occur, certain recessive characteristics (eg. grain size, colour, sweetness) must be selected preferentially, the plant must be harvested conversely to the natural seed dispersal process, and the harvested population must be sown regularly using grain from a seed stock (Harlan 1992). Hillman and Davies (1990, 1992) have used experimentally derived models to indicate that morphologically domesticated forms of Near Eastern cereals can become dominant within 20-200 years of the initial cultivation, if appropriate selection pressures are imposed.

The African Argument

Stemler (1980) argued that African cereals were being harvested from the mid-Holocene using a harvesting technology that emulated the natural seed-dispersal mechanism. Unlike harvesting mechanisms in other regions of the world, such as the Near East, this process did not select for certain recessive characteristics, such as a non-shattering rachis, that are characteristic of domesticated plants.

Ethnographic Evidence for Prehistoric Harvesting Technology

Ethnographic evidence from the Saharan and Sahelian zones of North Africa show that over 60 species of grass, including several staples, are harvested by beating or stripping the ripe grain from the fruiting head, or by sweeping the fallen grain from the ground (Jardin 1967). Several variations on this mechanism are found among modern African populations (Harlan 1989b). The desert grass, *Panicum turgidum*, for example, is harvested across the Saharan belt by beating the heads with sticks over deep, wide bowls made of wood and camel hide, or by rubbing the ears between the hands (figure 7.1) (Nicolaisen 1963). The swinging basket technique is used in harvesting savannah grasses of the Krebs complex (including *Eragrostis* and *Brachiaria* sp) or *Panicum* sp, or wild rice (*Oryza* sp). A large basket is swung through stands of wild grass so that the loose grains fall from the plant into the basket. Up to 10 kg of Krebs grasses can be collected by one man in a morning using this technique (Chevalier 1932). As these mechanisms emulate natural seed dispersal, they do not encourage the expression of recessive genetic characteristics, and therefore prevent the emergence of domesticated forms.

Populations of grasses gathered in this way can remain unchanged for thousands of years, even if they are gathered intensively and used as the seed stock for subsequent generations. Blumler and Byrne (1991) have suggested that intensive beating of seed heads in wild cereals, for example in drought conditions when the resources are under stress, could increase the frequency of indehiscence to produce effectively domesticated plants in the absence of agriculture. However there is no evidence to support this proposal, which may also be species specific, and the authors admit that it is only remotely possible. Harvesting techniques similar to those used today in arid North Africa may have been applied throughout the Holocene.

Stemler (1980) proposed that such harvesting technology was a response to the morphology of the tropical cereals. In contrast to thin-stemmed festucoid grasses such as wheat and barley, African cereals, such as sorghum and millet, generally have tough, thick stems that require a superior cutting tool to harvest the entire inflorescence. Stemler (1980) suggested that this technology was absent from Northern Africa, possibly until the introduction of metal sickles. The introduction of metal to Nubia in the 4th millennium BP would have had implications for improving harvesting technology, yet domesticated plants do not appear until a thousand years later. The work of Hillman and Davies (1990, 1992) has shown that phenotypic change can occur over roughly 20-200 years before morphologically domesticated species dominate the plant population, suggesting that certain selection pressures, such as storage or seed stock selection, were absent.

Archaeological Evidence for Prehistoric Harvesting Technology

Modern populations of domesticated African cereals are harvested intensively using metal technology and can offer little clue as to the prehistoric subsistence processing methods. The archaeological record can often be ambiguous in this respect and can result in misinterpretation. There are many examples throughout prehistoric Africa of stone tools with silica residues on the cutting edge, for example in the Maghreb and the Nile Valley. As the lithics are commonly associated with grinding stones, the 'sickle sheen' is cited as significant evidence that wild grasses were being harvested intensively using basic lithic technology (Clark 1976). Sheen on blades from Nilotic sites in Egypt dating between 15 000 and 10 000 years BP may indicate harvesting of grasses of the same subfamily as barley (Wendorf and Schild 1976, Hoffman 1980). Archaeobotanical and ethnobotanical data suggests that such species were indigenous to the Upper Nile Valley at this time and that they were exploited by resident human populations. Stemler (1980) argues that the formation of silica patina would differ significantly between tropical and festucoid grasses. The presence of silica gloss on prehistoric lithic tools from Africa implies that plants with siliceous skeletons, such as wild festucoid-type grasses or reeds, were being harvested rather than tropical grasses.

Another common assumption is that the presence of harvesting technology means that the harvested plants were used as food. Ethnographic data shows that topical grasses serve a variety of non-dietary functions. Wild *kram-kram* grass is harvested for cattle fodder in desert regions of North Africa, sorghum stalks are used for roofing and thatching in many areas and thin-stemmed Sahelian grasses serve a variety of functions such as bedding, basketry, building and ceramic temper (Barth 1857, Nicolaisen 1963, Harlan 1992).

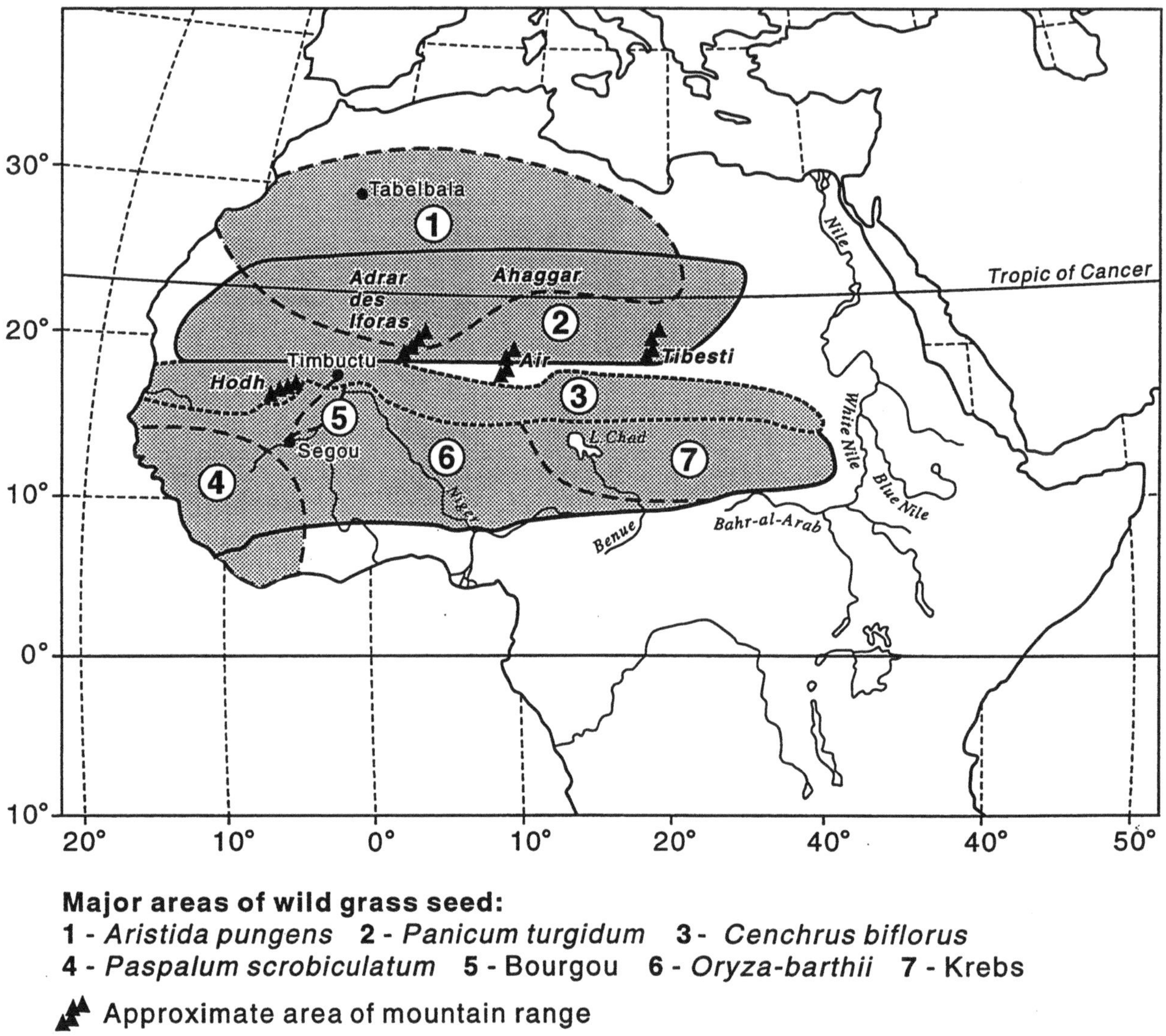

Figure 7.1. Major areas of wild grass-seed harvesting in Northern Africa (adapted from Harlan 1989).

Implications for the Domestication Process in Ethiopia

Although the genetic argument outlined above provides a viable explanation for the absence of morphologically domesticated plants in certain regions of Africa, this argument is less convincing when applied to the Ethiopian data.

Indigenous Highland cereals such as tef are thin-stemmed, and would be easier to harvest using the LSA obsidian technology that was available during the Holocene period. Potential harvesting tools such as blades and microlithic crescents are common to LSA sites. A silicated patina or 'sickle sheen' has been identified on lithics dating from 15 000 years BP onwards, suggesting that these may have been used for harvesting silicated grasses, such as *Eragrostis* and or possibly barley (Clark 1954, Clark and Prince 1978).

7.3. A THEORETICAL EXPLANATION FOR THE ABSENCE OF DOMESTICATED PLANTS IN ETHIOPIA

7.3.1. Background to the Emergence of Food Production

The Origins of Food Production in the Near East

The successful shift from food procurement to food production occurred independently in several regions of the world between c.10 000 and 5000 years BP (eg. Gebauer and Price 1992, Cowan and Watson 1992, Smith 1995). The major indicators of the transition, specifically domesticated plants and animals, were recorded earliest in the Near East from the start of the 10th millennium BP. In this region food production was neither a unique event nor limited to a single site, but emerged sporadically across the 'Fertile Crescent' - an area that stretches in an arc from the Levant and the Eastern Mediterranean into Syria and Southern Turkey along the foothills and the mountains of the Zagros mountains in Northern and Eastern Iraq. Domesticated plants[1] first appeared at sites such as Netiv Hagdud, Jericho, Abu Huryera and Mureybit in the Western section of the Fertile Crescent from the start of 10th millennium BP (eg. Zohary and Hopf 1993) where they were exploited as part of a hunting-gathering strategy. Conversely, in the more mountainous area of the Eastern Fertile Crescent, domesticated goats and sheep became economically dominant from the late 10th millennium BP (Legge 1996), while wild plant resources continued to be foraged as part of a transhumant subsistence strategy in this region until at least the 9th millennium BP.

Domesticated plant resources reliably sustained a 'village-based' farming economy during the 10th millennium BP in the Western and Central Fertile Crescent. Mixed agriculture only developed in this region after c.9000 BP following the expansion and assimilation of domesticated goats and sheep from the Eastern Fertile Crescent (Higgs and Jarman 1972). Important domesticated resources were added to the complex during the 9th and 8th millennia BP. These included the pig during the early 9th millennium BP (Clutton-Brock 1979), the cow during the 8th millennium BP (Grigson 1989), and flax from the late 9th millennium BP (Van Zeist and Bakker-Heeres 1975). The success and flexibility of this combination enabled agriculture to spread rapidly throughout the Near East, North Africa, the Mediterranean and Eurasia, so that this entire region was being successfully farmed within the next 3000 or 4000 years.

African Food Production from a Near Eastern Perspective

The emergence of food production in Africa has been interpreted through a Western perspective. It is generally thought to have derived from the Near East following introduction into the Nile Delta region in the late 7th millennium BP, and subsequent diffusion up the Nile Valley during the 6th and 5th millennia BP. When this 'moving frontier' of agriculture based on temperate plants reached its ecological limits in the African tropics, it became translated into agriculture based on tropical plants. The present evidence suggests that this process may not have occurred until the 3rd millennium BP.

Domesticated animals were more adaptable than temperate plants, and pastoral groups were able to spread into arid regions of tropical Northern Africa. In contrast to the absence of domesticated plants, food production based on domesticated livestock was practised from at least the late 7th millennium BP in tropical Northern Africa. Is the story of subsistence development in Africa primarily one of cattle and pastoralism in the absence of plant domestication?

The Emergence of Food Production from an Ethiopian Perspective

This emphasis on the Near Eastern as the centre for agriculture in Africa, Europe, the Mediterranean, Arabia and Central Asia has initiated extensive research in the Near East. This work has led to the construction of a dynamic theoretical framework to interpret the origins of agriculture from a Near Eastern perspective. However, Ethiopia is also considered as a possible centre for early plant domestication. If this assertion is valid, then the theoretical understanding derived from the Near Eastern data should have implications for understanding the emergence of food production in Ethiopia.

[1] Early domesticated plants include barley (*Hordeum vulgare*), emmer and einkorn wheat (*Triticum dicoccum* and *Triticum monococcum*), and pulses such as lentil (*Lens culinaris*), chick pea (*Cicer arietinum*), pea (*Pisum sativum*) and bitter vetch (*Vicia ervilia*).

The evidence discussed in this thesis suggests that a significantly different system of economic development prevailed in Ethiopia which was not consonant with the Near Eastern sequence. In this sense, Ethiopia provides a valuable tool for testing existing models of subsistence change, and for questioning the validity of current assumptions of human development.

The following section outlines the conceptual and historiographical framework for the emergence of food production (in the Near East), and identifies the current cognitive approach. This framework is then applied to the Ethiopian data. It is necessary to examine the potential for domestication in Ethiopia, and assess why this potential may ***not*** have been realised.

7.3.2. Ethiopian Food Production as a Failed Experiment in Cultural Evolution?

Are the anomalies in the evidence described here 'snapshots' of failed experiments in cultural evolution? Is it possible that similar patterns have characterised the course of human development? Could a dynamic evolutionary principal explain the absence of agriculture in Ethiopia? What drives the pattern of change? The following section explores Ethiopia's potential as a centre for the development of indigenous food production, from an ecological-evolutionary perspective.

Interactive Controlling Factors and the Ethiopian Database

Figure 7.2 illustrates the relative socio-economic developments in the three regions during the Holocene period. These developments display a striking chronological discrepancy. Cultural developments are roughly comparable in all regions, although the development of ceramic technology is more advanced in Africa. In the Near East, the emergence of permanent architecture is synchronous with the use of domestic resources during the 10th and 8th millennia BP. Conversely, in Africa and Ethiopia domesticated resources appear significantly later than cultural developments.

Three principal controlling factors determine the development of food production. These are environmental factors, climatic factors and cultural factors. The interaction between these three factors determines the nature of the human behaviour response.

The following approach attempts to use these factors comparatively to determine whether or not Ethiopia had the potential to develop food production indigenously, and therefore whether or not early evidence of domesticates would be expected. The Near East and Africa will be used as known variables against which the Ethiopian data can be compared and contrasted. Although it is extremely crude, this approach provides a basis for further investigation in the absence of viable data from Ethiopia.

A highly simplified equation is used, as follows:

- Certain factors were responsible for the development of agriculture in the Near East.
- The absence of domestication in Africa suggests that certain factors were lacking.
- By identifying which critical factors were also present or absent from Ethiopia during the Holocene, it is possible to broadly determine the potential for indigenous food production in Ethiopia.

Environmental Factors

A number of environmental factors have been recognised as necessary preconditions for the development of agriculture in the Near East (Binford 1968, Zohary 1970, Young et al 1982, Hole 1984, McCorriston and Hole 1991). If valid, these factors should also be generally applicable in other regions of the world. Table 7.1 compares the presence or absence of these preconditions in the Near East, Ethiopia and Africa.

Table 7.1 shows a close correlation between the presence of factors in the Near East and Ethiopia, but not in Africa. In other words, Ethiopia and the Near East had broadly analogous environmental contexts in the early Holocene. By inference, the potential for subsistence change may have existed in Ethiopia. This inference can be ratified through a closer examination of certain of these preconditions, as follows.

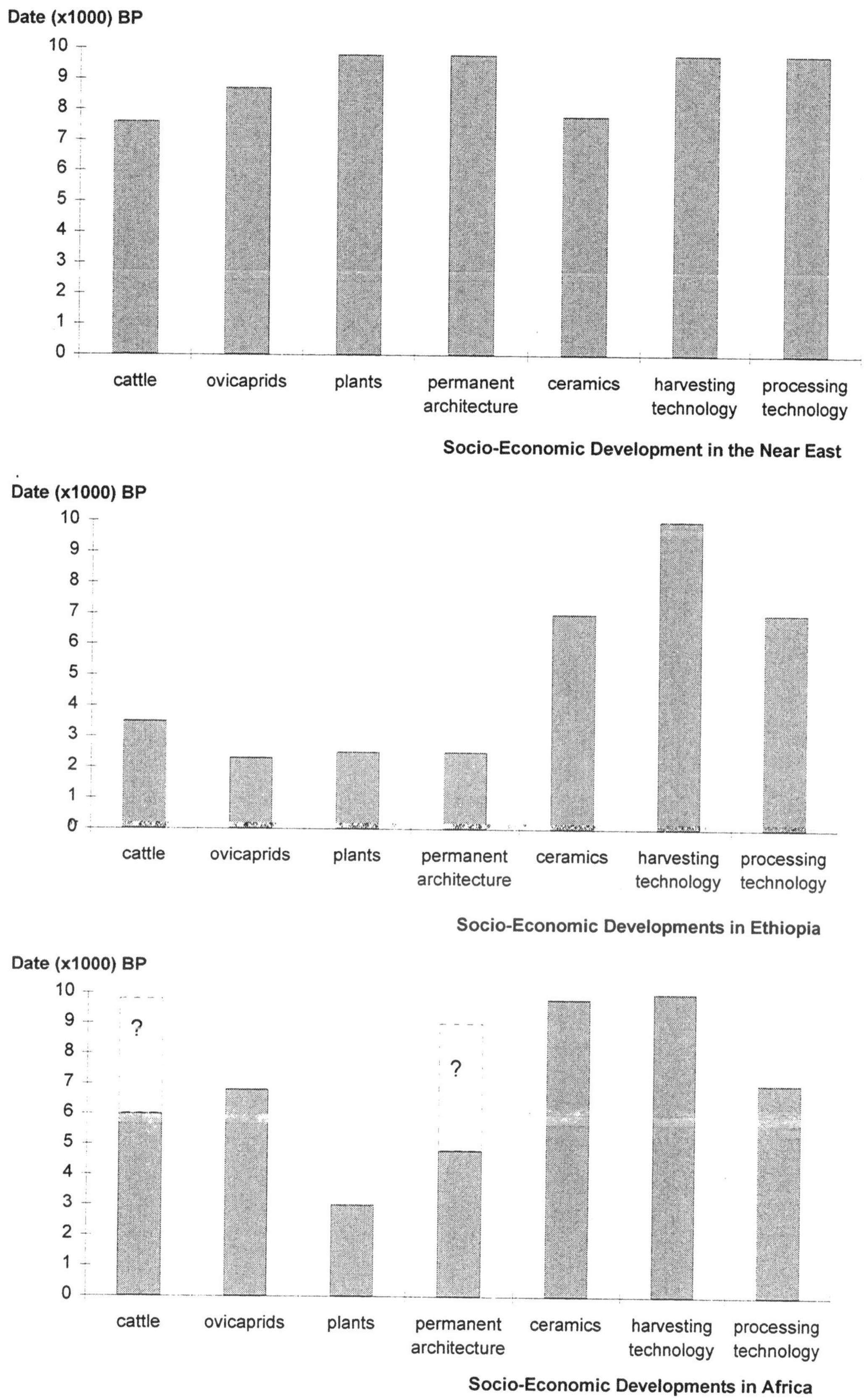

Figure 7.2. Comparative chronological framework of socio-economic developments and the origins of food producting societies in the Near East, Ethiopia and Africa.

Table 7.1. Early Holocene Environmental Preconditions for the Development of Agriculture

Environmental Variables	Near East	Ethiopia	Africa
Pronounced resource seasonality≠	YES	YES	YES
Concentration of wild progenitors*	YES	YES/POSSIBLE	NO
Closed ecosystems and circumscribed resource distribution*	YES	YES	NO
Vertical resource spectrum*	YES	YES	NO
'Centralised' and 'peripheral' resource zones	YES	YES	NO
High quality storable foods*	YES	YES	YES
Wide range of potentially exhaustible resources in small area*	YES	YES	NO
Potential for increasing yields through modification of territory or resources*	YES	YES	VARIABLE

≠ this precondition increases the need for storage and sedentism where group mobility is circumscribed.

- these preconditions reduce the necessity of adapting to altered conditions through increased group mobility.

Seasonal Resource Availability

As mentioned above, changes in seasonal climatic rhythms and the corresponding alterations in patterns of wild resource availability have been implicated as an 'ecological trigger' in the development of agriculture. McCorriston and Hole (1991) proposed that specific territorial and climatic circumstances following Pleistocene instability combined with the development of seasonality and its effects on human groups to create the conditions necessary for the development of agriculture. Although the Near Eastern evidence is still contentious (El-Moslimany 1984, Roberts 1984), recent work in Northern Africa suggests that changes in seasonality may have been a widespread phenomenon at the start of the Holocene (Lamb et al 1995, Roberts *pers.comm.*). If this was so - and future palaeoclimatic research should support or refute the hypothesis - then human groups across the world would have had to adapt to significant changes in resource availability at the start of the Holocene.

Resource Distribution and Circumscription

Zohary (1970) observed that domestication could only occur within the natural distribution zone of a wild progenitor. The potential for the development of food production is greater where there is a concentration of wild animal and plant progenitors as this ensures greater economic security through a broad resource base. Harlan (1971) extended this argument to propose that if the potential zone of domestication is ecologically defined by the natural habitat of the wild progenitor, then in areas where the progenitor has a diffuse distribution, the spatial parameters of its domestication zone would also be diffuse. He subsequently proposed a system of agricultural 'centres' and 'non-centres' defined by the relative distribution and concentration of wild resources, and suggested that the whole of the African Sahelian zone, including the Ethiopian Highlands and Western Africa, comprised a 'non-centre' for plant domestication.

Ethiopia, however, does not conform to Harlan's model. In contrast to the North African Sahel, Ethiopia's plant resources are circumscribed and are concentrated within a geographically limited area. Where Africa can be described as a diffuse, horizontal resource spectrum, the wild resources in Ethiopia form a concentrated vertical resource spectrum. A similar biogeographical pattern prevailed in the Near East at the start of the Holocene, where the contiguous arrangement of ecozones concentrated the diverse resources and reduced the need for group mobility[1]. Similar conditions may have prevailed in Ethiopia and encouraged more permanent settlement occupation with increased reliance on locally available resources.

[1] The main meat source in many areas of the Near East during the late Pleistocene and early Holocene was gazelle. Gazelle are relatively sedentary animals and their territory is reduced in areas with abundant flora. Other common species such as *Bos*, deer, wild boar and ibex have comparable home ranges. Dense and predictable wild food resources would therefore have been available without the need for long-range, mobile hunting strategies. Seasonal resource variations and deficits could then be overcome through small scale transhumance from a sedentary base, or adaptive strategies such as storage and food processing.

Wild Progenitors and the Potential for Agriculture
Zohary (1970, 1971) noted that the availability of wild progenitors was a universal precondition for the development of agriculture. Harris (1969) proposed that the indigenous resources must comprise substantial, balanced dietary components. All indigenous agricultural systems depend on a range of starchy staples, supplemented by oil and protein crops These include squash, maize and potato in South America; rice, foxtail millet and broomcorn millet in South East Asia, cereals, oil seeds and pulses in the Near East; with yams, manioc, rice and oil palm in West Africa. Alternatively, poor availability of vital metabolites, particularly carbohydrates, has been suggested as limiting the development of sustainable subsistence economies in tropical rainforests (Bailey et al 1989).

Resources with possible or probable Ethiopian ancestry are listed in Table 7.2. The crops which are endemic to Ethiopia are shown in the left hand column. These comprise one starchy staple, two species of oil seed and one source of protein. These resources alone do not constitute an adequate diet. They are too limited to provide a secure economic base for an emerging system of food production. The absence of a suitable range of wild progenitors would prevent the development of agriculture. However, if all the domesticates with ambiguous origins (shown in the right hand column) were to be included in this argument, the range of plant progenitors would be substantial enough to form the foundations of a mixed agricultural economy.

Table 7.2. Probable and Possible Indigenous Crops in the Traditional Farming System of the Ethiopian Highlands

Indigenous crops domesticated in Ethiopia	Some Ethiopian crops with uncertain origin of domestication in Ethiopia
tef	barley
noog	sorghum
cabbage seed	finger millet
cabbage	cow pea
	chick pea
	lentil
	castor
	sesame
	okra

Wild Faunal Resources
As discussed in chapter 2, there are no indigenous wild faunal progenitors in Ethiopia. Domesticated animals were probably unknown until the introduction of cattle and ovicaprids (see *Appendix A*). This provokes the question: *To what extent could agriculture have developed in the absence of domesticated animals?*

There is no global pattern for the chronological order of plant and animal domestication. The chronological relationship between the domestication of plants and animals in any region of the world must be considered within its specific bio-geographical framework and resource potential. In South America for example, the domestication of the llama and the alpaca was coeval with the domestication of the potato (Lynch 1983). In the Eastern section of the Fertile Crescent animal domestication preceded the appearance of domesticated plants by up to a thousand years, and by several millennia in Northern Africa (Smith 1995, Phillipson 1992). In the Western section of the Fertile Crescent, plant-based food production and hunting were successfully supporting permanent village settlements by the start of the 10th millennium BP. The inhabitants of sites such as Cayonu, Netiv Hagdud, Tell Abu Hureyra and Jericho were cultivating early domesticated forms of barley, wheat, lentils and chick peas for up to a thousand years before the assimilation of domesticated caprines (van Zeist 1972, Kislev et al 1986, Hillman et al 1989). Horticulture was possibly practised by at least the 5th millennium BP in the Melanesian archipelago (Bulmer 1975), and even today in Papua New Guinea the plant-dominated subsistence economy is supplemented by a single domesticated, introduced animal resource - the pig (Rappaport 1984).

Horticulture as a Viable Subsistence Strategy
A diet based entirely on animal products can lack vital nutrients, and is more conducive to physiological stress (Dennell 1979, Stryer 1995). Carbohydrate, particularly, is essential as a source of energy. Animal-product diets are carbohydrate-poor, and all pastoral groups, even 'extremists' such as the Dinka and Nuer of Southern Sudan, need to incorporate a minimum of plant-foods into their diet (Evans-Pritchard 1940). The majority acquire a substantial intake of plant resources through exchange, forage or agriculture (Wilding 1984, 1985a). Conversely, plant resources can provide a balanced diet containing all essential amino acids, carbohydrates, fatty acids, minerals and vitamins. The frequency of vegetarians and vegans in contemporary Western society is witness to the physiological viability of animal-free subsistence.

Although horticulture is feasible as an autonomous system of food production, its viability is enhanced by broadening the domestic resource base. Several researchers (eg. Boserup 1965, Sherratt 1980) have stressed that the development of agriculture required a shift from an extensive system of horticulture to an intensive form of land-use. The addition of domesticated animals increases the resource predictability and provides a mechanism for intensification of land use through draught animals, transhumance and increased mobility. This effectively overcomes an 'energy threshold' in the development of food production.

Implications of Environmental Factors for Ethiopian Food Production

The arguments outlined above indicate that more work is required to define the environmental factors for Ethiopia. The range and nature of indigenous Ethiopia resources available to a prehistoric population in the Holocene are especially important. These are unclear at present. The wild progenitors of modern crops were probably very limited in number, suggesting that they alone could not promote indigenous food production. A wider resource base and a means for intensification would have been required before agriculture could develop. In addition, the absence of domesticated animals in Ethiopia may have been a major factor in limiting the level of intensity of land use. This may have inhibited the domestication of wild plant foods.

Until a clearer picture of the prehistoric plant resources in Ethiopia can be produced, the question of whether or not Ethiopia was environmentally 'primed' for food production must remain unanswered. The present evidence suggests the following model:

- Indigenous food production would have been based on plant foods, supplemented by hunting and foraging.
- A number of environmental factors may have provided the context for intensification and specialisation of resource procurement.
- The nature and range of the available wild plant foods were not adequate to encourage the development of food production.
- Food production was therefore not feasible until the introduction of domesticated animals and/or foreign food plants.

Climatic Factors

Climatic factors provide a further possible explanation for the absence of indigenous food production in Ethiopia. In contrast to the comparative environmental patterns between Ethiopia and the Near East, and their divergence from North Africa, the Ethiopian palaeoclimate in the Holocene is distinct from the Near East. Instead, the Ethiopian palaeoclimate changes in parallel with North African palaeoclimate. *Could palaeoclimate be a primary factor in influencing the non-development of food production in Ethiopia?*

Figure 7.3 illustrates the cultural development patterns in the Near East, Ethiopia and Africa relative to simplified palaeoclimatic changes from these areas during the Holocene. In the Near East, alterations in vegetation patterns at the start of the Holocene provided a means for subsistence change, whilst the stress conditions created by the Younger Dryas event may have provided the motive for change. Subsequent climatic stability throughout the remainder of the Holocene enabled these adaptive responses to gain momentum in the Near East.

Conversely, in Africa and Ethiopia there were series of cultural developments during the Holocene that failed to become established. As figure 7.3 shows, the pattern of cultural development or cultural evolution oscillates in rhythm with the pattern of palaeoclimate change. This hypothesis corroborates the model of non-linear cultural evolution outlined above. This suggests that the proposed random trajectory of cultural development may in fact reflect a non-random relationship between human adaptation and climatic dynamism.

This agreement between environmentally and culturally disparate regions implies that there may be an integrating factor that overrides these disparities. The parallel cultural development pattern shown in figure 7.3 suggests that in Africa and Ethiopia the environmental variable caused a similar cultural response, despite the differences in the resource base.

Chapter 8 proposes a possible cultural-ecological sequence through which climatic instability in the Holocene may have influenced subsistence change in Ethiopia.

Cultural Factors

A cultural-theoretical perspective is inadequate for this thesis. The cultural component of the Ethiopian database is incomplete and cannot be used to test detailed cultural-development models. Much research is need in this area before the cultural development sequence can be interpreted from an archaeological perspective.

In the absence of comprehensive evidence for the cultural variable, this thesis has concentrated on the data for the Ethiopian palaeoclimate and palaeoenvironment. These factors have provided a dynamic context for understanding and interpreting the available cultural data.

Date x1000 (BP)	Development processes in Ethiopia	Development processes in tropical Northern Africa	Palaeoclimatic pattern in tropical Africa	Development processes in the Near East	Palaeoclimatic pattern in the Near East
12-11	harvesting wild grasses and plants using microliths and preadaptive technology	harvesting wild grasses and plants using microliths and preadaptive technology	**late Pleistocene aridity**	preadaptive technology (including storage, sickles, grindstones)	**late Pleistocene aridity**
11-10			*soil erosion*		**increased rainfall vegetation** *growth, soil formation, changes in seasonality*
10-9		ceramics ?domesticated cattle	**increased rainfall** *vegetation growth and soil formation*	plant domestication; permanent village architecture	
9-8				domesticated caprines	
8-7			**abrupt arid phase** *forest retreat*	domesticated cattle, ceramics	**gradual arid trend, especially from the mid Holocene**
7-6	reduced mobility, change in food procurement strategies and processing activities	reduced mobility, change in subsistence strategies and processing activities	**increased rainfall** *forest expansion, vegetation clearance?*		
6-5		domesticated animals appear, intensive exploitation of wild plants	**abrupt arid phase** *forest retreat*		
5-4			**increased rainfall, gradual drying trend** *forest retreat, modern vegetation pattern established*		
4-3	domesticated cattle and livestock	metal technology			
3-2		domesticated plants	*vegetation clearance?*		
2-1	domesticated plants Axumite culture		**abrupt cool phase**		

Figure 7.3. Proposed chronological pattern of cultural development and climate change in Ethiopia, Northern Africa and the Near East.

Chapter 8.

A SPECULATIVE MODEL FOR PREHISTORIC SUBSISTENCE CHANGE IN ETHIOPIA BETWEEN 10 000 AND 2500 YEARS BP

8.1. INTRODUCTION

Ethiopia has been described as one of the original 'centres ' for plant domestication and for agriculture, but traditional models to explain the emergence of food production in Ethiopia are confused, diverse and largely speculative. Little attempt has yet been made to explain how or why food production developed. At present, the incomplete and ambiguous nature of the database prevents the development of a definitive model for the emergence of food production in Ethiopia. This should not preclude attempts to construct an interpretative framework, and despite these constraints, a pattern can be detected in the available data. This pattern is narrated below as a speculative model for absence of evidence for domesticated plants in Ethiopia before the 3rd millennium BP.

8.2. SPECULATIVE MODEL FOR THE EMERGENCE OF FOOD PRODUCTION IN ETHIOPIA

Establishing a Pattern of Subsistence Change

There is no direct archaeological evidence to indicate early food production in Ethiopia, and no direct evidence to suggest that food production developed indigenously. What the available archaeological evidence does suggest is that domesticated cattle were introduced into the Rift Valley by the mid-4th millennium BP, and that foreign and indigenous food plants were being grown in the Northern Highlands by the mid-late 3rd millennium BP.

A range of evidence is available from other disciplines, including ethnography, ethnobotany, agronomy, botany, plant genetics, linguistics and rock art, which provides an insight into the pre-agricultural subsistence economy of Ethiopia. This evidence indicates that certain staple Ethiopian crops are of great antiquity and that these ancient resources include indigenous crops unique to Ethiopia. In addition there are several plants whose origins are uncertain. Although these are believed to derive from the Near East, this remains to be proven. It is unclear whether these plant resources were wild or domesticated before foreign domesticates were introduced, although domesticated cattle and livestock were introduced to Ethiopia at an early date.

An early system of subsistence is preserved in the traditional farming methods of Ethiopia. This system was based on the exploitation of local plant resources concentrated in the botanically rich and diverse *woina dega* and lower *dega* ecozones. Wild plants still comprise a vital part of the subsistence economy, and probably formed the basis of prehistoric subsistence prior to the introduction of domesticates. Domesticated animals were probably not an integral part of this subsistence system, suggesting that it was established before livestock were introduced.

A prehistoric subsistence economy can be tentatively proposed on the basis of these observations. This subsistence economy took advantage of the natural plant wealth and diversity found in contiguous ecotones of the Ethiopian Highlands. The economy was based principally on procurement of wild plant and animal resources and was supplemented by the cultivation of particular food plants, using a primitive form of land-management and preadaptive technology.

Establishing a Pattern of Climate Change

Detailed investigation and assimilation of recent palaeoenvironmental research in Ethiopia and North Africa shows that the climate changed significantly during the past 10 000 years under investigation. Three scales of change were identified:

Long-term change.

> Long-term climate change followed a precessional cycle, with a pattern of increased rainfall between c.10 000 and 5500 years BP changing to arid conditions between c.5500 and 4500 years BP, and becoming increasingly arid after c.4500 BP. This pattern would have allowed a long-term anthropogenic adaptation strategy.

Intermediate-term change.

> A separate pattern of change has been detected in more recent years. This pattern interrupts the long-term cycle to produce intense, abrupt climatic reversals. These events occur roughly every thousand years, and persist for 150-400 years. In Ethiopia, intense arid pulses have been detected in the early Holocene pluvial sequence at c.7500-7200 years BP and 6200-5800 years BP. A humid pulse interrupted the long-term arid trend at c.2500-1000 years BP. In addition, a change to more pronounced seasonality at the start of the Holocene is tentatively inferred from the available data.

These events are of immense significance to human survival. Their abrupt onset and relative brevity allows little time for long-term adaptive strategies. The existence of these climatic pulses in the tropics has only recently been recognised, and their implications for human development has yet to be determined.

Short-term change.

In addition to the above patterns of change, a low-amplitude, short-term rhythm exists. This affects the annual monsoon rainfall system every 10-30 years, and is responsible for drought conditions. The impact of this rhythm, and the extent to which it mitigates economic failure, depends on the socio-economic context in which it is translated. Over-use of the resources and potential of the land among extensive agricultural societies creates a more vulnerable environment that is more susceptible to drought-related catastrophe.

Short-term climatic change has a range of implications, including discouraging economic 'experimentation' and high risk subsistence, encouraging the construction of a system of exchange and trade to buffer the effects of localised economic catastrophe and encouraging the acquisition and maintenance of a 'fall-back' resource, such as cattle and livestock, with potential for trade-wealth and subsistence security.

Environmental change during the Holocene reflects the response of the vegetation to climate change. There was a gradual increase in vegetation and forest cover between c.10 000 and 8000 years BP which was briefly retarded by abrupt climate change in the 8th millennium BP. Climax vegetation was eventually established in the 7th millennium BP. The vegetation patterns altered again with climate change at the end of the 7th millennium BP, producing reduced forest cover and increased savannah woodland. By the end of the 6th millennium, increasingly arid conditions were encouraging the retreat of humid-type montane vegetation and the contraction of forest ecozones. Except for minor interruptions, particularly c.2000-1500 years BP, this process continued until relatively recently.

Using Integrated Evidence to Combine Patterns of Climate and Subsistence

Palaeoclimatic and palaeoenvironmental changes appear to have a resonance in the cultural development sequence in prehistoric Ethiopia. A contemporaneous pattern of behavioural change can be tentatively detected from the archaeological and palaeoenvironmental record:

Early Holocene Intensification

Indirect archaeological evidence shows a possible increase in the intensity of site use at the start of the Holocene, between the 10th and 8th millennia BP. This change is chronologically associated with a period of more extensive vegetation cover. Hypothetically, denser vegetation would discourage group mobility, whilst the increased availability of local resources would encourage sedentism. The primary step towards developing food production may have been taken at this stage.

Early-Mid Holocene Adaptation

Geomorphological and palynological evidence indicate possible anthropogenic clearance activities dating to the start of the 7th millennium BP in the Northern Highlands. This activity correlates to a period of maximum vegetation cover. By the 7th millennium BP, there was a shift in the subsistence economy. This can be detected in increased sedentism, a change in food processing activities and subsistence strategy with a concomitant shift in the human/environment relationship.

By expanding the range of local resources through food processing activities, human groups could exploit their immediate environment more efficiently. Such a move would encourage reduced group mobility, enhance population growth and stimulate changes in the cultural fabric. These adaptations represent a second step towards the development of food production.

This observation demands a rigorous investigation of human behavioural changes during the early Holocene. The relationship between humans and the environment during this period must be examined especially closely, particularly for the resource potential of the Ethiopian environment, and changes in this during the critical period.

Using the Combined Patterns to Speculate on Causality

Geomorphological and archaeological evidence suggest that there was no subsequent evolution of this initial development. The evidence suggests further that there was no major human impact on the environment in Ethiopia until the 3rd millennium BP. The first, tentative steps towards food production were unsuccessful. The critical question that we need to ask is why did food production *NOT* develop in Ethiopia? Two principal hypotheses can be proposed.

First, the rich diversity and distribution of the resources reduced the probability of environmental stress and the need for intensification. However, the intensity and abruptness of climate change would have significantly altered the vegetation pattern at certain stages during the first half of the Holocene, possibly inducing a stress situation. This would have necessitated a change in the prevailing subsistence strategies.

Second, the 7th millennium BP was a critical time in the cultural development of Ethiopia. This was a period of intense climatic and environmental instability. Unfortunately the poor resolution of the data does not allow more precise correlation of cultural and climatic events.

Hypothetically, increased sedentism and subsistence change during the 7th millennium may have been a response to increased forest cover in the early 7th millennium BP. It is possible that this was a reversion to the early Holocene development, that may have been interrupted by climate change. Following this argument further, we could propose that the subsistence change may have occurred earlier in the absence of the late 8th millennium BP arid interlude.

A second intense arid interval towards the end of the 7th millennium BP may have halted this development, which was then further repressed by more persistent aridity from the mid 6th millennium BP. These changes demanded an alternative adaptive human response, and they may have been intense enough to destabilise emerging social changes before a threshold of development could be reached.

Could it be, then, that climate change was actually preventing the development of food production in Ethiopia? This proposal has wider implications for the origins of agriculture. It must therefore be tested in regions such as Ethiopia that have the potential for developing food production, but where there is no evidence that food production arose indigenously.

An Alternative Explanation

After the 6th millennium BP, external cultural factors may have interfered with the pattern of indigenous cultural development in Ethiopia. The available evidence implicates two possible coincidental sources for this interference:

In the first, cattle and livestock may have been introduced from North East Africa, possibly as early as the 6th millennium BP. This contrasts with the 4th millennium date of traditional models. The mechanism by which cattle and livestock were introduced is unknown at present, but movement of pastoral groups, or exchange and assimilation by indigenous forage groups are both feasible and ethnographic analogues indicate that a combination of the two may have been most probable.

Second, domesticated plants of Near Eastern origin may have been introduced to the Ethiopian Highlands as early as the 5th millennium BP. This introduction arose as the result of trading contacts between Egypt and the land of Punt, in which the Northern Ethiopian Highlands may have actively participated.

The Role of Prehistoric Trade in Subsistence Change in Ethiopia

The movement of African plant resources has been used in this publication as a basis for discussion of prehistoric trade. This discussion has shown that long-range trade networks existed throughout the Holocene period. The nature of the contacts and the products that flowed along them changed over time. These networks interconnected all regions of Africa, and linked the African continent to Asia and the Mediterranean. Most probably, the entire Old World was broadly integrated through contact and movement between ancient trade centres.

The 5th millennium BP was a critical developmental period for Ethiopia, for Africa and for the Old World. It is possible this corresponds to a change in established trade networks between major centres such as Mesopotamia, Egypt and India, and a move to exploit the exotic resources of tropical Africa. This model is largely conjectural, and the motives behind such a re-orientation remain purely speculative. Demand for valuable raw materials, particularly gold and copper[1], timber, incense and exotica led to the establishment of new trade links. Tropical African food plants appear in Arabia, Egypt, India and possibly the Levant from at least the 5th millennium BP. It is feasible that an Africa 'package' was being transported immense distances along established communication routes. By implication, there was knowledge in tropical Africa of agricultural civilisations from at least the 5th millennium BP.

The most probable source of exported African plant resources was the Nile Valley. However, several of the exported crops were also indigenous to the Ethiopian Highlands, and the possibility of an African-Red Sea trade centre in the Horn cannot be discounted. More significantly, many of the exotic African resources exported to Egypt from the land of Punt are found in Ethiopia, suggesting that the Ethiopian Highlands may have formed part of the trading region of Punt from the 5th millennium BP. It is unclear whether plants and animals were exchanged for the products of Punt. However, contact with a developed agricultural society would have provided a mechanism for the eventual introduction of domesticated resources and agricultural technology. Close genetic similarity between Ethiopian and Egyptian wheats and barleys suggests close contact between these regions from an early date, whilst their divergence from Near Eastern crops

[1] For example, the need for copper, following the loss of the Anatolian source after the 6th millennium BP, caused the Sumerians to prospect widley for a new source. This was eventually found in the Oman (Mortensen 1986).

implies that there was genetic separation of either the wild progenitor or the primitive domesticate in prehistory.

Summary

Two significant phases of development can be tentatively identified in Ethiopian Holocene prehistory. First, in the 7th millennium BP there was a pre-adaptive, and possibly unsuccessful, shift towards food production. Subsequentl,y during the 6th/5th millennium BP, the introduction of domesticated plants and animals to Ethiopia precipitated the development of food production. The suggestion here is that communication networks, built on trade and exchange relations, may have been responsible for the introduction of domesticated animals to Ethiopia as early as the 6th millennium BP, and domesticated temperate crop plants from the 5th millennium BP. These introductions may have interfered with the indigenous cultural development processes.

The changes that occurred during these periods, and the contexts in which they arose, are critical to our understanding of cultural change in Ethiopia and the emergence of food production. A closer understanding of how and why this behavioural pattern developed also has wider implications for research in African prehistory, and for investigating the global origins of agriculture.

APPENDIX A: Ethiopian Resources

Section A.1.

Tables A.1.1.-A.1.8.

Natural plant resources in relation to elevation.

Section A.2.

Table A.2.1.

Domesticated plant resources in Ethiopia and their relative antiquity.

Section A.3.

Table A.3.1.

Wild faunal resources in Ethiopia.

Table A.3.2.

Domesticated faunal resources in Ethiopia and Africa and their relative antiquity.

References

Including Bedigan and Harlan (1986), Bilbert (1975), Burkhill (1953), Cheeseman (1956), Darby et al (1977), Doggett (1988, 1991), Edwards (1976, 1991), Farris (1965), Greenway (1945), Harlan (1969, 1989, 1992), Harlan et al (1976), Hayes (1905), Hedberg (1971), Hedberg and Edwards (1989), Hepper (1980), Huffnagel (1961), Meadow (1996), Mooney (1961), Pichi-Sermolli (1951, 1955), Porteres (1976), Potts (1993,1994), Purseglove (1976), Sebald (1968), Siegenthaler (1960), Simmonds (1976), Simmoons (1960), Tackholm (1956), Vishnu-Mittre (1969), von Breitenbach (1963), Westphal (1975), White (1983).

Table A.1.1. Lowland, Semi-Arid Vegetation of the *Kwolla* Zone in Ethiopia

Altitude-Environmental Zone	Vegetation Zone	Floral Components	Common Name	Uses
500 - 1000 m asl Sudanese/Ethiopian Border Region Lower *k'wolla*	Grassland savannah with dispersed trees and shrubs. Becomes open woodland savannah and shrub savannah away from the Sudanese border as altitude increases. Grasses can grow up to 80 cm, forming a continuous layer under which other plants grow. Many of these grasses have edible seeds. Xerophytic vegetation, such as *Acacia* spp, Baobab (*Adansonia digitaria*) and *Euphorbia* spp, increases towards the north as precipitation decreases, becoming thorn and scrub savannah.	**Trees and shrubs**		
		Acacia spp		certain *Acacia* spp are edible or useful
		Adansonia digitaria	baobab	edible fruits and leaves
		Anogeissus leiocarpus		
		Combretum hartmannianum		
		Hyphaene spp	dom palm	edible fruit, useful fibres
		Euphorbia spp		
		Gardenia lutea		edible fruits, timber
		Kilelia aethiopica	sausage tree	edible fruit
		Piliostigma thonningi		
		Terminalia spp		
		Grasses and herbs		
		Andropogon spp		edible seeds, animal fodder
		Aristida spp		animal fodder
		Cymbopogon spp		human and animal food, fibres, sources of oil
		Digitaria spp		edible seeds, animal fodder
		Eragrostis spp		edible seeds (human and animal), medicinal
		Hyparrhenia spp		thatching and construction
	Riparine gallery forest formations in lowland areas include certain common woody species.	*Panicum spp*	millet	edible (human and animal) seed and stem, fibres
		Sorghum spp	sorghum	edible seeds and stems, construction
		Sporobolus spp		
		Acacia spp		edible fruits
		Dracaena spp		
		Ficus spp	fig	edible fruits
		Oxytenanthera abysinnica	bamboo	edible fruits (cooked), edible young shoots, alcoholic beverages, construction
		Phoenix abyssinica	wild date	
		Syzygium guineense		edible fruits
		Tamarindus indica	tamarind	edible fruits
		Tamarix spp		edible fruits, medicinal, non-alcoholic beverages
		Zizyphus spp		edible fruits

Table A.1.2. Upper *Kwolla* Vegetation

Altitude-Environmental Zone	Vegetation Zone	Floral Componants	Common Name	Uses
		Trees and shrubs		
1000 - 1800 m asl	Open deciduous woodland.	*Anogeissus leiocarpus*		
kwolla zone		*Balanites aegyptica*	date	edible fruits, leaves and flowers
	Light tree cover with canopy 5-12 m high above short grass/herbaceous layer. Thickets of bamboo (*Oxythenanthera abyssinica*) among open woodland especially in the region 700-1200 m asl.	*Boswellia spp*	incense tree	resin, gum
		Combretum spp		edible fruits/seeds
		Commiphora spp		edible fruits (*C.africana*), resin, gum
		Dalbergia melanoxylon		
		Erythrina abyssinica		
		Gardenia lutea		edible fruits
		Lannea schimperi		edible fruits
		Lonchocerpus lexiflorus		
		Maytenus spp		edible fruits, fodder, construction
		Odina spp		
		Piliostigma thonningii		
		Stereospermum kunthianum		
		Terminalia browni		
		<u>At higher altitudes</u>		
		Euphorbia spp		
		Ficus spp	fig	edible fruits
		Olea chrysophylla	olive	edible fruits
		Grasses and herbs		
		Andropogon spp		edible seeds, fodder
		Heteropogon spp		fodder
		Hyparrhenia spp		thatching and construction
		Setaria spp	millet	edible seeds, alcoholic beverages, construction
		Themeda spp		

Table A.1.3. Vegetation of the *Woina Dega*

Altitude-Environmental Zone	Vegetation Zone	Floral Components	Common Name	Uses
		Trees and shrubs		
1800 - 2200 m asl	Montane evergreen thicket	*Brassica juncea*	Ethiopan cabbage	edible leaves, oil bearing seeds used in cooking
woina dega	and scrub interspersed with	*Buddleja polystachya*		medicinal
	shrub savannah.	*Calpurnia subdecandra*		leaves used medicinally
		Carissa spp		edible fruits
	Montane thicket comprises	*Coffea arabica*	coffee tree	coffee
	small trees, shrubs, big	*Cordia abyssinica*	white flowered *wanza*	divine and status associations
	bushes and some larger trees	*Ficus vasta*	fig	edible fruits
	over a herbaceous layer.	*Grewia spp*		edible fruits
		Gymnosporia spp		
		Hagenia abyssinica	*kosso*	medicinal/purgative
		Jasminum abyssinica		medicinal
		Olea spp	olive	edible fruits, medicinal, construction, firewood
		Osyris abyssinica		
		Otostegia spp		disinfectant
		Phoenix reclinata	wild date palm	edible fruits, fibres used in basketry
		Phytolacca dodecandra		edible leaves, fruit used as soap, roots used medicinally
		Pistacia chinensis		edible fruits
		Rhamnus spp	*gesho* and buckthorn	used in brewing beer (*t'ella*)
		Rhus spp		edible fruits
		Syzygium guineense		edible fruits/seeds
		Verbascum siaitium		
		Ximena americana		edible fruits and medicinal applications
		Zizyphus spp		edible fruits
		In well watered areas at higher altitudes		
		Juniperus procera	native Juniper/'pencil cedar'	medicinal, construction and firewood
		Podocarpus gracilior		construction
		Sapotaceae		

Table A.1.3. *(cont.)* Vegetation of the *Woina Dega* - Woody Plants and Grasses

Altitude-Environmental Zone	Vegetation Zone	Floral Components	Common Name	Uses
		Rampants and small woody plants		
1800 - 2200 m asl		*Asparagus racemosus*	wild asparagus	edible fruits/seeds
woina dega		*Chasmanthera dependens*		
		Cissus spp		edible roots
		Clematis spp		edible when cooked
		Coriandrum sativum	coriander	spice/flavouring
		Cucumis dispaceus		edible fruits
		Ensete ventricosum	wild banana	edible roots and leaves
		Gloriosa simplex		
		Glycine javanica		
		Hyperium lanceolatum		
		Pterolobium stellatum		
		Rosa abyssinica	rose	edible fruits
		Tragia mitis		
		Vernonia amydalina		flavouring for beer
		Grasses and herbs		
		Andropogon spp		edible seeds, animal fodder
		Aristida spp		animal fodder
		Crotalaria cylindrica		
		Eleusine spp	millet	edible seeds, fibres used for basketry
		Eragrostis spp		edible seeds, animal fodder
		Panicum spp	millet	edible seeds
		Pennisetum schimperi	millet	edible seeds, stems used for basketry
		Rhynchelytrum repens		edible seeds, animal fodder
		Setaria spp	millet	edible seeds,animal fodder
		Sporobolus spp		edible seeds, animal fodder
		Trifolium spp	clover	animal fodder

Table A.1.4. Localised Vegetation Mosaic of the *Woina Dega* - the Lake Tana Basin

Altitude-Environmental Zone	Vegetation Zone	Floral Components	Common Name	Uses
		Trees, shrubs and rampants		
c.2000 m asl Lake Tana Basin	Savannah woodland vegetation mosaic.	*Acacia spp*		
		Acanthus polystachius		edible leaves
		Cordia africana	*wanza*	edible fruits/seeds, construction, ideological signigficance
	Grassland on alluvial soils comprises many edible species. Interspersed with primary undifferentiated woodland of variously sized trees, bushes, climbers and scrub.	*Croton spp*		medicinal
		Ficus spp	fig	edible fruits
		Gardenia terrifolia		edible fruits, nail and hair dye
		Solanum campylacanthum		edible leaves/fruit
		Terminalia brownii		
		Grasses and herbs		
		Aristida spp		edible seeds, animal fodder
		Brassica juncea	Ethiopian cabbage	edible leaves, oil bearing seeds used in cooking
	Vegetation turns to shrub savannah towards the hills and is dominated by tall grass formations comprising *Hyparrhenia* and *Setaria* spp.	*Cyperus papyrus*	papyrus	construction and boat building
		Eleusine spp	millet	edible seeds
		Eragrostis spp		edible seeds
		Linum usitatissium		edible/oil bearing seeds
		Pennisetum spp	millet	edible seeds, animal fodder
		Setaria spp	millet	edible seeds
		Verbascum sinaiticum		edible seeds
Riparine formations in the Highlands	Plant communities along river valleys, especially the deep gorges of the Blue Nile the Angareb and the Takezze , are represented by isolated, heterogeous formations. Extensive stands of bamboo often occur amongst this vegetation type.	*Acacia spp*		
		Aphania senegalensis		edible fruits
		Arundinaria alpina	bamboo	construction
		Diospyrus mespiliformis		edible fruits
		Gossypium anomalum	includes cotton	useful fibres
		Phoenix spp	dates	edible fruits
		Salix spp	willow	
		Syzygium guineense		edible fruits
		Tamarindus indica	tamarind	edible fruits
		Tamarix spp		edible fruits, medicinal, non-alcoholic beverages
		Vernonia amygdalina		flavouring for beer
		Zizyphus spp		edible fruits/seeds

Table A.1.5. Examples of Montane Evergreen Forest in the Upper *Woina Dega* and Lower *Dega* Zones of the North Western Highlands

Altitude-Environmental Zone	Vegetation Zone	Floral Components	Common Name	Uses
2200 - 3500 m asl upper *woina dega* and lower *dega* zones	Montane dry evergreen forests. In the upper *woina dega* in the Western Highlands there are distinct forest types such as *Mimusops kummel* forest and *Acacia xiphocarpa* forest. *Acacia* forest for example comprises xerophilous montane vegetation with a 12 m high canopy over a stratum of small trees, bushes, climbers and lianas, above a herbaceous layer. The dry montane forest is dominated by juniper especially at higher altitutdes, with a lower stratum of smaller trees, wooded plants and shrubs. Grassy glades are common to upland formations. Montane grassland savannah forms a mosaic with montane forest formations. The two communities are mutually exclusive.	**Mimusops kummel forest**		
		Albizza schimperiana		
		Celtis kraussiana		edible fruits
		Diospyros abyssinica		
		Diphasia dainelli		
		Ficus thonningii	fig	edible fruits
		Millettia ferruginea		
		Mimusops kummel		
		Olea mildbraedii	olive	edible fruits
		Oxyanthus speciosus		
		Acacia xiphocarpa forest		
		Trees and shrubs		
		Acacia xiphocarpa		
		Asparagus asiaticus	wild asparagus	edible shoots and stems
		Bersama abyssinica		
		Ficus thonningii	fig	edible fruits
		Jasminium abyssinica		
		Maesa lanceolata		medicinal, construction
		Phytolacca dodcandra		edible leaves, soap, medicinal roots
		Salix spp	willow	
		Vernonia amygdalia		flavouring for beer
		Herbaceous layer		
		Cardamine africana		flavouring
		Commelina pyrroblepharis		
		Digitara abyssinica		
		Geranium simense		
		Guizotia villosa		

Table A.1.6. Vegetation of the *Dega* Zone of the North Western Highlands above 2400 m

Altitude-Environmental Zone	Vegetation Zone	Floral Components	Common Name	Uses
		Trees and shrubs		
2400 - 3400 m asl *dega*	Predominately Juniper forest forming a mosaic with montane savannah. This is interspersed with species such as *Rosa abyssinica, Erica arborea* and *Hypericum lanceolatum*	*Acacia abyssinica*		gum arabic
		Apodytes dimidiata		timber
		Arundiaria alpina	high altitude bamboo	construction
		Asparagus spp	wild asparagus	edible stems and shoots
		Carissa edulis		edible fruits
		Diospyros spp		edible fruits
		Dodonaea viscose		
		Ekebergia spp		edible fruits
		Euclea spp		edible fruits
		Ficus spp	fig	edible fruits
		Grewia spp		edible fruits
		Gymnsporia spp		
		Hagenia abyssinica	*kosso*	medicinal
		Juniperus procera	native juniper	medicinal and construction
		Maesa lanceolata		oil bearing fruits
		Myrsine spp		medicinal, construction
		Olea spp	olive	edible fruits
		Pistacia spp		edible fruits
		Pittosporum abyssinicum		
		Podocarpus gracilior		construction
		Pygeum africanum		timber
		Rapanea simensis		
		Rhus spp		edible fruits
		Grasses and herbs		
		Andropogon abyssinicum		edible seeds, animal fodder
		Commelina pyrrhoblepharis		edible leaves
		Echinops spp	giant thistle	
		Eragrostis spp		edible seeds
		Guizoitia spp		oil bearing seeds
		Heteropogon contortus		fodder
		Hyparrhenia spp		construction and thatching
		Pennisetum spp	millet	edible seeds, animal fodder

Table A.1.7. High Altitude *Urec* Vegetation

Altitude-Environmental Zone	Vegetation Zone	Floral Components	Common Name	Uses
Above 3500 m asl	High mountain vegetation	**Grasses, herbs and srcub**		
Urec zone	includes mountain scrub, high	*Agrostis spp*		edible seeds
	mountain steppe and	*Alchemilla haumannii*		
	afroalpine formations	*Asparagus spp*	wild asparagus	edible shoots and stems
		Blaeria spicata		
		Carex monostachya		
		Erica arborea	tree heather	
		Festuca spp		
		Helichrysum spp		
		Lobelia rhynchopetalum	giant lobelia	
		Papanea simensis		
		Poa spp		
		Swertia spp		

Table A.2.1. Relative Antiquity of Ethiopian Subsistence Resources

Crop	Botanical name	Country of origin	Earliest recorded date for use of domesticate in Ethiopia (BP)
Cereals			
tef	*Eragrotis tef* (Zucc.) Trotter	Ethiopia	late 3rd millennium
barley	*Hordeum vulgare* L.	South West Asia/ Egypt/Ethiopia	pre 2520
emmer wheat	*Triticum dicoccum* Schull.	South West Asia	c.2300
durum wheat	*Triticum durum* Desf.	South West Asia	c.2300
bread wheat	*Triticum aestivum* L.	South West Asia	c.2300
sorghum	*Sorghum bicolor* *Sorghum durra*	Sudan/Egypt/Ethiopa	2nd millennium. In Arabia by 6th/5th millennium
finger millet	*Eleusine coracana* (L.) Gaertn.	North East Africa/Ethiopia	2nd millennium. In India by 5th millennium
bullrush millet	*Pennisetum americanum*	Tropical Africa	In India by c.3700
oats	*Avena abyssinca*	South West Asia/Ethiopia	2nd millennium. In Arabia by 5th millennium
maize	*Zea mays*	Central America	post AD 1500
Pulses and legumes			
field/horse bean	*Vicia faba*	South West Asia	2nd millennium?
chick pea	*Cicer arietinum* L.	Ethiopia?/South West Asia	pre 2520
lentil	*Lens esculenta* Moenan	South West Asia/Ethiopia?	c.2300
cow pea	*Vigna unguiculata* (L.) Walp	Ethiopia/East Africa/West Africa	spread with sorghum to Asia c.4th millennnium
pigeon pea	*Cajanus cajun* (L.) Millsp.	India/North Africa	cultivated in Egypt since 2000 BC(XII Dynasty)
pea	*Pisum sativum* L.	West Asia	2nd millennium
grass pea	*Lathyrus sativus*	West Asia	2nd millennium
common bean	*Phaseolus vulgaris* L	Central America	post AD 1500
hyacinth bean	*Dolichos lablab* L.	India/Asia	?
Oil seed			
noog	*Guizotia abyssinica* (L) Cass.	Ethiopia	c.1600
castor	*Ricinus communis*	Ethiopia/East Africa	early records of use in PreDynastic Egypt and India
sesame	*Sesamum indicum*	Ethiopia/Horn/India/ Far East	cultivated from c.6000 in Asia and Egypt?
safflower	*Carthaus tinctorius*	South West Asia/Ethiopia/Central Africa	spread to Egypt by 4th millennium
linseed	*Linum usticitissimum*	South West Asia	2nd/1st millennium
melon seed	*Cucumis melo*	Ethiopia/East Africa	exploited from wild from early date
sunflower seed	*Helianthus annus*	North America	post AD 1500
pumpkin seed	*Curcurbita maxima*	Central/South America Ethiopia	post AD 1500
Ethiopian cabbage	*Brassica juncea* (L.) Czern.+Cass.	Ethiopia	exploited from wild from early date
Other			
cotton	*Gossypium herbaceum*	North, East and South Africa	found in 7th millennium contexts in India and 5th millennium contexts in the Near East and central Nile Valley. Earliest date in Ethiopia is c.2000 BP.

Crop	Botanical name	Country of origin	Earliest recorded date for use of the domesticate in Ethiopia (BP)
Roots and tubers			
ensete	*Ensete edulis*	Ethiopia	depictions on Gerzean pottery suggest use in Egpyt c.5000 post AD 1500
potato	*Solanum tuberosum* (L.)	South America	pre 19th century
kaffir potato	*Coleus edulis*	Ethiopia/Arabia	post AD 1300
onion	*Allium cepa*	West Asia	1st millennium AD?
garlic	*Allium satium*	Asia	
Vegetables			
Ethiopian cabbage	*Brassica juncea* (L.) Czern.+Cass.	Ethiopia	?
okra	*Hibiscus esculentus (L.)*	Ethiopia/Tropical East Africa	evidence from Ancient Egypt still being debated
Ethiopian kale	*Amaranthus caudatus*	Tropical Africa	*Amaranthus* sp in 5th millennium contexts in Ethiopian Highlands
tomato	*Lycopersicum esculentum* Mill.	South America	post AD 1500
Spices, drugs and flavourings			
coriander	*Coriander sativum* (L.)	Ethiopia?	cultivated in Egypt by 3rd millennium
fennel	*Nigella sativa*	Ethiopia	?
buckthorn	*Rhamun prinoides* Her.	Ethiopia	?
fenugreek	*Trigonella foenum-graecum* (L.)	South west Asia/Mediterranean/ Ethiopia?	?
tumeric	*Curcuma domestica* Val.	South East Asia	post AD 1100?
ginger	*Zingiber officinale*	South East Asia	post AD 1100?
basil	*Ocinum basilicum* (L.)	Southern Asia, also wild *Ocinum* sp in Ethiopia used as herbs	post AD 1500
caraway	*Carum copticum*	South East Asia	post AD 1500
chilli peppers	*Capsicum annum* (L.)	CentralAmerica	post AD 1500
Fruit			
fig	*Ficus carica*	Ethiopia	not cultivated
blackberry	*Rubus spp*	Ethiopia	not cultivated
prickly pear	*Ficus opuntia indica*	Ethiopia	not cutivated
rosehip	*Rosa abyssinica*	Ethiopia	not culitvated
tamarind	*Tamarindus indica* (L.)	Ethiopia/East Africa	not cultivated in Ethiopia but cultivated in India in prehistory
melon	*Cucumis melo*	Ethiopia	?
'gaba'	*Zizyphus spinachristi*	Ethiopia/N.Africa	not cultivated, but exploited since pre-2500 in Ethiopia and c.7000 in Central Sudan

Wild Mammalian Resources of Ethiopia

Ethiopia has a rich diversity of wild fauna including over 103 mamalian species of which 7 are endemic, and 832 bird species. Until earlier this century many large mammals roamed the slopes and plateaux of the Highlands of Ethiopia. Widespread ecological destruction, the introduction of firearms and the recent warfare has obliterated many indigenous species or driven them to neighbouring countries. Large game animals such as elephant, rhinoceros, giraffe and gazelle inhabited the *kwolla* and lower *woina dega* lowlands to c.2000 m, and the lion, symbol of Ethiopia, proliferated in the Sudanese border region. Agile herbivores such as red duiker, greater kudu, klipspringer and mountain reedbuck browsed on lush vegetation along with bush pigs, monkeys, baboon and rabbit. Hippos, now possibly extinct, were endemic to Lake Tana as recently as the 1960s (Simoons 1960). Mammals such as walia ibex and mountain nyala inhabited the high mountain zone (Mloszewski and Mahaney 1989).

The Highlands are especially rich in bird life. Guinea hens are common, also partridge, quail, pigeon, parrots, ibis, eagles, kingfishers and storks. Historical accounts (Bruce 1790, vol 3) document the availability of game birds and the alacriaty with which the Highlanders embarked on hunting expeditions. Bees also represent an important commodity to modern Highland economy and honey is used for sweetening, flavoring and for brewing Ethiopian mead (*T'ej*).

Table A.3.1. Origins of Domesticated Faunal Resources in North and East Africa and their Relationship to Ethiopia

Animal Species	Proposed Date of Introduction to Ethiopia	Background and Evidence
Donkey *Equus asinus*	?4th-3rd millennium BP	Domesticated from the African wild ass *Equus africanus* which was widespread across the Sahara and the Horn in prehistory (Clutton-Brock 1987). Very important for transport and trade, especially in mountainous regions such as the Ethiopian Highlands. Known in Egypt from the 6th millennium BP, from Shaqadud in central Sudan from the 5th millennium BP and from Mahal Teglinos in the Gash Delta from the 4th millennium BP (Krzyzaniak 1978, Fattovich 1990, Blench 1993). No evidence from Ethiopia until the historical period, but depictions of domesticated donkeys at Deir el-Bahari indicate their presence in Punt. It is possible that donkeys were known in Ethiopia by the mid-4th millennium BP. Ethiopia is unusual in sub-Saharan Africa for using domesticated donkeys and oxen for traction from an early date (i.e. at least the start of the 2nd millennium BP). Recent linguistic studies suggest that the donkey preceded the introduction of Ethio-Semitic into Ethiopia, and may therefore date as early as the 4th or 5th millennium BP (Blench 1995; see chapter 2)
Horse *Equus caballus*	?late 3rd millennium BP	Not indigenous to Africa. Known in Egypt and the Nile Valley from the mid-4th millennium BP onwards, and may have reached Ethiopia from Nubia. Became of widespread importance in trade, transport and warfare, particularly in association with emerging urbanism. With >1.5 million horses, Ethiopia now has the highest equine population in Africa (Smith 1992).
Camel *Camelus dromedarius*	?5th millennium BP	Origins in Africa and Ethiopia unknown. Proposed domestication in Arabia, where dromedary remains have been identified at the Neolithic site of Sihi on the Arabian coast c.8200±200 BP (Grigson et al 1989). Possibly introduced to North Africa via the Sinai, and to the Horn via the Strait of Bab-el-Mandeb during the 3rd millennium BP or earlier (Kholer-Rollefson 1990, Wilson 1984). Camel hair rope from the Fayum in Egypt dates to c.4600 BP (Caton-Thompson 1934) and camel dung from Qasr Ibrim has been dated to c.2690 BP (Rowley-Conwy 1988). A domesticated camel tooth from Gobedra in the Ethiopian Highlands has been tentatively dated to the 5th millennium BP, suggesting a possible early use in Ethiopia (Phillipson 1977b). Camels are not adapted to the present Highland climate and are virtually unknown in the area today, but may have been more widespread in the past during more arid climatic intervals.

		Camels are extremely important for transport in arid regions and formed the backbone of historic - and possibly prehistoric - trade networks in Africa and Arabia. Camels are mentioned in Axumite inscriptions from the early-mid 2nd millennium BP and are represented in Ethiopian and Eritrean rock art dating possibly to the 2nd or 3rd millennia.
Pig *Sus domesticus*	?2nd/3rd millennium BP	Wild pigs (*Sus scrofa*) are indigenous to Ethiopia, North Africa and the Nile Valley (Clutton-Brock 1987), although domesticated pigs were probably introduced from the Near East via North Africa between the 4th and 2nd millennia. Pigs are of little dietary importance in Ethiopia and there is even a mild taboo against eating pigs in many regions, which may have derived from Egypt or from the Near East.
Sheep *Ovis aries*	?4th millennium BP	Derived from the South West Asian mouflon (*Ovis orientalis*), although it is possible that wild Barbary sheep may have been domesticated in North Africa in the 7th millennium (eg. at Haua Fteah in North Cyrenica) and taken to the Nile Valley by the 6th millennium BP (Higgs 1967, Shaw 1977, Klein and Scott 1986). Barbary sheep also comprised 74% of the identifiable faunal remains at Ti-n-Torha in the Northern Sahara, suggesting increased resource selectivity by c.8600 BP (Barich 1992). Sheep/goat remains have been recovered in Central Sudan from the start of the 6th millennium BP and from Lake Turkana from the late 5th/early 6th millennium BP (Arkell 1953, Krzyzaniak 1978, Barthelme 1985). Ovicaprid remains have been tentatively identified in late 3rd millennium BP contexts at Lalibela cave in the Northern Highlands (Dombrowski 1971). Rock art of fat-tailed sheep and goats in Ethiopia and Eritrea have been dated comparatively to the 4th millennium. The date and route of introduction to Ethiopia are unknown, although Clark (1976, 1980) has suggested that pastoralists moved from the north west into the fringes of the Ethiopian plateau during the 4th millennium BP.
Goat *Capra hircus*	?4th millennium BP	Derived from the scimitar-horned goat (*Capra aegagrus*) of South West Asia (Clutton-Brock 1987). A species of wild goat (*Capra walia ibex*) is endemic to the Northern Ethiopian Highlands, but has never been domesticated. This is now an endangered species. Remains of goats from Neolithic Capsian levels in Algeria date to at least 6500 BP, and domesticated goats are known from the Central Nile Valley by the 6th millennium BP (Roubet 1978, Peters 1985/6). Dwarf goats are known in West Africa by the 4th millennium BP (Carter and Flight 1972). Very little evidence for goats exists in Ethiopia or East Africa prior to the 2nd millennium BP, although rock art depictions may date substantially earlier than the 4th millennium BP. Possible caprid remains have been identified in late 3rd millennium levels at Lalibela cave.
Cattle *Bos* sp	?5th/6th millennium BP	Morphological and genetic diversity of cattle in Africa has led to much debate over the origins of the domesticate. African cattle have at least two independent ancestors in humpless European *Bos taurus* and humped Asian *Bos indicus* (Clutton-Brock 1987). More recent genetic evidence suggests an indigenous African breed may have been domesticated separately. Recurrent droughts and cattle epidemics have caused massive loss of cattle in Ethiopia, and have extinguished the original gene pool (for example 90% of all cattle died between 1888 and 1890. Pankhurst 1965). However a relict population of humpless bovids surviving in Southern Ethiopia may represent an ancient, prehistoric breed (Epstein 1971). Although Ethiopia currently has the highest cattle population of any country in Africa, with over 26 million animals, these are derived from zebu or zebu-hybrids.

The earliest, unequivocal evidence of domesticated cattle in Africa dates from the early 6th millennium BP in the Sahara (Gautier 1982, Carter and Clark 1976). Domesticated cattle were widespread across the Sahara and Nile Valley during the 6th and 5th millennia, as documented by archaeological and pictorial evidence. It is unclear at present whether these cattle were domesticated indigenously during the early Holocene, as suggested by the evidence from Bir Kiseiba and Nabta Playa (Wendorf et al 1984), or whether they were introduced from the Near East during the 7th millennium BP. The possibility that domesticated cattle were introduced to the Near East from North Africa cannot be entirely ignored.

The early North African cattle have been identified as short-horned breeds, but the horn size is age- and gender-dependent. Short-horned cattle are occasionally represented in Saharan rock art, but do not appear in Egyptian art until the 4th millennium BP. These may be a later introduction to North Africa. However short-horned cattle are an ancient breed in the Sahel and in West Africa and areas of Ethiopia, and their resistance to tsetse fly suggests that they are of considerable antiquity in these regions and may even have originated there. Depictions of the Land of Punt from Deir el-Bahari show that both long-horned and short-horned humpless cattle were known in that region by the mid-4th millennium BP (Naville 1898).

Humped zebu cattle are not unequivocally represented in Africa before the 3rd millennium BP, although undated rock art depictions may be older. Zebu are better adapted to more arid conditions. Their introduction may have been a response to climatic deterioration in the late Holocene, and may have facilitated the spread of pastoralists through East Africa during the 3rd and 2nd millennia BP (Marshall 1989).

Domesticated bovids have been tentatively identified in Ethiopia from fragmented remains at Laga Oda and Lake Besaka in the Southern Highlands and Afar Rift Valley respectively, dating to the mid-4th millennium BP (Clark and Williams 1978, Brandt 1982). Rock art depictions of cattle in Ethiopia and the Horn have been tentatively dated to the 4th millennium, but may be considerably earlier (see Chapter 3). Domesticated cattle are attested at Lake Turkana on the Kenyan-Ethiopian border from the mid-5th millennium BP (Barthelme 1977). As pastoral groups are thought to have spread from Ethiopia to East Africa, this implies that domesticated cattle must have been known in Ethiopia before the mid-5th millennium BP (Ehret 1974, 1979).

APPENDIX B. Pollen Diagrams from Ethiopia

Figures B.i - B.viii show pollen diagrams from Lake Turkana and from seven Ethiopian sample sites:
Mount Badda
Lake Abijata
Danka Valley
Dega Sala
Lake Langano
Wenchi
Lake Awassa
The distribution of these sites is given in figure 4.4, chapter 4.

General vegetation spectrum:
savannah woodland at lower altitudes to c.1400 m asl
evergreen thicket c.1500 - 2000 m asl
dry montane forest (*Podocarpus* dominated) c.2000 - 2600 m asl
Juniperus/Hagenia dominated montane forest c.2700 - 3300 m asl
Ericaceous and afro-alpine scrub above c.3300 m asl.

Predominant pollen types:
Podocarpus, Olea, Juniperus, Dodonea and *Myrica* are characteristic of dry montane forest at higher altitudes.
Acacia and Gramineae indicate dry savannah woodland at lower altitudes.
Increased *Alchemilla, Hypericum Hagenia, Macaranga, Rapanea,* Urticaceae and Ericaceae indicate wetter conditions at higher altitudes, particularly an increased moist component of montane forest.

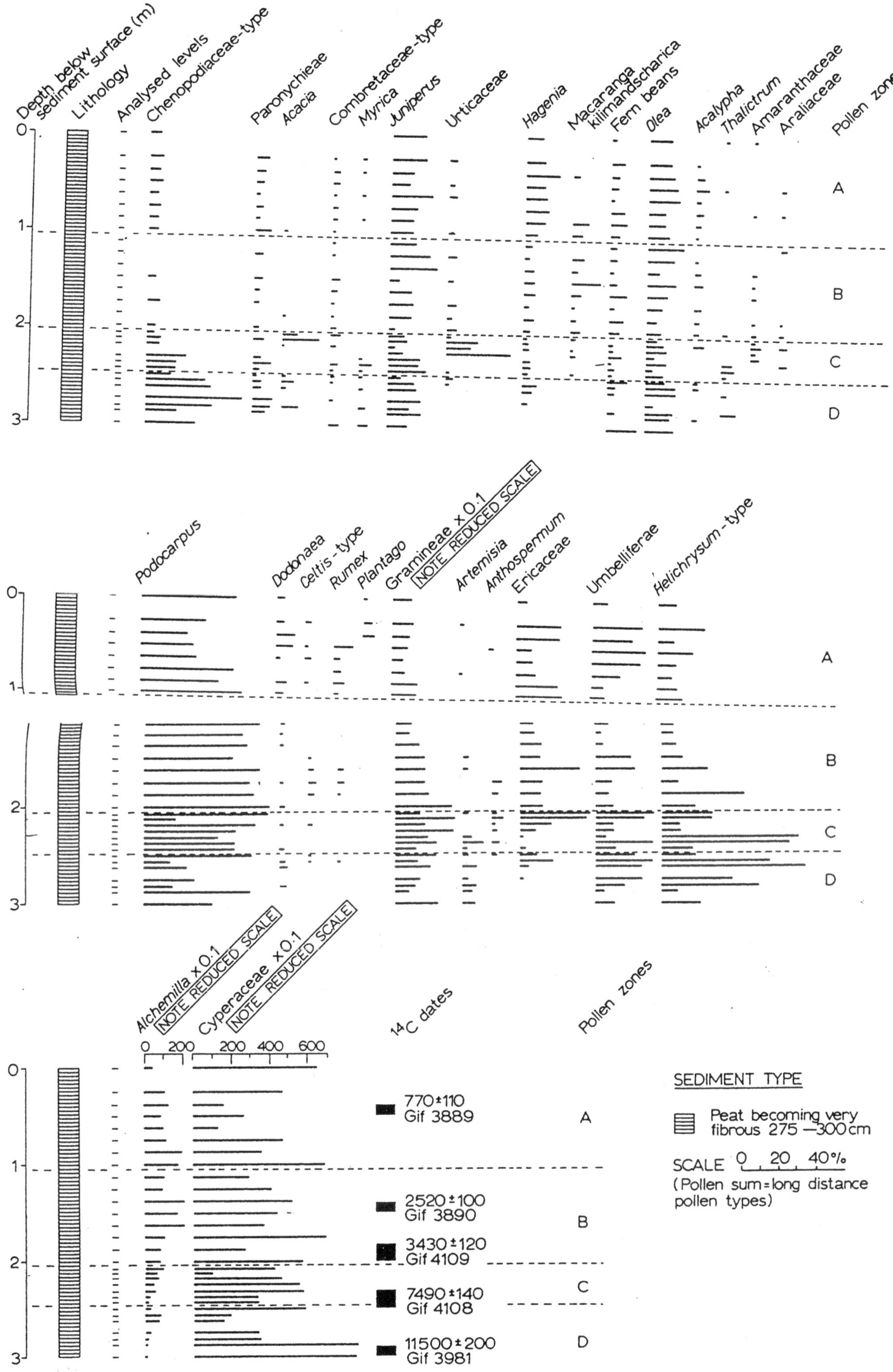

Figure B.i. Mount Badda pollen diagram (after Hamilton 1982).

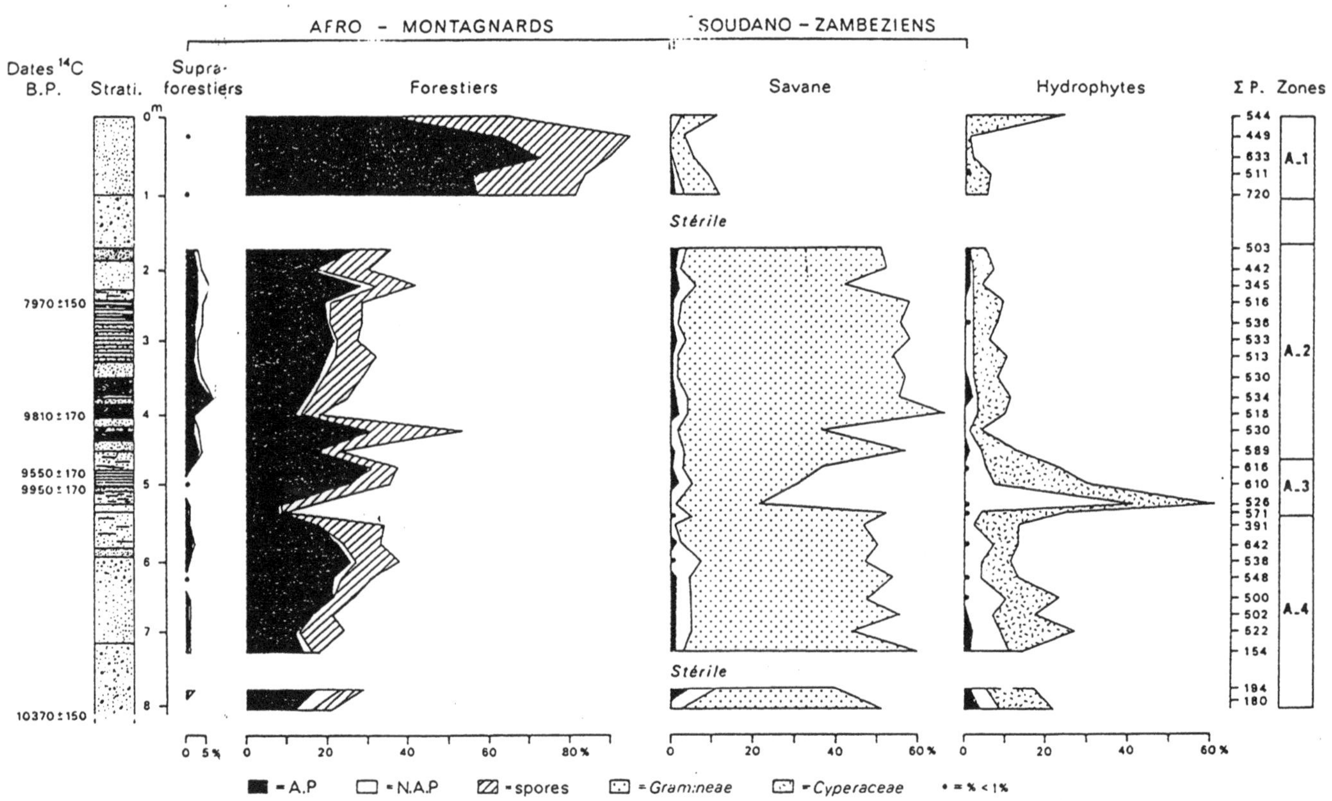

Figure B.ii. Lake Abiyata pollen diagram (after Lezine and Bonnefille 1982).

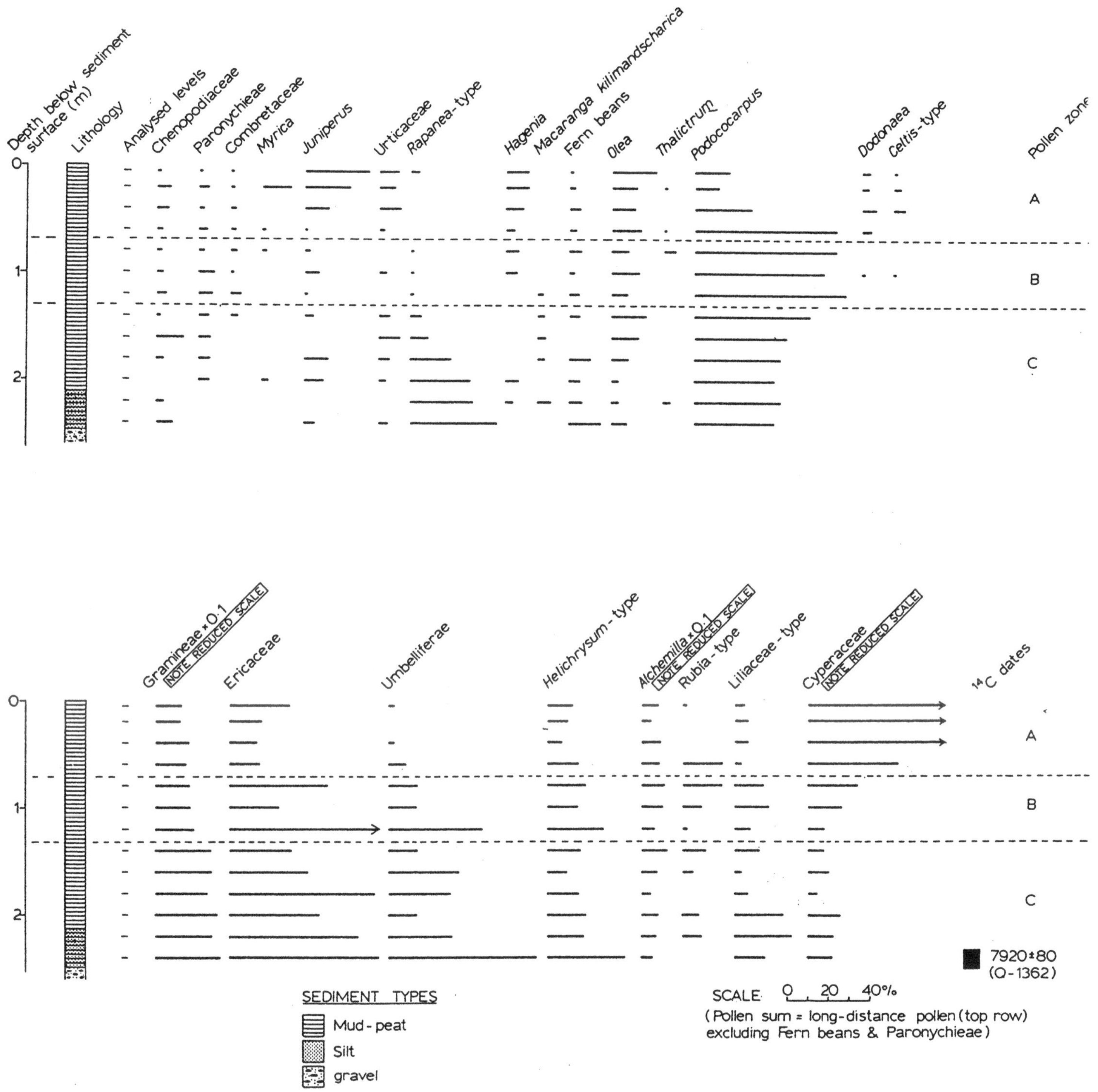

Figure B.iii. Danka Valley pollen diagram (after Hamilton 1982).

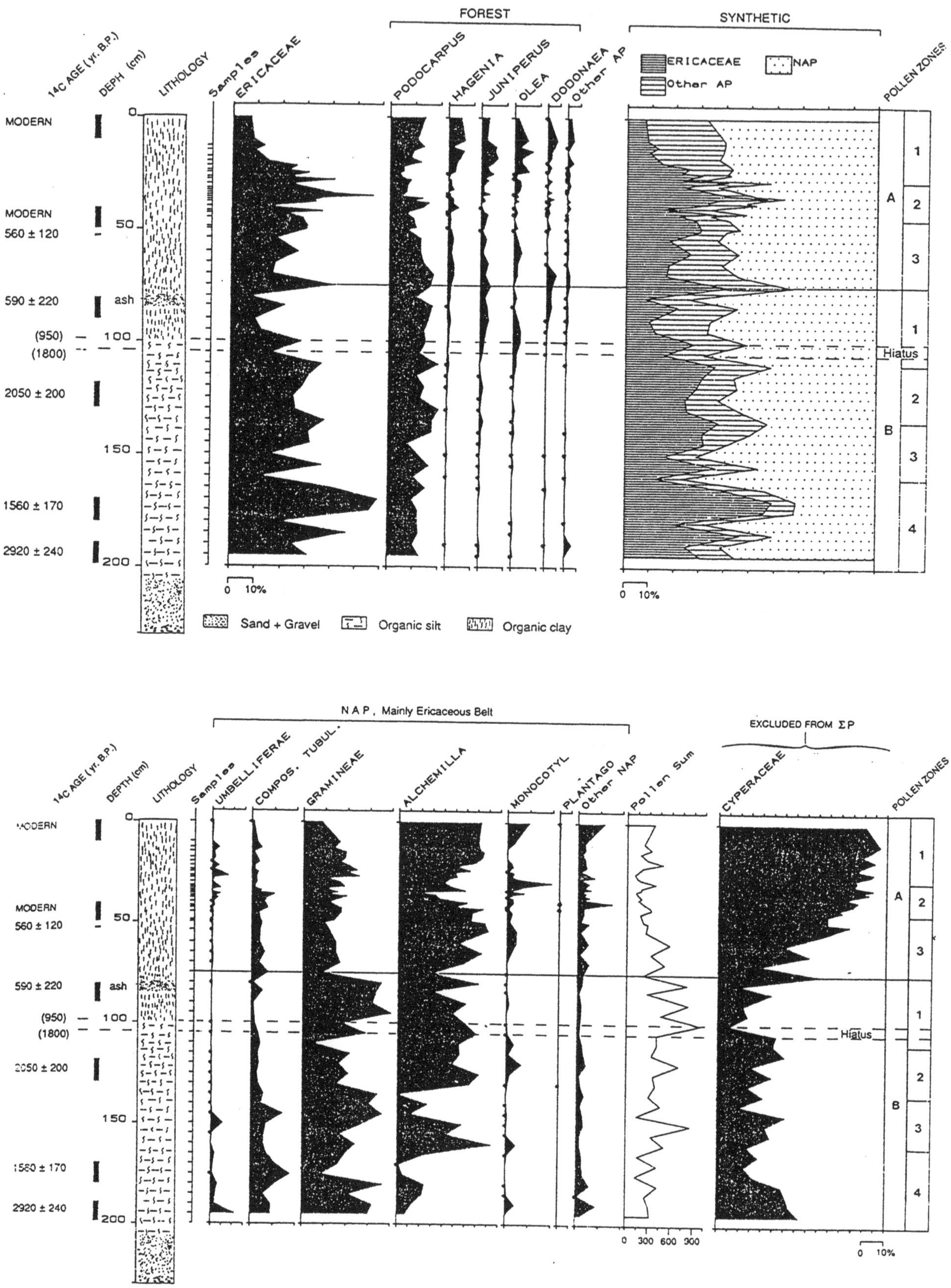

Figure B.iv. Dega Sala pollen diagram (after Mohammed and Bonnefille 1994).

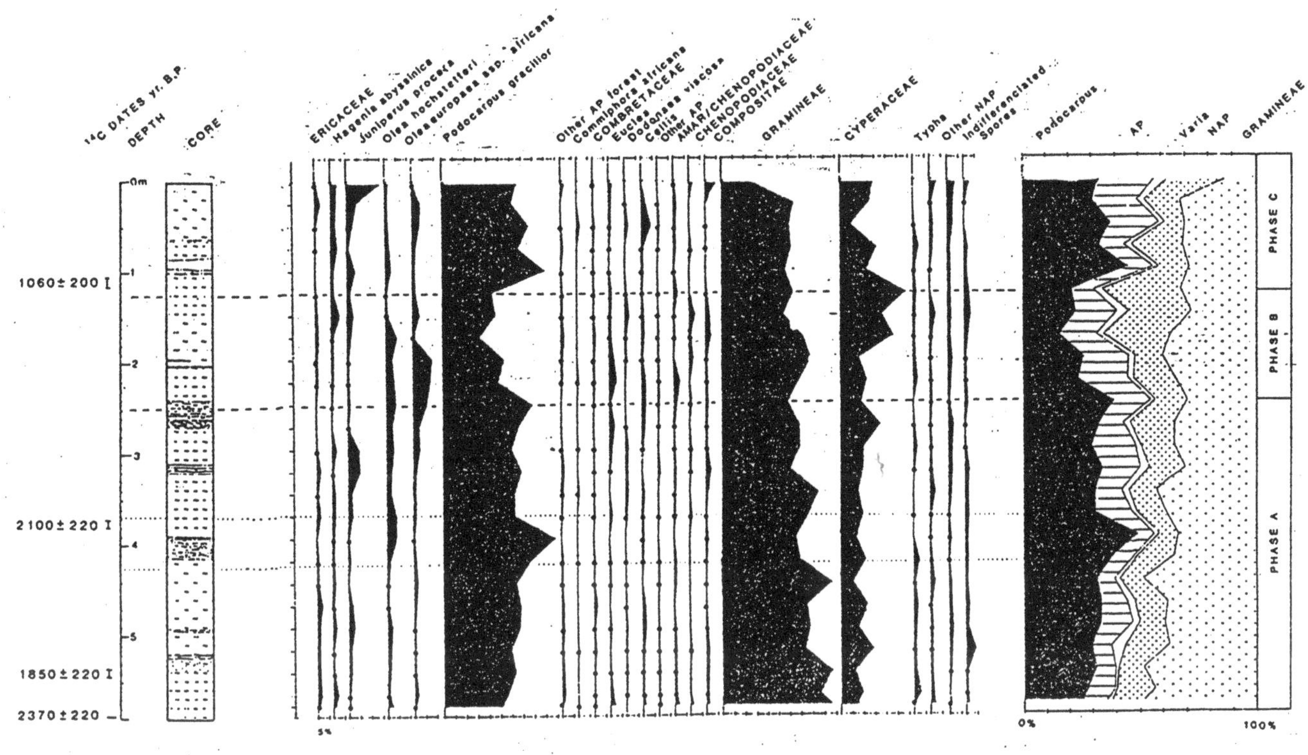

Figure B.v. Lake Langano pollen diagram (after Mohammed and Bonnefille 1991).

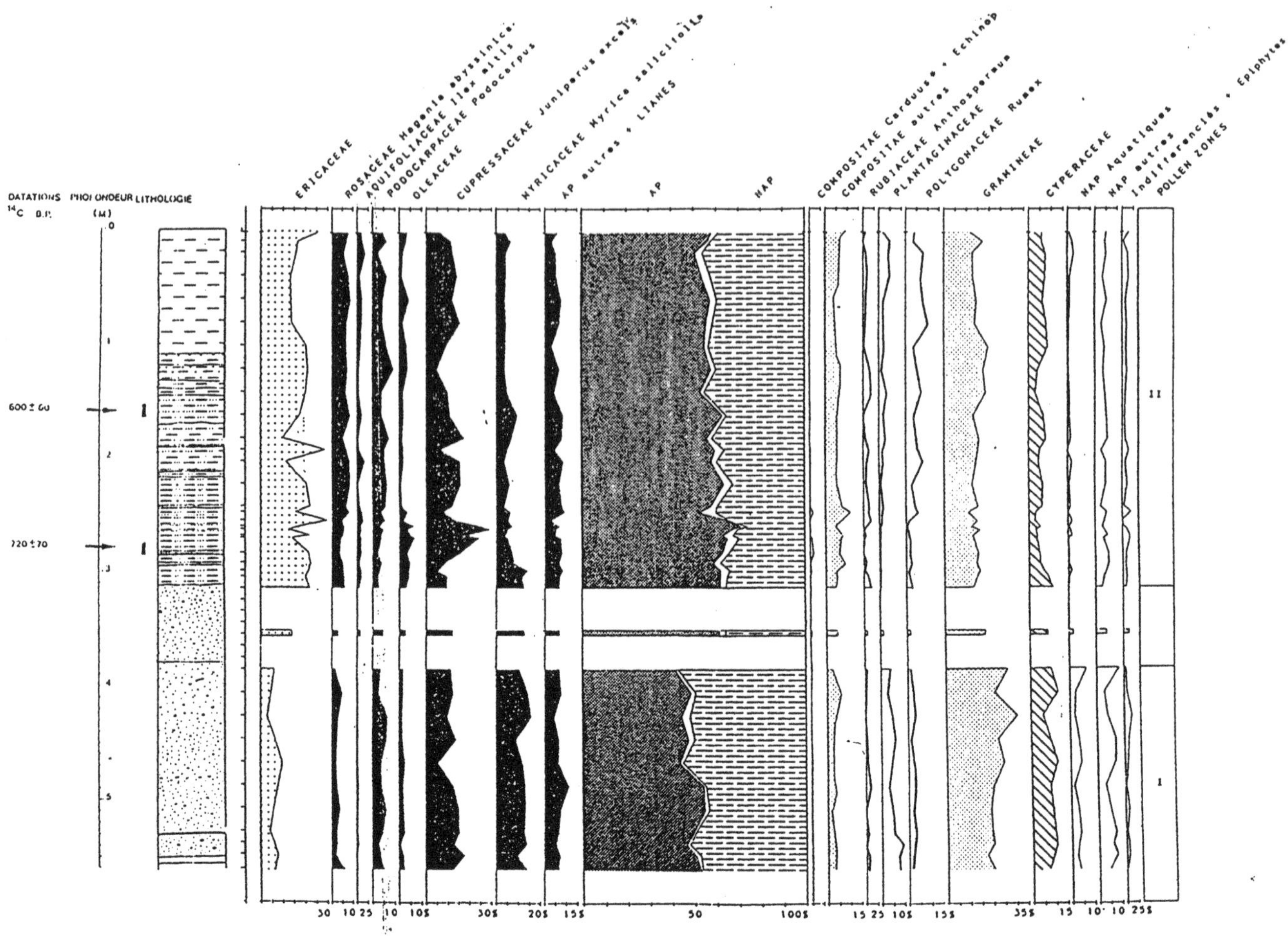

Figure B.vi. Wenchi pollen diagram (after Bonnefille and Buchet 1986).

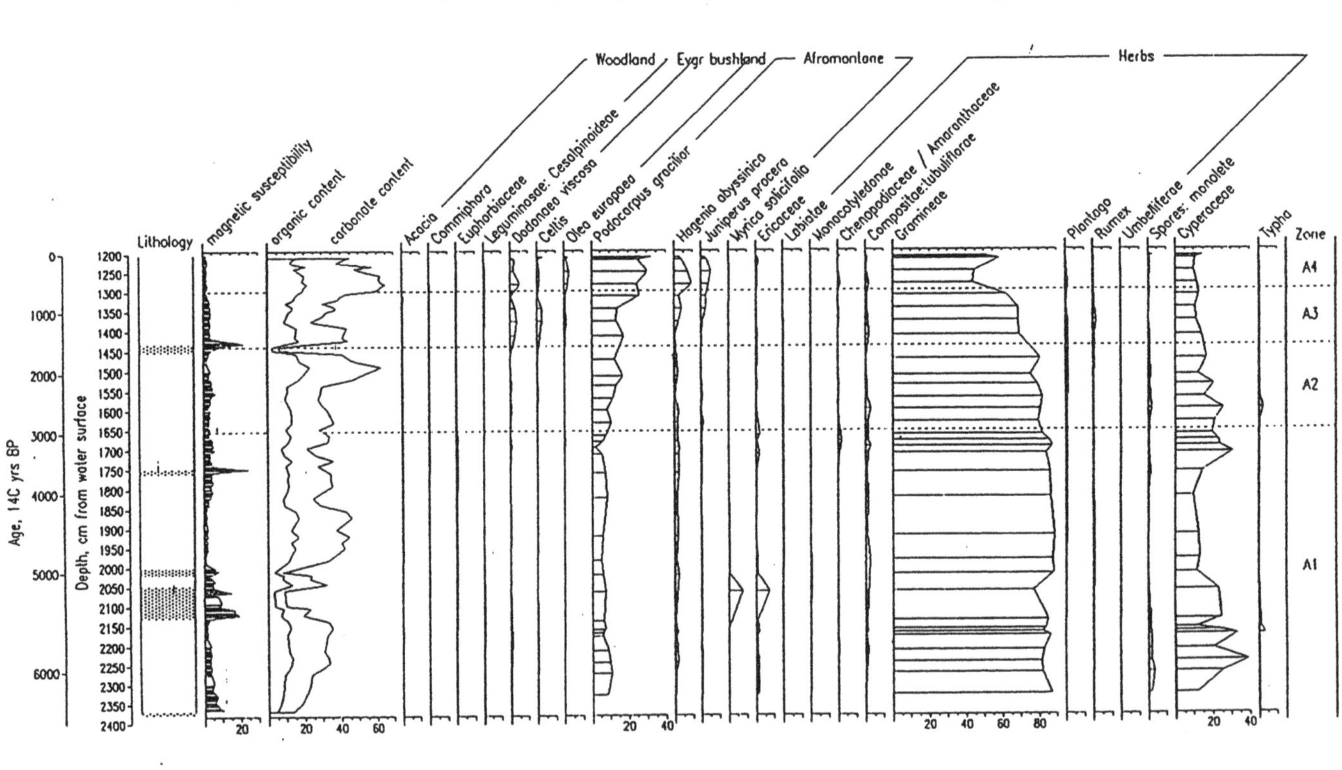

Figure B.vii. Lake Awassa pollen diagran (after Lamb *pers.comm.*).

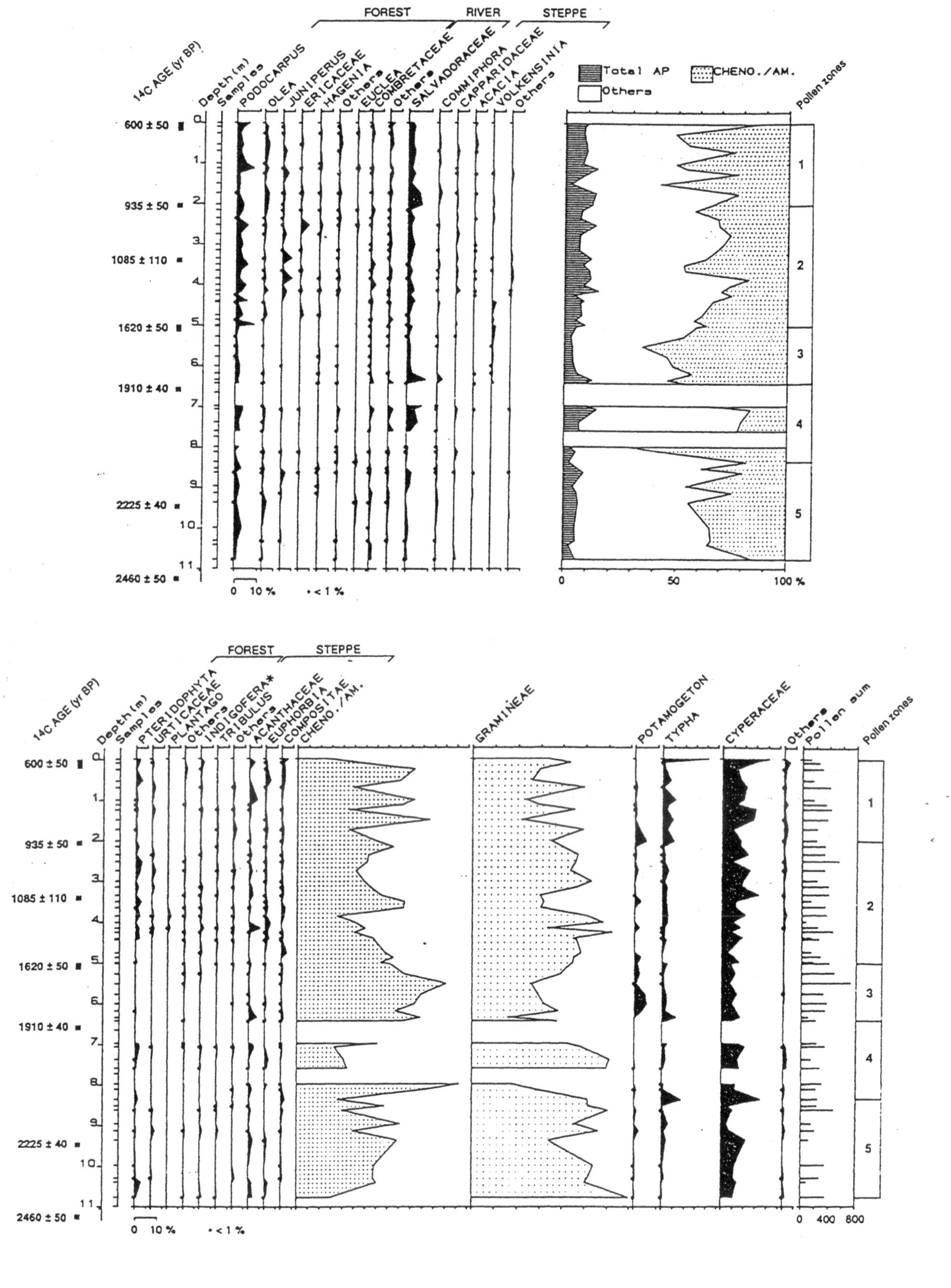

Figure B.viii. Lake Turkana pollen diagram (after Mohammed et al 1995).

APPENDIX C: Archaeological Evidence from LSA Ethiopia

Table C.1.

Summary of Archaeological Evidence and Geographical Locations of Prehistoric Sites in Ethiopia

Table C.2.

Summary of Cultural Remains from Archaeological Sites in Ethiopia

Table C.1. Summary of Archaeological Evidence and Geographical Features of Prehistoric Sites in Ethiopia

Site	Location	Nature of Site	Date of Use	Type of Site	Altitude (m asl)	Total Excavated Area	Environmental zone
Laga Oda (Clark and Williams 1978)	SE escarpment 25 km S of Dire Dawa.	Double rock shelter in limestone cliff overlooking stream.	c.15 000 BP to recent (5 radiocarbon dates) ?hiatus c.8000 BP.	Limited/specialised activity site during LSA rather than habitation site. Long-term, intermittent use.	2000-2300	3 m^2 excavated in lower shelter to a depth of 80 cm. Below this only 1 m^2 dug to a depth of 140 cm.	Ecotone between evergreen scrub woodlai and montane woodland savannah.
Porc Epic (Clark and Williamson 1984)	SE Rift Valley at the foot of the escarpment, 2 km S of Dire Dawa.	Cave in limestone outcrop, 140 m above wadi floor, difficult access.	Extensive MSA occupation overlying breccia, sealed by stalagmite. Date of 5700±110 BP from charcoal in sealed deposit may be valid, although rejected by excavators. Date of c.75 000 BP, obtained from obsidian hydration dating, on artefacts from earlier excavation, should be treated with extreme caution. MSA/LSA ceramic-bearing deposit near entrance thought to derive from stratigraphical disturbance. Cave also contains a recent shallow LSA occupation deposit.	MSA/LSA hunting/pastoral camp, seaonally occupied.	c.1800	6m^2	Ecotone between open, xerophytic vegetat of the Rift Valley plains and montane and evergreen vegetation of the Southern Plateau. The wadi is a major access route the Southern Highlands.
Aladi Springs (Clark and Williams 1978)	Near base of ecsarpment 120 km W of Dire Dawa.	Deposits sealed by spring sediment.	At least 2 occupation phases - MSA-LSA ?pre 17000 BP. Hiatus 17-12 000 BP. Second activity phase 11 070±160 BP dated from molluscs in contemporary tufa deposits.	MSA/LSA ?special purpose site used during humid intervals. Hiatus coincides with hyper-arid phase.	c.1600	-	Semi-arid woodland savannah, with limit aquatic-type flora aroud the spring itself.
Lake Besaka (Brandt 1980, 1982)	Rift Valley lake near N escarpment edge.	4 open air sites on or close to scarp of lake edge.	3 stratified occupation phases: *Phase A* c.22-19 000 BP (3 dates on ostrich egg-shell). *Phase B* c.11-5 000 BP comprises aceramic and ceramic levels dating c.11-8/7 000 and ?7/6 000 BP respectively (dates related to comparable geological deposits). *Phase C* 3460±280 BP (^{14}C date on charcoal from hearth).	LSA -?seasonal settlement sites.	c.950	(FeJx1 and 2) >50 m^2 (FeJx3) 4 m^2 (FeJx4) 6 m^2	Semi-arid/arid wooded savannah. Lake Besaka is currently saline, but would havi been less alkaline during lacustrine phase

K'one (Clark and Williams 1978, Kurashina 1978)	Southern Afar Rift 150 km E of Addis Ababa. 20-25 km E of Lake Besaka.	Crater complex with several ephemeral open air sites on the caldera floor and nearby slopes.	1 date on ostrich egg shell near upper levels at single site c.14 500 BP for MSA/LSA 'transitional' industries. MSA levels may extend back to 70 000 BP. Discontinuous use relating to environmental context.	MSA and LSA lithic workshops or dumps. LSA use known principally from undated surface deposits and a 1 m^2 test pit.	c.1200	Total of $27m^2$	Semi-arid wooded savannah.
Lake Ziway Area (Gasse and Street 1978, Humphries 1978, Gallagher 1973)	Central Ethiopian Rift 130 km S of Addis Ababa.	Deposits around freshwater lake in extensive lake basin.	*Bulbula River*: limited occupation horizons c.27 500 BP and c.11 870 BP dated on charcoal from geological section. *Macho and Waso:* 3 occupation sites excavated. Single date of c.10 330 on charcoal. 1 surface deposit investigated. 230 BP date on charcoal from associated hearth. *Gademotta*: MSA material and surface LSA deposits.	Early LSA and LSA occupation sites on lake shore. Continuous activity from Early Holocene LSA to recent. Occupation sites on lake shore.	c.1636	- Total of 32 m^2 at Waso. Total of 25 m^2 at Macho.	Lacustrine - surrounded by extensive belt dense aquatic vegetation, beyond which is the wooded savannah of the Rift floor.
Gobedra (Phillipson 1977b)	Northern Ethiopian Highlands 5 km from Axum.	Rock shelter in rocky ridge.	c.12 000 BP - post c.800 AD.	LSA special activity/occupation site.	c.2250	4 m^2	Highland - evergreen thicket and montane forest/savannah mosaic.
Gorgora Rock-shelter (Moysey 1943, Leakey 1943, *personal observations 1994*)	Northern shore of Lake Tana, Northern Highlands.	Rock shelter 50 m up a small hill 5 km from present lake shore.	MSA and LSA levels undated. Pleistocene and Holocene occupation.	MSA and LSA intermittent occupation site. Fairly good typological continuity.	c.2000	c.7.5 m^2	Lacustrine - changing lake levels mean th the site was alternately an island, a shore site, or surrounded by evergreen forest and savannah woodland.
Gorgora Peninsula (Moysey 1943, Dombrowski 1971, *personal observations 1994*)	Northern shore of Lake Tana.	Small, rocky peninsula.	LSA undated surface material.	-	c.2000	-	Lacustrine - evergreen forest and savanna woodland.
Quiha (Clark 1954, *personal observations 1994*)	Northern Ethiopian Highlands, close to Makelle.	Rock shelter.	LSA levels, undated. Intermittent occupation over extended period ?possibly c.6th-3rd/2nd millennium BP. Upper levels probably PreAxumite (3rd millennium BP) to recent.	-	c.2200	*	Ecotone between evergreen forest and montane woodland/savannah mosaic.
Zergnat (Clark 1954)	Northern Ethiopian Highlands, close to Makelle.	Surface scatter - open-air site?	LSA undated.	-	c.2200	-	Ecotone between evergreen forest and montane woodland/savannah mosiac.

Amba Sel (Clark 1954)	Northern Ethiopian Plateau, close to Makelle.	Surface scatter - open air site?	LSA undated.	-	?	-	*
Lalibela and Natchabiet (Dombrowski 1971)	Northern Highlands, on escarpment 50 km east of Lake Tana.	Caves on escarpment overlooking lacustrine plain.	LSA occupation from at least 2520 ±80 BP.	LSA occupation spanning c.2000 years.	c.2300	38.5 m^2 at Natchabiet 20 m^2 at Lalibela.	Ecotone between evergreen forest and montane woodland/savannah mosaic, with access to the wooded savannah of the lacustrine plain. Rolling hills at top of escarpment are now under cultivation.
Melka Konture (Chavaillon 1976)	50 km S Addis Ababa on southern edge of Northern Highlands.	Extensive surface scatters of lithic material.	ESA, MSA and LSA activity. ESA and MSA excavated but LSA only known from undated aceramic surface collections at 3 sites (Kella, Wofi and Balchit).	Dynamic continuity over the past 1.8 million years. Probable obsidian quarrying and working at at least one of the sites (Balchit).	c.2200	-	Ecotone between evergreen forest and montane woodland-savannah.
Agordat (Arkell 1954)	Western Eritrea.	4 surface scatters of prolific ceramic and lithic material with limited copper artefacts.	LSA ?4th millennium BP date suggested on basis of comparison with Nile Valley material.	?Occupation sites.	c.1400	-	Semi-arid wooded savannah/evergreen forest.

* Not known

Table C.2. Summary of Cultural Remains from Archaeological Sites in Ethiopia

Site Name	Lithics	Inorganic resources	Ceramics	Faunal Remains	Floral Remains	Art	Burials	Imports/Other
Laga Oda	Esp. backed bladelets. A few crescents and lunates. Edge damage and gloss present on all types throughout sequence. EBTT.	Mostly local chert. Some imported obsidian.	Only in upper levels, late 1st millennium BP.	Cattle bones c.80-70 cm depth (c.4000-3500 BP). Camel bones in upper levels only. No caprid remains found, but faunal sample very small.	-	Paintings of cattle, fat-tailed sheep and humans.	-	Obsidian.
Porc Epic	MSA esp. points and knives [inc. retouched points,scrapers, burins and backed blades]. LSA esp. microliths, small scrapers and *outils ecailles*. EBTT.	Local basalt, shale, chert. Obsidian? imported.	Ceramics from LSA deposit, mixed with MSA lithics near cave entrance. No other ceramics known from this cave.	Limited: mostly bovids butchered away from the site, also bushbuck, duiker, pig and zebra in MSA levels. Similar game available to LSA occupants.	LSA levels inc *Zizyphus* sp, cowpea (*Vigna unguiculata*) and *Pisum sativum*.	Poorly preserved paintings of wild animals and human figures.	Human mandible fragment.	LSA and MSA hearths. MSA hearths used for the preparation of pigments from various rocks, for cooking and for heat treating chert and pigments. MSA deposit contained pieces of rubbed haematite. Grindstone fragments from MSA levels showed no pigment stains.
Aladi Springs	MSA points and scrapers. LSA microblades and heavy duty scrapers. Comparable to Besaka Phase B. EBTT.	Obsidian with some chert and chalcedony.	-	-	-	-	-	Pierced ostrich egg shell.
K'one	Esp. end scrapers, burins, awls, points, but shaped tools represent only 0.5% of total artefacts. Large quantities of flakes and cores with conjoining flakes. EBTT, and Elmeneitan affinities in long blades and backed bladelets.	Pitchstone and obsidian.	-	-	-	-	-	-

Lake Besaka	*Phase A:* MSA/LSA transitional - macro and micro blades, end scrapers and burins. EBTT *Phase B*: Typical blade and burin technology of the EBTT. Large and microlithic backed blades, end-scrapers, burins and awls. Upper and lower grindstone fragments. *Phase C*: convex scrapers, few blades or burins. Increased frequency of grindstone fragments. Shaped lava fragment from stone bowl?	Good quality green-banded obsidian, also some silicates and basalt, all from 30 km radius of site, inlcuding Fentale volcano and K'one crater complex.	*Phase A:* - *Phase B*: aceramic (METAHARA) and ceramic (ABADIR) levels. Small number of decorated and undecorated sherds from ceramic (Abdair) phase. *Phase C*: large quantities of decorated and undecorated pottery. Reconstructible fragments of 1 large, conical bottomed vessel with exterior burnish.	*Phase A:* burnt bone inc. small-large bovids, hippo, reptiles, birds, warthog, fish (inc. catfish). *Phase B*: medium sized bovids, fish. *Phase C*: increased concentration and range of faunal remains - oryx, hartebeest, warthog, wild boar, zebra, lizards, birds. Domesticated cattle tentatively identified from fragments of 3 bovid teeth.	-	-	*Phase B*: ? truncated, partially burnt burials of 5 individuals, associated with cairns and burial items.	*Phase B*: Red Sea shells, pierced ostrich egg shell and carnelian beads. Grindstones in Abdair Phase. *Phase C:* stone bowls from animal burrow - possible connection with pastoral 'Stone Bowl Culture' of East Africa?
Gobedra	Esp. scrapers, blades, backed flakes and geometrics, points and fragments. EBTT.	Esp jasper, quartz, chalcedony, chert and opal.	3 types represented: Thick, fine ware and decorated or plain thin, coarse ware (local developments?). 1 sherd of thin, fine ware, probably imported. Decoration includes rows of comb stamping or vertical incisions. Fragment of handle also present.	Fragmented bone found from 10 000 BP onwards, including bovids, alcelaphine, tree hyrax, ?domesticated camel from c.5000 BP and possibly domesticated bovid from 1st millennium.	*Amaranthus* sp in 7th millennium BP level and above. *Eleusine coracana* in 1st millennium level.	-	Fragmented remains of human skull in 7th-1st millennium level.	2 soapstone artefacts and several grindstones in 7th millennium BP levels. Iron working evidence from 1st millennium BP.

Gorgora Rock-shelter	Esp. points in the MSA level with some scrapers and backed blades. 'Transitional' industry in Level 3 with MSA points and LSA microliths. Increased tool variety in LSA (Levels 2 and 1) with esp. microliths, scrapers, backed blades and crescents and hammerstones. Points are absent. Tools are crude and belong to EBTT.	Local materials including chert, quartz, chalcedony and jasperite.	Total of 36 potsherds from upper 1.4 m of deposit (early Holocene?). 3 main types: crude, coarse ware (plain and decorated) in lower levels; smoothed, coarse ware (plain and decorated) in all levels; thin, fine ware (plain) in upper level only. Vessels of local manufacture (except fine ware?). Decoration with simple linear incisions or finger-nail impressions.	-	-	-	-	Blue-grey and haematite pigments in MSA and LSA deposits. Grindstones appear from Level 3 ('transitional' level). 1 ovoid grindstone in Level 3 stained with blue-grey pigment, others abraided by grinding plant material. Part of ground stone object in Level 3 comparable to soapstone rods from Gobedra?
Gorgora Peninsula	Peninsula is rich in archaeological material and warrents systematic survey. Discrete scatters of LSA material in eroded open area sites, small caves and rock-shelters. Types include scrapers, bladelets and microliths.	Chert, quartz, obsidian.	1 plain potsherd in association with LSA material	-	-	-	-	-
Quiha	Esp. scrapers, blades, lunates and broken lunates. Industry is distinct from EBTT and has affinities with Elmenteitan. Similar industry also at Amba Sel and Zergnat.	Obsidian and some chert.	Majority of sherds from well fired, decorated vessels. Wide range of decoration, esp. comb stamping, scraping, grooving, polishing, woven mat impression and incision. Handle from decorated pot. Upper level contains PreAxumite pottery, pot bases, and foreign wares.	14 bovid teeth of which at least 9 were from domesticated cattle (8 from Level 3, 1 from Level 4). Upper part of bovid ulna and 4 equid teeth from Levels 3 and 4.	-	-	1 human incisor.	'Ritual object' (lip plug?) Large ovoid grindstone with surface abrasion, polish and pitting.
***Zergna*t**	Esp. long blades and sub-trapezoidal flake blades.	Obsidian.	-	-	-	-	-	Blades and flakes show use-wear and edge gloss.
Amba Sel	4 end or side scrapers 5 *outils ecailles* 1 square *outil ecaille* worked on 4 edges 8 micro-flakes and blades.	Obsidian and chert.	-	-	-	-	-	-

Lake Ziway	*Bulbula River*: esp. backed blades, end scrapers, burins, blade cores. EBTT.	Obsidian.	-	Baboons, bovids and numerous fragments of animal bones.	-	-	Human parietal	-
	Macho and Waso: esp. blades, non-geometric microliths, scrapers, burins and blade cores. EBTT. Surface deposit esp scrapers similar to Phase C Besaka.	Obsidian.	Only in recent surface deposit.	Unidentified bone in surface deposit.	-	-	-	possible ovoid hut circle associated with concentration of surface material.
Lalibela and Natchabiet	Wide variety of tool types in lowest levels. Variety decreases in upper levels. Grinding stones present throughout.	Obsidian and chert.	Wide variety of decorated and undecorated vessels including large water jars, storage jars, coffee pots, open bowls, injera cooking plates (*mugogo*) and incense burners. Many similar to modern vessels	Domesticated cattle and ovicaprids. Cattle are more common in earlier levels, becoming gradually replaced by ovicaprids. Wide variety of wild fauna present throughout - savannah and woodland fauna - as well as fish and birds.	Earliest levels contain domesticated barley, chickpea, legumes, horsebean and wild bitter vetch and *Zizyphus* sp. Tef recovered from post 2000 BP levels.	-	-	Quantities of worked bone, shell and, in later levels, iron.
Melka Konture	Scrapers, flakes and mircoliths esp at Wofi. Flakes and waste predominate at Balchit quarry site.	Obsidian.	-	None in surficial LSA material. Large quantities of faunal remains in excavated ESA and MSA deposits.	-	-	Hominid remains in ESA and MSA contexts.	-

BIBLIOGRAPHY

Abdel-Magid A (1989) *Plant Domestication in the Middle Nile Basin. An archaeoethnobotanical case study.* Cambridge Monographs in African Archaeology 35.

Aberg E (1948) Cereals and peas from Eastern Tibet and their importance for the knowledge of the origins of cultivated plants. *Annales of the Royal Agricultural College of Sweden 15*:235-250.

Abir M (1965) Salt, trade and politics in Ethiopia in the Zamara Mesafint. *Journal of Ethiopian Studies 4(2).*

Abir M (1968) *Ethiopia, the Era of Princes 1796-1855.* London.

Adams R M (1966) *The Evolution of Urban Society: Early Mesopotamia and Prehistoric Mexico.* Chicago.

Adams W Y (1977) *Nubia Corridor to Africa.* London and Princeton, Allen Lane and Princeton University Press.

Adamson D A, Clark J D and Willliams M A J (1974) Barbed bone points from Central Sudan and the age of the Early Khartoum Tradition. *Nature 249*: 120-123.

Adamson D A, Gasse F, Street F A and Willimas M A J (1980) Late Quaternary history of the Nile. *Nature 287:* 50-55.

Addison F (1949) *Jebel Moya.* Oxford University Press, London.

Agren G and Gibson R (1969) Food composition table for use in Ethiopia. *CNU Reoprt 16.*

Ahmed M M M (1994*) Indigenous Knowledge for Sustainable Development in the Sudan.* Sudan Library Series 20, Institute of African and Asian Studies, University of Khartoum.

Aitken M J (1990) *Science-based Dating in Archaeology.* Longman, London.

Alberro M and Haile-Mariam S (1982) The indigenous cattle of Ethiopia. *World Animal Review 41*:2-10/*42*:27-34.

Alexander J (1969) The indirect evidence for domestication. In Ucko P J and Dimbleby G W (eds) *The Domestication and Exploitation of Plants and Animals.* Duckworth.

Allan W (1972) Ecology, techniques and settlement patterns. In Ucko and Dimbleby:211-226.

Allchin B and Allchin R (1982) *The Rise of Civilization in India and Pakistan.* Cambridge University Press, Cambridge.

Allen J (1977) The hunting Neolithic: adaptations to the food quest in prehistoric Papua New Guinea. In Megaw (ed):167-188.

Abbreviations used in the text

MSA - Middle Stone Age.

The approximate period 200 000-30 000 years BP

LSA - Late Stone Age.

The approximate period from 30 000 years BP until recently.

EBTT - Ethiopian Blade Tool Tradition.

SPN - Savannah Pastoral Neolithic.

ITCZ - Inter-Tropical Convergence Zone.

asl - Above sea level.

Alley R B, Mayewski P A, Sowers T, Stuiver M, Taylor K C and Clark P U (1997) Holocene climate instability: a prominent, widespread event 8200 years ago. *Geology 25(6)*: 483-486.

Almeida M de (1954) *Some records of Ethiopia.* Beckingham C F and Huntingford G W B (translators).

Alvarez F (1881) *Narrative of the Portuguese Embassy to Abyssinia during the years 1520-1527.* Translated by Lord Stanley of Alderley.

Amblard S and Pernes J (1987) The identification of cultivated Pearl Millet (*Pennisetum*) amongst the plant impressions on pottery from Oued Chebbi (Dhar Oualat, Mauritania). *African Archaeological Reveiw 7*:117-126.

Ambrose S H (1980) Elmenteitan and other late Pastoral Neolithic adaptations in the Central Highlands of East Africa. In Leakey and Ogot (eds):279-282.

Ambrose S H (1984) The introduction of Pastoral adaptations to the Highlands of East Africa. In Clark and Brandt (eds):212-239.

Ames K M (1985) Hierarchies, stress and logistic strategies among hunter-gatherers in North West North America. In Price and Brown (eds):155-180.

Amirkhanov H (1994) Research on the Palaeolithic and Neolithic of Hadramant and Matra. *Arabian Archaeology and Epigraphy 5(4)*:217-228.

Anfray F (1963) Une campagne de fouilles a Yeha (Fevrier-Mars 1960). *Annales d'Ethiopie 5*:171-192.

Anfray F (1967a) Les sculptures rupestres de Chabbe, Sidamo. *Annales d'Ethiopie 7:*19-32.

Anfray F (1967b) Matara. *Annales d'Ethiopie 7:*33-35.

Appleyard D L (1977) Linguistic evidence of non-Semitic influence in the history of Ethiopian Semitic: lexical borrowing in Ge'ez and other Ethiopian Semitic languages. *Conference on Ethiopian Origins, SOAS, June 1977.*

Arkell A J (1947) Early Khartoum. *Antiquity 21*: 172-181.

Arkell A J (1949) *Early Khartoum.* Oxford, Oxford University Press.

Arkell A J (1953) *Shaheinab.* London

Arkell A J (1954) Four occupation sites at Agordat. *Kush 2*:33-62.

Arnold D E (1985) *Ceramic Theroy and Cultural Process.* Cambridge University Press, Cambridge.

Artzy M (1994) Incense, camels and collared rim jars: desert trade routes and maritime outlets in the 2nd millennium. *Oxford Journal of Archaeology 13(2)*:121-147.

Asfaw Z (1990) An ethnobotanical study of barley in the Central Highlands of Ethiopia. *Biologisches Zentralblatt 109*: 51-62.

Asfaw Z (1991) Hybrids between *spontaneum* and Ethiopian barley. *Barley Genetics 6*: 56-61.

Aumassip G (1984) Ti-n-Hanakaten, Tassili-n-Ajjer, Algerie, Bilan de 6 campagnes de fouilles. *Libyca 28/9*:115-127.

Bailey R C, Head G, Jeinke M, Owen B and Rechtman R (1989) Hunting and gathering in tropical rain forests: is it possible?. *American Anthropologist 91*:59-82.

Baker C M A and Manwell C (1980) Chemical classification of cattle I. Breed groups. *Animal Blood Groups and Biochemical Genetics 11*:127-150.

Bakhteyev F Kh (1971) Some experimental aspects of the problem of origin and pylogeny of barley. *Barley Genetics 2*: 36-44.

Banks K M (1984) *Climates, cultures and cattle: the Holocene archaeology of the Eastern Sahara*. Dallas, Southern Methodist University.

Bar-Yosef O (1975) The Epi-Palaeolithic in Palestine and Sinai. In *Problems in Prehistory: North Africa and the Levant*. Wendorf F and Marks A E (eds) Dallas. Southern University Press: 363-378.

Bar-Yosef O (1981) Pre-pottery Neolithic sites in Southern Sinai. *Biblical Archaeologist 45(1)*: 9-12.

Bar-Yosef O (1983) The Natufian of the Southern Levant. In Young C T, Smith P E L and Mortensen P (eds) *The Hilly Flanks and Beyond*. University of Chicago Press, Chicago:11-42.

Bar-Yosef O and Belfer Cohen A (1989) The origins of sedentism and farming communities in the Levant. *Journal of World Prehistory 3(4)*:447-498.

Bar-Yosef O and Belfer Cohen A (1991) From sedentary hunter-gatherers to territoral farmers in the Levant. In Gregg S A (ed) *Between Bands and States*. Centre for Archaeological Investigations, Carbondale:181-202.

Bar-Yosef O and Belfer Cohen A (1992) From foraging to farming in the Mediterranean Levant. In Gebauer and Price (eds):21-48.

Bar-Yosef O and Phillips J L (1977) *Prehistoric Investigations in Gebel Meghara, Northern Sinai*. Monographs of the Institute of Archaeology, Hebrew University, Jerusalem.

Barakat H N (1996) Human impact on the vegetation during the Middle Holocene period in the Eastern Sahara and Central Sudan. *Paper presented at the 13th Biennial Congress of SAFA, Posnan, September 1996.*

Bard K A, Coltorti M, Diblasi M C, Dramis F and Fattovich R (*in press*) The environmental history of Tigray (Northern Ethiopia) during the Holocene: a preliminary outline. Paper submitted to The African Archaeological Review 1996.

Bard K A and Fattovich R (1993) The 1993 excavations at Ona Enda Aboi Zague (Axum, Tigre). *Nyame Akuma 40*:14-17.

Bard K A and Fattovich R (1995) The IUO/BU excavation at Bieta Giyorgis (Axum): an interim Report. *Nyame Akuma 44*: 25-27.

Barich B E (1987) *Archaeology and the Environment in the Libyan Sahara: the Excavations in the Tadrat Acacus, 1978-1983*. Oxford. BAR International Series 368.

Barich B E (1992) The botanical collections from Ti-n-Torha/Two Caves and Uan Muhuggiag (Tadrart Acaucus, Libya) - an archaeological commentary. *Origini 16*:109-123.

Barth H (1857) *Travels and Discoveries in North and Central Africa*. New York, Harper and Bros.

Barthelme J W (1977) Holocene sites from north-east of Lake Turkana: a preliminary report. *Azania 12*:33-41.

Barthelme J W (1984) Early evidence for animal domestication in Eastern Africa. In Clark and Brandt (eds):200-205.

Barthelme J W (1985) *Fisher-hunters and Neolithic Pastoralists in East Turkana, Kenya*. BAR, Oxford.

Bass G F (1986) A bronze age shipwreck at Ulu Burun (Kas): 1984 campaign. *American Jounal of Archaeology 90*:269-296.

Bayliss-Smith T (1996) People-plant interactions in in the New Guinea Highlands: agricultural hearthland or horticultural backwater? In Harris (ed):499-523.

Beauvisage G (1897) Le bois d'ebere. *Recheil de Travaux 19*:77-83.

Bedigian D and Harlan J R (1986) Evidence for the cultivation of sesame in the Ancient World. *Economic Botany 40(2)*:137-154.

Bednarik R G (1996) Only time will tell: a review of the methodology of direct rock art dating. *Archaeometry 38(1)*:1-13.

Behre K E (1986) Anthropogenic Indicators in Pollen Diagrams. Balkema, Rotterdam.

Beke L T (1852) *Letters on the Commerce of Abyssinian and other parts of East Africa Addressed to the Foreign Office and Board of Trade.*

Bekele E (1983) A differential rate of regional distribution of barley flavonoid patterns in Ethiopia, and a view of the centre of origin of barley. *Hereditas 98*:269-308.

Bekele E (1983) Some measures of gene diversity analysis on landrace populations of Ethiopian barley. *Hereditas 98*:127-143.

Bekele E and Lester R N (1981) Biochemical assesment of the relationships of *Eragrostis tef* (Zucc.) Trotter with wild *Eragrostis* species. *Annals of Botany 48*: 717-25.

Bell B (1970) The oldest records of the Nile Floods. *Geographical Journal 136*:569-573.

Bell B (1971) The dark ages in ancient history I. The first dark age in Egypt. *American Journal of Archaeology 75*:1-26.

Bell M and Walker M J C (1992) *Late Quaternary Environmental Change*. Longman, New York.

Bender B (1978) Hunter-gatherer to farmer: a social perspective. *World Archaeology 10*:204-222.

Bender M L (1971) The languages of Ethiopian: a new lexicostatistic classification and some problems of diffusion. Anthropological Linguistics 13(5): 175.

Bender M L (1975) *Omotic: a New Afroasiatic Language Family*. Carbondale, South Illinois University Museum.

Bender M L (1976) *The non-semitic languages of Ethiopia*. African Studies Centre, East Lansing, Michigan.

Bender M L (1982) Livestock and linguistics in North East African ethnohistory. *Current Anthropology 23*:316-317.

Bent J T (1893) The ancient trade route across Ethiopia. *Geographical Journal 2*:140-146.

Beraki O, Brancaccio L, Calderoni G, Coltorti M, Dramis F and Umer M (1994) Geomorphic and stratigraphic evidences of Holocene landscape evolution in the Highlands of Tigre (Ethiopia). *Proceedings of the 12th Conference of Ethiopian Studies, East Lansing. September 1994.*

Berger A and Loutre M F (1991) Insolation values for the climate of the last 1 000 000 years. *Quaternary Science Reviews 10*:297-317.

Berger W H and Labeyrie L D (1987*) Abrupt Climate Changes: Evidence and Implications.* Reidel, Dordrecht.

Bernal M (in prep.) *Moses and Muses: the African and Asian Roots of European Civilization.*

Berndt R and Berndt C (1951) Discovery of pottery in North East Arnhem Land. *Journal of the Royal Anthropological Institute 77.*

Bethke S (1976) Basic zonal rainfall patterns in Ethiopia. In Hussein (ed):97-103.

Betts A, van der Borg K, de Jong A, McClintock C and van Strydonck M (1994) Early cotton in Northern Arabia. *Journal of Archaeological Science 21*:489-499.

Biagi P, Maggi M and Nisbet R (1985) Excavation at the aceramic coastal settlement of RH5 (Muscat, Sultanate of Oman) 1983-5. In Frifelt K and Sorenson P (eds) *South Asian Arcaheology 1985.* Scandanavian Institute of Archaeology Occasional Papers 4:1989:1

Bietak M (1979) Ceramics of the C-Group culture. *Meroitica 5*:146.

Bilbert M (1975) Proceedings of the Ethiopian wildlife and Natural History Society on the study of the fauna and flora of Ethiopia. *Walia 6*:1-23.

Binford L R (1968) Post Pleistocene adaptations. In *New Perspectives in Archaeology.* Binford L (ed) Aldine, Chicago: 313-341.

Bintliff J L and van Zeist W (1982) *Palaeoclimate, Palaeoenvironment and Human Communities in the Eastern Mediterranean Region in Later Prehistory.* BAR International Series 133. Oxfrod.

Birch H F (1967) Soil map of Ethiopia and description of agricultural value of the soil groups. *Institute of Agricultural Research, Addis Ababa.*

Bisschopp J H R (1937) Parent stock and derived types of cattle, with particular reference to tyhe importance of confrmational characteristics in the study of their origin. *South African Journal of Science 33*:852-870.

Blackburn R H (1974) The Okiek and their history. *Azania 9*:139-158.

Blench R (1991) Connections between Ancient Egypt and sub-Saharan Africa: the evidence of cultivated plants. In Davies W V (ed):54-56.

Blench R (1993) Ethnographic and linguistic evidence for the prehistory of African ruminant livestock, horses and ponies. In Shaw et al (eds):71-103.

Blench R (1993) Recent developments in African language classification and their implications for prehistory. In Shaw et al: 126-138.

Blench R (1995) Linguistic evidence for the history of asses, donkeys and mules in Africa. *Paper presented at the Institute of Archaeology Conference on African Livestock: Archaeology, Linguistics and DNA. September 22-24, 1995. London.*

Blench R (1996) Linguistic evidence for cultivated plants in the Bantu borderland. In Sutton J E G (ed) *The Growth of Farming Communities in Africa from the Equator Southwards.* British Institute in Eastern Africa, Nairobi:83-102.

Blumler M A (1996) Ecology, evolutionary theory and agricultural origins. In Harris (ed):25-50.

Blumler M A and Byrne R (1991) The ecological genetics of domestication and the origins of agriculture. *Current Anthropology 32:*23-54.

Boardman S and Jones G (1990) Experiments on the effects of charring on cereal plant components. *Journal of Archaeological Science 17:*1-11.

Boessneck J (1988) *Die Tierwelt des Alten Agypten.* C H Beck, Munich.

Bondestam L (1974) People and capitalism in the northeastern lowlands of Ethiopia. *Journal of Modern African Studies 12(3):*423-439.

Bonnefille R and Buchet G (1986) Contribution palynologique a l'histoire recente de la foret de Wenchi (Ethiopie). *Memoir de Travaille EPHE, Institute of Montpellier 17*:143-158.

Bonnefille R and Hamilton A (1986) Quaternary and late Tertiary history of Ethiopian vegetation. *Acta Universitet Uppsala Symposium Botanica Uppsala 26(2):*47-63.

Bonnefille R and Mohammed U (1994) Pollen-inferred climatic fluctuations in Ethiopia during the last 3000 years. *Palaeogeography, Palaeoclimatology, Palaeoecology 109:*331-343.

Boserup E (1965) *The Conditions of Agricultural Growth.* London.

Bower J (1991) The Pastoral Neolithic in East Africa. *Journal of World Prehistory 5*:49-104.

Bower J and Lubell D (1988) *Prehistoric Cultures and Environments in the Late Quaternary of Africa.* Cambridge Monographs in African Archaeology 26. BAR International Series 405.

Bower J R F, Nelson C M, Waibel A F and Wandibba S (1977) The University of Massachusetts' Later Stone Age/Pastoral Neolithic comparative study in Kenya. *Azania 12:*119-146.

Bower J R F and Nelson C M (1978) Early pottery and pastoral cultures of the Central Rift Valley, Kenya. *Man 13*:554-66.

Braidwood R J (1960) The agricultural revolution. *Scientific American 203:*130-148.

Braidwood R J and Braidwood L (1950) Jarmo:a village of early farmers in Iraq. *Antiquity 24*:189-195.

Braidwood R J and Howe B (1960) *Prehistoric Investigations in Iraqui Kurdistan.* Studies in Ancient Oriental Civilzations 31. University of Chicago Press, Chicago.

Braithwaite M (1982) Decoration as ritual symbol: a theoretical perspective and an ethnographic study in Southern Sudan. In Hodder I (ed) *Symbolic and Structural Archaeology.* Cambridge University Press, Cambridge:80-88.

Brancaccio L, Caleroni G, Coltorti M, Dramis F and Ogbaghebriel B (1994) Phases of soil erosion during the Holocene in the Highlands of Western Tigray (Northern Ethiopia): a preliminary report. *Paper presented at the 12th Congress of Ethiopian Studies, East Lansing, Michigan.*

Brandt S A (1980) Archaeological investigations at Lake Besaka, Ethiopia. In *Proceedings 8th PanAfrican Congress of Prehistory* *and Quaternary Studies, Nairobi.* Leakey L S B and Ogot B A (eds):239-243.

Brandt S A (1982) *A Late Quaternary Environmental Sequence from Lake Besaka, Southern Afar, Ethiopia.* PhD thesis, University of California, Berkeley. Ann Arbor.

Brandt S A (1984) New perspectives on the origins of food production in Ethiopia. In Clark J D and Brandt S A (eds):173-90.

Brandt S A (1986) The Upper Pleistocene and Early Holocene prehistory of the Horn of Africa. *African Archaeological Review 4*: 1-82.

Brandt S A (1996) Inter-ethnic variability in flaked stone tool use among the contemporary hide workers of South Ethiopia. *Paper presented at the 13th Biennial Congress of SAFA, Posnan, September 1996.*

Brandt S A and Brook G A (1984) Archaeology and palaeoenvironmental research in North Somalia. *Current Anthropology 25*:119-121.

Brandt S A and Carder N (1987) Pastoral rock art in the Horn of Africa: making sense of udder chaos. *World Archaeology 19*:194-223.

Breasted J H (1952) *A History of Egypt.*

Breuil l'A H (1934) Peintures rupestre prehistorique du Harar (Abyssinie*). L'Anthropologie 44(5-6)*:473-483.

Brewer D J (1989) A Model for Resource Exploitation in the Prehistoric Fayum. In Krzyzaniak L and Kobusiewicz M (eds): 127-38.

Brogan E (1954) The camel in Roman Tripolitania. *Papers of the British School at Rome 9:*126-131.

Brown F H (1975) Barbed bone points from the Lower Omo Valley, Ethiopia. *Azania 10*:144-148.

Brown J (1990) Horn-shaping ground stone axe hammers. *Azania 25*:57-67.

Bruce J (1790) *Travels to Discover the Source of the Nile in the Years 1768, 1769, 1770, 1771, 1772 and 1773.* London. G G J and J Robinson.

Bulliet R W (1975) *The Camel and the Wheel.* Harvard University Press, Cambridge, Massachusetts.

Bulmer S (1975) Settlement and economy in prehistoric Papua New Guinea: a review of the archaeological evidence. *Journal de la Societe des Oceanistes 31*:7-75.

Burkill I H (1953) Habits of man and the origins of the cultivated plants of the Old World. *Proceedings of the Linnean Society, London 164:*12-42.

Butler A (1995) The small-seeded legumes: an enigmatic prehistoric resource. *Acta Botanica 35(1)*:105-116.

Butzer K W, Brown F H and Thurber D L (1969) Horizontal sediments of the lower Omo Valley: the Kibish formation. *Quatermaria 11*:15-29.

Butzer K W, Isaac G Ll, Richardson J L and Washburn-Kamau G K (1972) Radiocarbon dating of East African lake levels. *Science 175*:1069-1076.

Butzer K W (1971) *Environment and Archaeology: an Ecological Approach toPrehistory.* Methuen, London.

Butzer K W (1971) Recent history of an Ethiopian Delta. *University of Chigaco Dpeartment of Geography Research Papers No.136.*

Butzer K W (1975) Patterns of environment change in the Near East duirng the late Pleistocene and early Holocene times. In Wendorf F and Marks A E (eds) *Problems in Prehistory: North Africa and the Levant.* Southern Methodist University Press:389-410.

Butzer K W (1976) *Early Hydraulic Civilization in Egypt.* University of Chicago Press, Chicago.

Butzer K W (1981) Rise and fall of Axum, Ethiopia: a geo-archaeological interpretation. *American Antiquity 46(3)*:471-495.

Butzer K W (1982) Empires, capitals and landscapes of Ancient Ethiopia. *Archaeology 35(5)*:30-37.

Butzer K W (1984) Long-term Nile flood variation and political discontinuities in Pharonic Egypt. In Clrak and Brandt (eds):102-112.

Byrne R (1987) Climatic change and the origins of agriculture. Manzanilla L (ed) *Studies in the Neolithic and Urban Revolutions: the V. Gordon Childe Colloquium.* BAR International Series 349:21-34.

Calegari G (1996) I testimioni impassibili. *Archaelogia African – Saggi Occasionali 2*, Centro Studio Archeologia Africana.

Campell J (1991) Land or peasants? The dilemma confronting Ethiopian resource conservation. *African Affairs 90*:5-21.

Candolle A de (1886) *Origins of Cutivated Plants.* New York.

Cane S (1989) Australian Aboriginal seed grinding and its archaeological record: a case study from the Western Desert. In Harris and Hillman (eds):99-119.

Caneva I (1983) Pottery using hunter-gatherers at Saggai (Sudan): preconditions for food production. *Origini 12*:7-29.

Caneva I (1987) Pottery decoration in prehistoric Sahara and Upper Nile Valley: a new perspective. In Barich B E (ed) *Archaeology and Environment in the Libyan Sahara.* Cambridge Monographs in African Archaeology 23: BAR International Series 368:231-254.

Caneva I (1988) Prehisatoric settlements along the Nile between Geili and Kabbashi. In Geili E (ed) *The History of a Middle Nile Environment 7000 BC-AD 1500.* Oxford, BAR 424:321-344.

Caneva I and Marks A (1990) More on the Shaqadud pottery: evidence for Saharao-Nilotic connections during the 6th-4th millennia BP. *Archeologie du Nil Moyen 4*:11-35.

Carter P L and Clark J D G (1976) Adrar Bous and African cattle. *Proceedings of the Pan-African Prehistory Congress 7:*487-493.

Carter P L and Flight C (1972) A report on the fauna from the sites of Ntereso and Kintampo Rock Shelter Six in Ghana: with evidence for the pracitce of animal husbandry during the second millennium BC. *Man 7*:277-282.

Carter P L and Higgs E S (1979) A study of the faunal remians from 'La Grotte Capeletti Khanguet si Mohamed Tahar', Aures, Algeria. In Roubet C (ed) *Economie Pastorale Preagricole en Algerie Orientale: le Neolithique de Tradition Capsienne.* CNRS, Paris:411-414.

Caton-Thompson G and Gardener E W (1934) *The Desert Fayum.* London.

Central Statistics Office (1984) *Time series data on area, production and yield of principal crops by regions 1979/80 - 1983/4.* Addis Ababa.

Cervick P (1971) Rock paintings of Laga Oda (Ethiopia*). Paideuma 17*:121-136.

Cervick P (1979) Some African affinities of Arabian rock art. *Rassegna di Studi Etiopici 27*:5-12.

Chapman T (1992) *Desertified Grasslands: their Biology and Management.* Academic Press, London.

Chang K C (1970) The beginnings of Agriculture in the Far East. *Antiquity 44*:175-85.

Chaplin J H (1974) The prehistoric rock art of the Lake Victoria region. *Azania 9*:1-50.

Charters S, Evershed R P, Goad L J, Leyden A, Blinkhorn P W and Denham V (1993) Quantification and distribution of lipid in archaeological ceramics: implications for sampling potsherds for organic residue analysis and the classification of vessel use. *Archaeometry 35(2)*:211-223.

Chavaillon J, Chavaillon N, Hours F and Piperno M (1979) From the Oldowan to the MSA at Melka-Konture (Ethiopia). Understanding cultural changes. *Quaternaria 21*:87-114.

Chavaillon J (1976) Evidence for the technical practices of early Pleistocene hominids, Shungura Formation, Lower Omo Valley, Ethiopia. In Coppens Y, Howell F C, Isaac G Ll and Leakey R E F (eds) *Earliest Man and Environments in the Lake Rudolf Basin.* Univeristy of Chicago Press, Chicago:565-573.

Chavaillon J (1976) Mission archaeologique Franco-Ethiopienne de Melka Konture, repport preliminaire: 1972-1975. In *L'Ethiopie Avant L'Histoire. Cahier No.1.*

Cheeseman R E (1936) *Lake Tana and the Blue Nile.* London, Macmillan and Co.

Chevalier A (1932) *Resources Vegetales du Sahara et de ses Confins Nord et Sud.* Paris, Musee d'Historie Naturelle.

Childe V G (1928) *The Most Ancient East: the Oriental Prelude to European Prehistory.* Kegan Paul, Trench, Trubner, London.

Childe V G (1956) *Piecing Together the Past.* Routledge and Kegan Paul, London.

Chilodnicki M (1984) Pottery from the Neolithic settlement at Kadero (Central Sudan). In Krzyzaniak and Kobusiewicz (eds):337-342.

Chowdury K A and Buth G M (1970) 4500 year old seed suggests that true cotton is indigenous to Nubia. *Nature 227*:85-86.

Churcher C S (1984) Zoological study of the ivory handle from Abu Zaiden. In Needler E (ed) *PreDynastic and Archaic Egypt in the Brooklyn Museum.* Wilbour Monograph 9, Brooklyn Museum, New York:152-168.

Clark J D (1945) Short notes on Stone Age sties at Yavello, Southern Abysinnia. *Transactions of the Royal Society of South Africa 31(1)*:29-37.

Clark J D (1954) *The Prehistoric Cultures of the Horn of Africa.* Cambridge University Press, Cambridge.

Clark J D (1962) The spread of food production in sub-Saharan Africa. *Journal of African History 3*: 211-28.

Clark J D (1967a) *Atlas of African Prehistory.* University of Chicago Press, Chicago.

Clark J D (1967b) Prehistoric populations and pressures favouring plant domestication in Africa. In Harlan et al (eds):67-105.

Clark J D (1972) Mobility and settlement patterns in sub-Saharan Africa: a comparison of late Prehistoric hunter-gatherer and early agricultural occupation units. In Ucko et al (eds):127-48.

Clark J D (1976) Prehistoric populations and pressures favouring plant domestication in Africa. In Harlan et al (eds):67-105.

Clark J D (1980) The origins of domestication in Ethiopia. In Leakey and Ogot (eds): 268-70.

Clark J D (1982) *The Cambridge History of Africa. Vol.1.* Cambridge, Cambridge University Press.

Clark J D and Brandt S A (1984) *From Hunters to Farmers: The Causes and Consequences of Food Production in Africa.* University of Califiornia Press. Berkeley.

Clark J D and Prince G R (1978) Excavations at Laga Oda, Haraghi, Ethiopia. *Azania 13*: 109.

Clark J D and Stemler A B L (1975) Early domesticated sorghum from central Sudan. *Nature 254*: 588-91.

Clark J D and Williams M A J (1978) Recent archaeological results in South Eastern Ethiopia: some preliminary results. *Annales d'Ethiopie 11*: 19-45.

Clark J D and Williamson M A J (1984) A Middle Stone Age occupation site at Porc Epic Cave, Dire Dawa (East Central Ethiopia). *African Archaeological Review 2*:37-71.

Cleuziou S and Constantini L (1980) Premiers elements sur l'agriculture protohistorique de l'Arabie Orientale. *Paleoorient 6*:245-251.

Cleuziou S and Constantini L (1982) A l'origine des oasis. *La Recherche 13:* 1181-9.

Close A E and Wendorf F (1992) The beginnings of food production in the Eastern Sahara. In *Transitions to Agriculture in Prehistory.* Gebauer and Price (eds): 63-72.

Clutton-Brock J (1979) The mammalian remains from Jericho Tell. *Proceedings of the Prehistoric Society 45*:135-158.

Clutton-Brock J (1987) *A Natural History of Domesticated Animals.* Cambridge University Press, British Museum.

Clutton-Brock J (1989) Cattle in ancient North Africa. In Clutton-Brock (ed):200-206.

Clutton-Brock J (1989) *The Walking Larder: Patterns of Domestication, Predation and Pastoralism.* Unwin Hyman, London.

Coe M (1989) Biogeographic affinities of the high mountains of tropical Africa. In Mahaney (ed):257-278.

Coetzee J A (1967) Pollen analytical studies in East and Southern Africa. *Palaeoecology of Africa and the Surrounding Islands 3:* 1-146.

Cohen M N (1970) A reassesment of the Stone Bowl cultures of the Rift Valley, Kenya. *Azania 5*:27-38.

Cohen M N (1975) Population pressure and the origins of agriculture. an archaeological example from the Coast of Peru. In *Population, Ecology and Social Evolution.* Polgar S (ed) The Hague, Mouton: 79-122.

Cohen M N (1977) *The Food Crisis in Prehistory*. Yale University Press, New Haven.

Cohen R (1981) Evolutionary epistemology and human values. *Current Anthropology 22*:201-218.

Collett D and Robertshaw P (1981) Early Iron Age pottery: finds from Gogo Falls, South Nyanza. *Azania 16*:133-145.

Collet D and Robertshaw P (1983a) Pottery traditions of early pastoral communities in Kenya. *Azania 18*:107-125.

Collet D and Robertshaw P (1983b) Problems in the interpretation of radiocarbon dates: the Pastoral Neolithic of East Africa. *African Archaelogical Review 1*:57-74.

Collier S and White JP (1976) Get them young? Age and sex inferences on animal domestication in archaeology. *American Antiquity 41(1)*:96-102.

Connah G (1981) *Three Thousand Years in Africa: Man and His Environment in the Lake Chad Region of Nigeria*. Cambridge University Press, Cambridge.

Constantini L (1979) Plant remains from Pirak. In Jarrige J -F and Samtoni M (eds) *Texte Fouilles des Pirak I*. Publications de la Commission des Fouilles Archaeologiques.

Constantini L (1984) The beginnings of agriculture in the Kachi Plain: the evidence of Mehrgarh. In Allchin B (ed) *South Asian Archaeology 1981*. Cambridge University Press, Cambridge:29-33.

Constantini L (1990) Ecology and farming of the protohistoric communities in the Central Yemeni Highlands. In De Maiget A (ed) *The Bronze Age Cultures of Hawlan At-Tyial and Al-Hada*. Ismeo, Rome:187-204.

Constantini P, Fattovich R and Piperno M (1982) Preliminary report of archaeological investigations at the site of Mahal Teglinos (Kassala) November 1981. *Journal of Field Archaeology 13*:30-33.

Constantini L, Fattovich R, Piperno M and Sadr K (1983) Gash Delta Archaeological Project 1982 Field Season. *Nyame Akuma 23*:17-19.

Constantini L and Constantini Biasini L (1985) Agriculture in Baluchistan between the seventh and the 3rd millennium BC. *Newsletter of Baluchistan Studies 2*:16-30.

Constanza S H, de Wet J M J and Harlan J R (1979) Literature review and numerical taxonomy of *Eragrostis tef* (tef). *Economic Botany 33*: 413-424.

Constanza S H (1974) *Literature and Numerical Taxonomy of Tef (*Eragrostis tef*)*. MSc Thesis, Cornell University, Illinois.

Conti Rossini C (1928) *Storia d'Ethiopia*. London.

Corbett J E M (1988) Famine and household coping strategies. *World Development 16(9)*:1099-1112.

Cowan C W and Watson P J (1992) *The Origins of Agriculture: an International Perspective*. Smithsonian Institution Press, London.

Coursey D G (1976) The origins and domestication of yams in Africa. In Harlan et al (eds):383-408.

Crawford O G S (1958) *Ethiopian Itineraries c.1400-1524*.

Cremaschi M, d'Alessandro A, Fattovich R and Piperno M (1986) Gash Delta archaeologicalproject: 1985 field season. *Nyame Akume 27*:45-48.

Cribb R (1991) *Nomads in Archaeology*. Cambridge University Press, Cambridge.

Cufodontis G (1953-1972) Enumeratio plantarum Aethiopiae Spermatophyta. *Bull.Jard.Bot.Etak.Brux. 23-42*.

Cziesla E (1980) Excavations at Wadi Sahal. In Leakey and Ogot (eds):143-145.

D'Souza F (1988) Famine:social security and an analysis of vulnerability. In Harrison G A (ed) *Famine*. Oxford University Press, Oxford:1-56.

Darby W J, Ghalioungui P and Grivetti L (1977) *Food: the Gift of Osiris*. Academic Press, London.

David N, Harvey P and Goudie C J (1981) Excavations in Eastern Equatoria, South Sudan, 1980. *Azania 16*:7-54.

Davies W V (1991) *Egypt and Africa: Nubia from Prehistory to Islam*. British Museum Press, London.

Dawson A G (1992) *Ice Age Earth; Late Quaternary Geology and Climate*. Routledge.

de Candolle A (1882) *Origine des Plantes Cultivees*. Germer Bailliere, Paris.

de Contensen H (1963) Les fouilles de Haoulti en 1959. *Annales d'Ethiopie 5*:41-52.

de Contensen H (1981) PreAxumite culture. In Mokhtar G (ed) *General History of Africa II. Ancient Civilizations of Africa*. Heinemann, California. UNESCO:360.

de Maret P (1996) Pits, pots and far-west streams. In Sutton (ed):318-323.

Dekker G and Selassie T C (1972) Rock engravings from Ch'ew Bahir. *Annales d'Ethiopie 9*:19.

Delano-Smith C (1979) *Western Mediterranean Europe*. Academic Press, London.

Demoz A (1978) Ethiopian origins: a survey. *Abbay 9*:11-14.

Dennell R W (1979) Prehistoric diet and nutrition: some food for thought. *World Archaeology 11(2)*:121-135.

Dirar H A (1994) Indigenous fermented foods and beverages of rural areas of the Sudan. In Ahmed (ed):43-80.

Dirks R (1980) Social response during severe food shortages and famine. *Current Anthropology 21(1)*:21-44.

Doe B (1971) *Southern Arabia*. Thames and Hudson, London.

Doggett H (1965) The development of cultivated sorghums. In Hutchinson J B (ed) *Essays on Crop Plant Evolution*. Cambridge University Press, Cambridge: 50-69.

Doggett H (1970) *Sorghum*. 1st Edition. Longman, London.

Doggett H (1988) Sorghum history in relation to Ethiopia. In Engels J M M (ed) *The Conservation and Utilization of Ethiopian Germplasm*. Proceedings of an International Symposium, Addis Ababa 13-16 October 1986:97-115.

Doggett H (1988) *Sorghum*. 2nd Edition, Longmans, London.

Doggett H (1991) Sorghum history in relation to Ethiopia. In *Hawkes et al* (eds):140-59.

Dombrowski J C (1970) Preliminary report on excavations in Lalibela and Natcheibet Caves. *Annales d'Ethiopie 8*:21-29.

Dombrowski J C (1971) *Excavations in Ethiopia: Lalibela and Natcheibet Caves, Begemder Province*. Ph.D. dissertation, Boston University, Ann Arbor.

Doutrelepont H, Vrydaghs L, Langhe E de, Janssens B, Swennen R, Mbida Ch and Maret P de (1996) Banana in Africa. *Paper Presented at the 13th Biennial Congress of SAFA, Posnan, September 1996.*

Dove K (1890) *Kulturzonen von Nord-Abessinien*. Petermanns Mitt. ErganzHft.97.

Drew S F (1952) The mysterious rock carvings of the Eritrean Highlands. *Illustrated London News (Aug 9)*: 225-227.

Drew S F (1954) Notes from the Red Sea Hills. *South African Archaeological Bulletin 9*:101-102.

Ebba T (1969) *Tef* (Eragrostis tef*). The Cultivation, Useage and Some of the Known Disease and Insect Pests*. Part 1, Hailie Sellassie University College of Agriculture, Experimental Station Bulletin 60.

Ebba T (1975) *Tef* (Eragrostis tef) *Cultivars: Morphology and Classification*. Addis Ababa University College of Agriculture, Dire Dawa Experimental Station Bulletin 66.

Edwards S (1976) *Some Wild Plants of Ethiopia*. Addis Ababa University Press, Addis Ababa.

Edwards S (1991) Crops with wild relatives found in Ethiopia. In Hawkes et al (eds):42-74.

Eggert M K H (1994) Pots, farming and analogy: implication of early ceramics in the Equatorial Rainforest. *Cambridge BIEA Conference on the Spread of Farming in SubEquatorial Africa*. June 1994.

Ehret C (1971) *Southern Nilotic History*. Evanston, Northwest University Press.

Ehret C (1974) *Ethiopians and East Africans: the Problem of Contacts*. Nairobi.

Ehret C (1976) Cushitic prehistory. In Bender M L (ed) *The Non-Semitic Languages of Ethiopia*. East Lansing.

Ehret C (1979) On the antiquity of agriculture in Ethiopia. *Journal of African History 20*:161-177.

Ehret C (1983) Nilotic and the limits of Eastern Sudanic. In Vossen R and Beckhaus-Gerst M (eds) *Nilotic Studies, Part Two*. Dietrich Reimer, Berlin:377-421.

Ehret C (1984) Historical/linguistic evidence for early African food production. In Clark and Brandt (eds):26-35.

Ehret C (1993) Nilo-Saharans and the Saharo-Sudanese Neolithic. In Shaw et al (eds):104-125.

El Mahi A T (1988) *Zooarchhaeology in the Middle Nile Valley. A study of four Neolithic sites near Khartoum*. Cambridge Monographs in African Archaeology 27.

El-Moslimany A P (1984) Comment on 'Age, palaeoenvironments and climatic significance of late Pleistocene Konya Lake, Turkey' by Neil Roberts. *Quaternary Research 21*:115-116.

Elamin Y M and Khabir A M (1987) Neolithic pottery from survey sites around Shaqadud cave, Western Butana., Sudan. *Archeologie du Nil Moyen 2*:175-184.

Eldredge N and Gould S J (1972) Punctuated equilibria: an alternative to phyletic gradualism. In Schopf T J M (ed) *Models in Palaeobiology*. Freeman, Cooper, San Francisco:82-115.

Engels J M M (1990) The genetic diversity in Ethiopian barley in relation to altitude. In Tyama S and Takede G (eds) *Proceedings of the 6th International Congress of the Society for Advancement of Breeding Research in Asia and Oceana 1989, SABRAO, Japan*: 107-110.

Engels J M M (1991) A diversity study in Ethiopian barley. In Hawkes et al: 131-40.

Engels J M M and Goettsch E (1991) Konso agriculture and its plant genetic resources. In Hawkes et al (eds):167-186.

Engels J M M and Hawkes J (1991) The Ethiopian gene centre and its genetic diversity. In Hawkes et al (eds):23-41.

Epstein H (1971) *The Origin of Domesticated Animals of Africa*. Africana Publishing Corporation, New York.

Epstein H and Mason I L (1984) Cattle. In Mason I L (ed) *Evolution of Domesticated Animals*. Longman, London:6-27.

Evans-Pritchard E E (1940) *The Nuer*. Clarendon Press, Oxford.

Evernden J F and Curtis G H (1965) The K/Ar dating of late Cenozoic rocks in East Africa and Italy. *Current Anthropology 5*:343-364.

Evershed R P, Heron C, Charters S and Goad L J (1992) The survival of food residues: new methods of analysis, interpretation and application. *Proceedings of the British Academy 77*:187-208.

Evershed R P, Heron C and Goad L J (1991) Epicutilar wax components preserved in potsherds as chemical indicators of leafy vegetables in ancient diets. *Antiquity 65*:540-544.

Faber J and Petit Maire N (1988) Holocene climatic evolution at 22-23°N from two palaeolakes in the Taoudenni area (North Mali). *Palaeogeography, Palaeoclimatology, Palaeoecology 65*:133-148.

Faris D G (1965) The origin and evolution of the cultivated forms of *Vigna sinensis. Canadian Journal of Genetic Cytology 7*: 433-452.

Fattovich R (1975) The contribution of the Nile Valley cultures to the rising of Ethiopian civilzation. *Meriotic Newsletter 16*:2-8.

Fattovich R (1977) Some data for the study of cultural history in Ancient Northern Ethiopia. *Nyame Akuma 10*:6-18.

Fattovich R (1982) The problem of Sudanese-Ethiopian contacts in antiquity: *Status Quaestionis* and current trends of research. *Proceedings of the Symposium for Nubian Studies, Cambridge 1978*: 76-86.

Fattovich R (1988) The stelae of Kassala: a new type of funerary monuments in the Eastern Sudan. *ANM 3*:55-63.

Fattovich R (1989) The late prehistory of the Gash Delta (Eastern Sudan). In Krzyzaniak and Kobusiewicz (eds):481-498.

Fattovich R (1990a) The peopling of the Northern Ethiopian-Sudanese borderland between 7000 and 1000 BP: a preliminary model. *Nubica 1/2*: 3-45.

Fattovich R (1990b) Remarks on the PreAxumite period in Northern Ethiopia. *Journal of Ethiopian Studies 23*: 1-33.

Fattovich R (1991) The problem of Punt in the light of recent field work in the Eastern Sudan. In Schoskes S (ed) *Akten des Vierten*

Internationalen Agyptologen Kongresses, Munchen 1985, 4:257-272.

Fattovich R (1993) Punt: the archaeological perspective. Sesto Congresso Internationale di Egittologia. *Atti 2*:399-404.

Fattovich R (1993) The Gash Group of the Eastern Sudan: an outline. In *Environment Change and Human Culture in the Nile Basin and North Africa until the 2nd Millennium BC*. Pozan, Poland.

Fattovich R and Piperno M (1982) Archaeological research in the Gash Delta, Kassala Province (1980-1981 field seasons). In Krause M (ed) *Nubische Studien. Proceedings of the Symposium of International Society of Nubian Studies*. Heidleberg 1982:47-53.

Fattovich R, Sadr K and Vitagliano S (1988) Societe e territorio nel delta del Gash (Kassala, Sudan Orientale) 3000 aCr. 300/400 d.Cr. *Africa (Roma) 43*:394-453.

Fattovich R, Sadr K and Vitagliano S (1989) Society and territoy in the Gash Delta (Kassala, E. Sudan) 3000 BC-300/400 AD. *Origini 14*: 329-57.

Fein J S and Stephens P L (1987) *Monsoons*. John Wiley and Sons, New York.

Flannery K V (1968) Archaeological systems theory and early Mesoamerica. In Meggers (ed) *Anthropological Archaeology in the Americas*. Anthropological Society of Washington, Washington:67-87.

Flannery K V (1969) Origins and ecological effects of early domestication in Iran and the Near East. In Ucko and Dimbleby (eds): 73-100.

Flannery K V (1973) The origins of agriculture. *Annual Review of Anthropology 2*: 271-310.

Ford J (1975) *African Trypanosomiasis: an assesment of the tsetse fly problem today*. African Environment: Problems and Perspectives. IAI, Animals and Fisheries Resources Developement Authority (Vetinary Department).

Friedman J and Rowlands M J (1977) *The Evolution of Social Systems*. Duckworth, London.

Frost S, Holm G and Asker S (1975) Flavonoid patterns and the phylogeny of barley. *Hereditas 79*:133-142.

Frost S, Harbourne J B and King L (1977) Identification of the flavonoids in five chemical races of cultivated barley. *Hereditas 85*:163-168.

Frost S and Holm G (1973) A new phylogenetic approach in barley. *Hereditas 74*:304-308.

Frost S and Holm G (1975) Variation of flavonoid patterns in *Hordeum spontaneum* and *H.argriocrithon*. *Hereditas 80*:167-172.

Gabriel B (1984) Great plains and mountain areas as habitats for the Neolithic man in the Sahara. In Krzyzaniak L and Kobusiewicz M (eds): 391-398.

Gabriel B (1987) Palaecological evidence from Neolithic fireplaces in the Sahara. *African Archaeological Reveiw 5*: 93-103.

Gallagher J P (1973) Preliminary report on archaeological research near Lake Zuai, Ethiopia. *Annales d'Ethiopie 9*:64-80.

Gamst F C (1965) *Travel and research in North West Ethiopia: notes for anthropologists and other field workers in Ethiopia*. Haile Sellassie I University.

Gamst F C (1969) *The Qemant: a Pagan-Hebraic Peasantry of Ethiopia*. Holt, Rinehart and Winston.

Garcia R V and Escudero J C (1982) *Drought and Man. Volume 2. The Constant Catastrophe: Malnutrition, Famines and Drought*. Pergamon Press, Oxford.

Gardiner A H, Peet T E and Cerny J (1952-1955) *The Inscriptions of Sinai*. London.

Gasse F, Rognon P and Street F A (1980) Quaternary history of the Afar and Ethiopian Rift lakes. In Williams and Faure (eds): 361-400.

Gasse F, Tehet R, Durand A, Gibert E and Fontes J-C (1990) The arid-humid transition in the Sahara and the Sahel during the last deglaciation. *Nature 346*:141-146.

Gasse F (1978) Les diatoms Holocenees d'une tourbiere (4040 m) d'une montagne Ethiopienne: le Mont Badda. *Revue Alyologine 13*:105-149.

Gasse F (1980) Late Quaternary changes in lake levels and diatom assemblages on the South East margin of the Sahara. *Palaeoecology of Africa and the Surrounding Islands 12*:333-350.

Gasse F and Derbyshire E (1996) Environmental changes in the Tibetan Plateau and surrounding areas. *Palaeogeography, Palaeoclimatology, Palaeoecology:120* (special issue).

Gasse F and Street F A (1978) Late Quaternary lake level fluctuations and environments of the Northern Rift Valley and Afar region (Ethiopia and Djibouti*)*. *Palaeogeography, Palaeoclimatology, Palaeoecology 24*:279-325.

Gautier A (1982) Prehistoric fauna from Ti-n-Torha (Tadrat Acacus, Libya). *Origini 11*:87-127.

Gautier A (1987) Prehistoric men and cattle in North Africa: a dearth of data and a surfeit of models. In Close A E (ed) *Prehistory of Arid North Africa: Essays in Honour of Fred Wendorf.* Southern Methodist University Press, Dallas:163-187.

Gebauer A B and Price T D (1992) *Transitions to Agriculture in Prehistory*. Monographs in World Archaeology 4. Prehistoric Press, Madison.

Geological Map of Ethiopia and Somalia (1979) 1:2 000 000. Ethiopian Mapping Agency.

Geraads D (1983) Faunal remains from the Gash Delta, Sudan. *Nyame Akuma 23*:22.

Getahun A (1973) Root and tuber crops grown and used in Ethiopia. *Paper presented at 3rd International Symposium on Tropical Root Crops*. Ibadan, Nigeria.

Getahun A (1974) The role of wild plants in the native diet of Ethiopia. *Agro-Ecosystems 1*: 45-56.

Getahun A (1980) Agro-climate and agricultural systems in Ethiopia. *Agricultural Systems Journal 5(1)*: 35-50.

Getahun A (1984) Stability and instability of mountain ecosystems in Ethiopia. *Mountain Research and Development 4(1)*: 39-44.

Getahun A (1990) Tropical African mountains and their farming systems. In Riley et al (eds): 105-124.

Gibb H A R (1929) *Ibn Battzta: Travels in Asia and Africa 1325-1354*. Translated and selected by Gibb H A R.

Gillespie R, Street-Perrott F A and Switsur R (1983) Post-galcial arid episodes in Ethiopia have implications for climate prediction. *Nature 306*:680-682.

Gillet J B (1955) The relationship between the Highland floras of Ethiopia and British East Africa. *Webbia 11*:459-466.

Girad A (1873) *Souveniers d'un Voyage en Abyssinie*. Paris.

Goettsch E, EngelsJ M M and Demissie A (1984) Crop diversity in Konso agriculture. *PGRC/E-ILCA Germplasm Newsletter 7*:18-26.

Goodfriend G A (1991) Holocene trends in ^{18}O in land snail shells from the Negev Desert and their implications for changes in rainfall source areas. *Quaternary Research 35*:417-426.

Goudie A S (1981) *The Human Impact: Man's Role in Environmental Change*. Basil Blackwell, Oxford.

Graziosi P (1940) *L'Eta della Pietra in Somalia*. R.Universitas degli Studi di Firenze 18. Firenze: Centro Studi Coloniali.

Graziosi P (1964) New discoveries of rock art paintings in Ethiopia. *Antiquity 38*: 91-101.

Greenberg J H (1964) Historical inferences from liguistic research in sub-Saharan Africa. In *Boston University Papers in African History 1*. Butler J (ed) Boston, Boston University Press: 1-15.

Greenberg J H (1966) *The Languages of Africa*. The Hague, Mouton.

Greenway P J (1945) Origins of some East African food plants. *East African Agricultural Journal 11*:34-119, 177-225; *12*:56-63.

Grigson C, Gowlett J A J and Zarins J (1989) The camel in Arabia - a direct radiocarbon date, calibrated to about 7000 BC. *Journal of Archaeological Science 16*:355-362.

Grigson C (1989) Size and sex: evidence for the domestication of cattle in the Near East. In Milles A, Williams D and Gardner N (eds) *The Beginnings of Agriculture*. BAR International Series 496, Oxford:77-109.

Grigson C (1991) An African origin for African cattle - some archaeological evidence. *African Archaeologcial Reveiw 9*:119-144.

Grigson C (1995) *Bos africanus* (Brehm?). Notes on the archaeology of native cattle of Africa. *Paper presented at the Institute of Archaeology Conference on African Livestock: Archaeology, Linguistics and DNA. September 22-24, 1995, London.*

Grove A T (1978) *Africa*. Oxford, Oxford University Press.

Grove A T (1993) Africa's climate in the Holocene. In *Shaw T. et al*: 32-42.

Grove A T and Goudie A S (1971) Late Quaternary lake levels in the Rift Valley of southern Ethiopia and elsewhere in Tropical Africa. *Nature 234*: 403-5.

Gryseels G and Anderson F (1983) Research on farm and livestock producitivity in the Ethiopian Highlands: Initial results 1977-1980. *ILCA Research Report No 4*, ILCA, Addis Ababa.

Haaland R (1981) *Migratory Herdsmen and Cultivating Women: the Structure of Neolithic Seasonal Adaptation in the Khartoum Nile Environment*. Bergen, Universitet i Bergen.

Haaland R (1984) Continuity and discontinuity. *Norweigan Archaeological Review 17(1)*:39-52.

Haaland R (1991) Specialised pastoralism and the use of secondary products in prehistoric Central Sudan. *Archeologie du Nil Moyen 5*:149-155.

Haaland R (1992) Fish, pots and grain: early and mid-Holocene adaptations in the Central Sudan. *African Archaeologcial Reveiw 10*:43-64.

Haaland R (1995) Sedentism, cultivation and plant domestication in the Holocene Middle Nile region. *Journal of Field Archaeology 22*:157-174.

Haaland R (*in press*) Theory and evidence in archaeological interpretation of the transition from gathering to domestication. In Hathu J and Gosden C (eds) *Changes in Subsistence Systems, Social Theory and Biological Processes*.

Haaland R and Magid A A (1991) Atbara research project: the field seasons of 1985, 1987, 1989 and 1990. *Nyame Akuma 35*:36-41.

Haberland E (1963) *Volker Sud-Athiopiens 3: Galla, Sud-Athiopiens*. Stuttgart.

Haile Selassie Y, Hagos T and Hundie G (1990) An overview of the Stone Age sites of Gadeb, Melka Konture and Besaka. *Proceedings of the 1st International Conference of Ethiopian Studies, Institute of Ethiopian Studies, Addis Ababa.*

Hallam B, Warrren S and Renfrew C (1976) Obsidian in the Western Mediterranean:characterisation by neutron activation analysis and optical emission spectroscopy. *Proceedings of the Prehistoric Society 42*:85-110.

Hallpike C R (1970) Konso agriculture. *Journal of Ethiopian Studies 8*:31-43.

Hamilton A, Taylor D and Vogel J C (1989) Neolithic forest clearence at Ahakagyezi, Western Uganda. In Mahaney (ed):435-464.

Hamilton A (1976) The significance of patterns of distribution shown by forest plants and animals in tropical Africa for reconstruction of Upper Pleistocene palaeoenvironments: a review. *Palaeoecology of Africa and the Surrounding Islands 9*:63-97.

Hamilton A C (1982) *Environmental History of East Africa*. Academic Press, London.

Hamilton A C and Perrott R A (1980) Modern pollen deposition on a tropical mountain. *Pollen et Spores 22*:437-468.

Hamilton A C and Taylor D (1986) Mire sediments in East Africa. In Forstick L E et al (eds) *Sedimentation in the African Rifts.Geological Society Special Publication 25*:211-217.

Hancock G (1985) *The Challenge of Hunger*. Victor Gollancz Ltd, London.

Handwerker W (1983) The first demographic transition: an analysis of subsistence choices and reproductive consequences. *American Anthropologist 85*:5-27.

Harlan J R, de Wet J R J and Stemler A B L (1976) *Origins of African Plant Domestication*. The Hague, Mouton.

Harlan J R (1957) *One Man's Life with Barley*. Exposition Press, New York.

Harlan J R (1967) A wild wheat harvest in Turkey. *Archaeology 20:*197-201.

Harlan J R (1968) On the origin of barley. In *Barley, Origin, Botany, Culture, Winter-hardiness, Genetics, Utilization and Pests.* USDA Agricultural Handbook 338: 12-34.

Harlan J R (1969) Ethiopia: a centre of diversity. *Economic Botany 23(4)*: 309-14.

Harlan J R (1971a) Agricultural origins: centres and non-centres. *Science 174*: 468-74.

Harlan J R (1971b) On the origins of barley: a second look. In *Barley Genetics II.* Proceedings of the International Barley Genetics Symposium. Pullman 1971, Washington State University Press:45-62.

Harlan J R (1979) Wild grass seeds as food sources in the Sahara and sub-Sahara. *Sahara 2:* 69-74.

Harlan J R (1989a) The tropical African cereals. In Harris and Hillman (eds): 335-343.

Harlan J R (1989b) Wild-grass seed harvesting in the Saharan and Sub-Sahara of Africa. In Harris and Hillman (eds): 79-98.

Harlan J R (1992) *Crops and Man.* 2nd Edition. American Society of Agronomy Inc, Madison.

Harlan J R (1993) The tropical African cereals. In Shaw et al (eds): 53-60.

Harlan J R amd Stemler A B L (1976) The races of sorghum in Africa. In Harlan et al (eds):465-478.

Harlan J R and de Wet J M J (1972) A simple classification of cultivated sorghum. *Crop Science 12:* 172-176.

Harlan J R and Zohary D (1966) Distribution of wild wheats and barley. *Science 153*:1074-1080.

Harpending H and Davies H (1977) Some implications for hunter-gather ecology derived from the spatial structure of resources. *World Archaeology 8*:275-286.

Harris D R (1969) Agricultural systems, ecosystems and the origins of agrculture. In Ucko and Dimbleby (eds):3-16.

Harris D R (1972) The origins of agriculture in the tropics. *American Scientist 60*: 180-193.

Harris D R (1973) The prehistory of tropical agriculture: an ethnoarchaeological model. In Renfrew C (ed) *The Explanation of Culture Change: Models in Prehistory.* London, Duckworth: 391-417.

Harris D R (1977a) Alternative pathways toward agriculture. In Reed (ed): 179-243.

Harris D R (1977b) Settling down: an evolutionary model for the transformation of mobile bands into sedentary communities. In Friedman and Rowlands (eds): 401-17.

Harris D R (1989) An evolutionary continuum of people-plant interaction. In Haris and Hillman (eds):11-26.

Harris D R (1996) *The Origins and Spread of Agriculture and Pastoralism in Eurasia.* University College Press, London.

Harris D R and Hillman G C (1989) *From Foraging to Farming.* Unwin, London.

Hassan F A (1980) Prehistoric settlements along the Nile. In Williams and Faure (eds): 421-50.

Hassan F A (1984) Environment and subsistence in Predynastic Egypt. In Clark and Brandt (eds):57-64.

Hassan F A (1986) Chronology of the Khartoum Mesolithic and Neolithic and related sites in the Sudan: a statistical analysis and comparison. *African Archaeological Reveiw 4*:83-102.

Hassan F A (1986) Desert environments and origins of agriculture. *Norweigan Archaeological Reveiw 19*: 63-76.

Hassan F A (1988) Holocene Nile floods and their implications for the origins of Egyptian agriculture. In Bower and Lubbel (eds):1-18.

Hassan F A (1995) Early origins for domestic cattle in North Africa: the climatic factor. *Paper presented at the Institute of Archaeology Conference on African Livestock: Archaeology, Linguistics and DNA. September 1995.*

Hassen M (1990) *The Oromo of Ethiopia: a History 1570-1860.* African Studies Series 66, Cambridge University Press, Cambridge.

Haudricourt A G and Delamarre M J B (1955) *L'Homme et la Charrue a Traverse le Monde.* Paris.

Hawkes J, Engels J M J and Wordede T (1991) *Plant Genetic Resources of Ethiopia.* Cambridge, Cambridge University Press.

Hawkes J G (1983) *The Diversity of Crop Plants.* Harvard University Press, Cambridge, Massachusetts.

Hayden B (1990) Nimrods, piscators, pluckers and planters: the emergence of food production. *Journal of Anthropological Research 9*: 31-69.

Hayden B (1992) Models of domestication. In Gebauer and Price (eds):11-19.

Hayes A J (1905) *The Source of the Blue Nile . A Record of a Journey through the Soudan to Lake Tsana in Western Abyssinia, and of the Return to Egypt by the Valley of the Atbara with a note on the Religion, Customs etc of Abyssinina and an Ethnobotanical Appendix by E B Poulton.* Smith, Elder, London.

Haynes G (1996) Prehistoric human impacts on the environment of North Western Zimbabwe. *Paper presented at the 13th Biennial Congress of SAFA, Posnan, September 1996.*

Hedberg O (1961) The phytogeographical position of afro-alpine vegetation. *Recerche Adr. Botanique 1*:914-919.

Hedberg O (1971) The high mountain flora of the Galama Mountain in Arussi Province, Ethiopia. *Webbia 26*:101-124.

Hedberg O and Edwards S (1989) *Flora of Ethiopia, Volume 3.* The Hebarium, Addis Ababa University, Ethiopia and Department of Systematic Botany, Uppsala Univeristy, Sweden.

Helbaek A (1960) The palaeoethnobotany of the Near East and Europe. In Braidwood R J and Howe B (eds):99-118.

Helbaek A (1965) Archaeological evidence for genetic changes in wheat and barley. *Plant Introduction Review (CSIRO) Canberra 2(19)*:10-16.

Helbaek A (1966) Commentary on the phylogenesis of Triticum and Hordeum. *Economic Botany 20*:350-360.

Helbaek A (1970) The plant husbandry of Hacilar. In Mellart J (ed) *Excavations at Hacilar.* Volume 1. Edinburgh University Press, Edinburgh:189-244.

Henry D O (1983) Adaptive evolution within the Epi-Palaeolithic of the Near East. *Advances in World Archaeology 2*: 99-160.

Henry D O (1985) Preagricultural sedentism: the Natufian example. In Price and Brown (eds):365-384.

Hepper F M (1980) On the botany of James Bruce's expedition to the source of the Blue Nile 1768-1773. *Journal of the Society of Bibliography of Natural History 9(4)*:527-537.

Herring R S (1979) Hydrology and chronology: the Rodah Nilometer as an aid in dating interlacustrine history. In Webtser J B (ed).

Herzog R (1968) *Punt.* Gluckstadt.

Hess R L (1970) *Ethiopia - the Modernization of Autocracy.* Cornell University Press, London.

Higgs E S (1967) Environment and chronology: evidence from mammalian fauna. In McBurney C M (ed): 16-44.

Higgs E S (1972) *Papers in Economic Prehistory.* Cambridge University Press, Cambridge.

Higgs E S and Jarman M R (1972) The origins of animal and plant husbandry. In Higgs E S (ed) *Papers in Economic Prehsitory.* Cambridge Univeristy Press, Cambridge:3-14.

Hillman G C (1975) The plant remains from Tell Abu Hureyra: a preliminary report. *Proceedings on the Prehistoric Society 41*:70-73.

Hillman G C (1989) Late Palaeolithic plant foods from Wadi Kubbaniya in Upper Egypt: dietary diversity, infant weaning and seasonality in a riverine environment. In Harris and Hillman (eds):207-239.

Hillman G C (1996) Late Pleistocene changes in wild plant-foods available to hunter-gatherers of the Northern Fertile Crescent: possible preludes to cereal cultivation. In Harris (ed):159-203.

Hillman C G, Colledge S M and Harris D R (1989) Plant-food economy during the Epi-Palaeolithic period at Tell Abu Hureyra, Syria: dietary diversity, seasonality and modes of exploitation. In Harris and Hillman (eds):240-268.

Hillman G C and Davies M S (1990) Domestication in wild-type wheats and barley under primitive cultivation. *Biology Journal of the Linnean Society 39(1)*:39-78.

Hillman G C and Davies M S (1992) Domestication rate in wild wheats and barley under primitive cultivation: preliminary results and archaeological implications of field measurements of selection coefficients. *Prehistoire de l'Agriculture: Nouvelles Approches Experimentales et Ethnographiques. Monographie du CRA 6. CNRS*:113-157.

Hilu K M, de Wet J M J and Seigler S (1976) Domestication of *Eleusine coracana. Economic Botany 30*: 199-208.

Hilu K W, de Wet J M J and Harlan J R (1979) Archaeobotanical studies of *Eleusine coracana* subsp *coracana* (finger millet). *American Journal of Botany 66*: 330-335.

Hilu K W and de Wet J M J (1976) Racial evolution in *Eleusine coracana* subsp *coracana* (finger millet). *American Journal of Botany 63*: 1311.

Hodder I (1977) The distribution of material culture items in the Baringo district, Western Kenya. *Man 12(2)*:239-269.

Hodder I (1979) Economic stress and material culture patterning. *American Antiquity 44(3)*:446-454.

Hodder I (1982) *Symbols in Action.* Cambridge University Press, Cambridge.

Hodder I (1982) Theoretical archaeology: a reactionary view. In *Symbolic and Structural Archaeology.* Hodder I (ed) Cambridge University Press, Cambridge: 1-16.

Hodder I (1987) The meaning of discard: ash and domestic space in Baringo. In Kent (ed): 424-48.

Hodder I (1990) *The Domestication of Europe.* Blackwell, Oxford.

Hoffman M A (1980) *Egypt Before the Pharaohs.* Routledge and Kegan Paul, London.

Hogg R (1980) Pastoralism and impoverishment: the case of the Isiolo Borana of Northern Kenya. *Disastes V 4(3)*:299-310.

Hole F A, Flannery K A and Neeley J A (1969) *Prehistory and Human Ecology on the Deh Luran Plain.* Museum of Anthropology, University of Michigan.

Hole F A (1984) A reassesment of the Neolithic Revolution. *Palaeorient 10*:49-60.

Holl A (1985) Subsistence patterns of the Dhar Tichett Neolithic, Mauretania. *African Archaeological Review 3*:151-162.

Holm G and Frost S (1983) Genetical and biochemical aspects on the phylogeny of cultivated barley. *Barley Genetics 6*: 45-50.

Holmes L L (1989) *The Predynastic Lithic Industries of Upper Egypt.* Cambridge Monographs in African Archaeology 33.

Hopfen H J and Biesalski E (1953) *Small Farm Implements.* FAO, Rome.

Hornblower G D (1928) An humped bull of ivory. *Journal of Egyptian Archaeology 13*:222-225.

Horowitz A (1971) Climatic and vegetational developments in North West Israel during Upper Pleistocene-Holocene times. *Pollen et Spores 13*:255-278.

Hudson G (1978) Geolinguistic evidence for Ethiopian Semitic prehistory. *Abbay 9*:71-85.

Huffnagel H P (1961) *Agriculture in Ethiopia.* FAO, Rome.

Hughen K A, Overpeck J T, Peterson L C and Trumbore S (1996) Rapid climate change in the tropical Atlantic region during the last deglaciation. *Nature 380*:51-54.

Hugot H S (1968) The origins of agriculture in the Sahara. *Current Anthropology 9*:483-489.

Humphreys G K (1978) A preliminary report of some LSA occurences in the Lake Ziway area of the central Ethiopian Rift Valley. *Annales d'Ethiopie 11*:45-53.

Hunri H (1983) Soil formation rates in Ethiopia. FAO, Rome.

Hunri H (1986) *Soil Conservation in Ethiopia.* Community Forests and Soil Conservation Development Department, Ministry of Agriculture and Livestock Development, Addis Ababa.

Hunri H (1989) Late Quaternary of Simien and other mountains in Ethiopia. In Mahaney (ed):105-120.

Huntingford G W B (1989) *The Historical Geography of Ethiopia from the 1st Century AD to 1704.* Oxford University Press, Oxford.

Hurst H E (1952) *The Nile: a General Account of the River and the Utilization of its Water.* Constable, London.

Hurst H E and Phillips V (1931) *The Nile Basin. I. General Description of the Basin.* Cairo, Physical Department Paper, Government Press.

Hussein A M (1976) *Drought and Famine in Ethiopia.* African Environment Special Report 2. International African Institute, London.

Hutchinson J B (1974) *Evolutionary Studies in World Crops: Diversity and Changes in the Indian SubContinent.* Cambridge University Press, Cambridge.

Hutchinson J B (1976) India: local and introduced crops. *Phil.Transactions of the Royal Society of London B 275*:129-141.

Jaques-Felix H (1971) Grain Impressions. In *The Titchett Tradition: A Late Prehistoric Occupation of the South Western Sahara.* Munson P J PhD Thesis, University of Illinois:355-61.

Jardin C (1967) *List of Foods Used in Africa.* FAO, Rome.

Jarman M R, Bailey G N and Jarman H N (1982) *Early European Agriculture: its Foundations and Development.* Cambridge Unviersity Press, Cambridge.

Jersnimo Lobo (1984) *The Itineries of Jersinmo Lobo.* Translated by Lockhardt D M. Hakluyt Society, London.

Johanson D C (1975) Newly discovered Plio-Pleistocene hominid remains from Central Afar, Ethiopia. *American Journal of Physical Anthropology 42(2)*:309-310.

Johanson D C and Taieb M (1976) Plio-Pleistocene hominid discoveries in Hadar, Ethiopia. *Nature 260*:293-297.

Jones R and Meehan B (1989) Plant foods of the Gidjingali: ethnographic and archaeological perspectives from North Australia on tuber and seed exploitation. In Harris and Hillman (eds): 120-35.

Joussaume R (1981) L'art rupestre de l'Ethiopie. In Roubet C, Hugot H J and Souville G (eds) *Prehistoire Africaine: Melanges Offerts au Doyen Lionel Balout.* ADPF, Paris:159-175.

Joussaume R (1995) *Tiya: l'Ethiopie des Megalithes.* CNRS, Paris

Kajale M D (1974) Ancient grains from India. *Bulletin of the Deccan College Research Institute 34*:55-74.

Kajale M D (1977) On the botanical findings from the excavations at Dainabad, a chalcolthic site in Western Maharashtra, India. *Current Science 46*: 818-827.

Kajale M D (1991) Current status of Indian palaeoethnobotany: Introduced and indigenous food plants with a disscussion of the historical and evolutionary development of Indian agriculture and agricultural systems in general. In Renfrew J (ed) *New Light on Early Farming.* Edinburgh, Edinburgh University Press:155-191.

Kamm A (1954) The discovery of wild 6-row barley and wild *Hordeum intermedium* in Israel. *Lantbrukshogsk Ann.21*:287-320.

Kazanas N and Fields M L (1981) Nutritional improvement of sorghum by fermentation. *Journal of Food Science 46*:819-821.

Keeley L H (1980) Technique and methodology in microwear studies: a critical review. *World Archaeology 5*:323-336.

Kent S (1989) *Farmers as Hunters - the Implications of Sedentism.* Cambridge University Press, Cambridge.

Ketema S (1987) Research recommendation for production and brief outline of strategy for the improvement of tef (*Eragrostis tef* (Zucc.)Trotter). *9th National Crop Improvement Conference, 22-26 April, Addis Ababa.*

Ketema S (1991) Germplasm evaluation and breeding work on tef (*Eragrostis tef*) in Ethiopia. In Hawkes et al (eds): 323-328.

Kim S O (1994) Burials, pigs and political prestige in Neolthic China. *Current Anthropology 35(2)*:119-142.

Kinfe B and Megenasa T (1984) Handweeding of tef (*Eragrostis tef* (Zucc.)Trotter). *Ethiopian Journal of Agricultural Sciences 6(2):* 88-101.

Kirkbride D (1966) Five seasons at the PPN village of Beidha in Jordan. *Palestine Exploration Quaterly. Jan-June 1966.*

Kislev M E, Bar-Yosef O and Gofer A (1986) Early Neolithic domesticated and wild barley from Netiv Hagdud region in the Jordan Valley. *Israel Journal of Botany 35:*197-201.

Kislev M E (1989) Pre-domesticated cereals in the Pre-Pottery Neolithic A period. In Hershkovitz I (ed) *People and Culture Change.* BAR International Series 508 (i), Oxford:147-152.

Kitchen K A (1971) Punt and how to get there. *Orientalia 40*:184-207.

Kitchen K A (1993) The Land of Punt. In Shaw et al (eds):587-608.

Klees K and Kuper R (1992) *New light on the North East African past.* Monographs on African Archaeology and Environment No.5, Koln

Klein R G and Scott K (1986) Re-analysis of faunal assemblages from the Haua Fteah and other late Quaternary archaeological sites in Cyrenican Libya. *Journal of Archaeological Science 13*:515-542.

Klichowska A (1984) Plants of the Neolithic Kadero (Central Sudan): a palaeoethnobotanical study of the plant impressions on pottery. In Krzyzaniak and Kobusiewicz (eds):321-326.

Klichowska M. (1978) Preliminery results of palaeoethnobotanical studies of plant impressions on pot sherd from the Neolithic settlemnet of Kadero. *Nyame Akuma 12*:42-43.

Klima B (1957) Ubersicht uber die jungsten Palaeolithischen Forscungen in Mahren. *Quartar 9:*85-130.

Klima B (1963) *Dolni Vestonice.* Czechoslovak Academy of Sciences, Prague.

Knutson K E (1969) *Plowlands and Swidden: a Dual System of Agriculture in Western Ethiopia.* Addis Ababa.

Kobischanov Y M (1966) The sea voyages of Ancient Ethiopians in the Indian Ocean. *Proceedings of the International Congress of Ethiopian Studies 1*:19-22.

Kobischanov Y M (1981) Axum: political system, economics and culture, first to fourth century. In Mokhtar G (ed) *General History of Africa II. Ancient Civilizations of Africa.* University of California Press, Berkeley:381-400.

Koch C P (1996) Neolithic mortuary behaviour in the Lake Turkana basin, Kenya: Il Lokeridede. *Paper presented at the 13th Biennial Conress of SAFA, Posnan, Septenmber 1996.*

Kohler-Rollefson I (1990) An unpdate on domestication and dispersal of the dromedary. In *Proceedings of the International Conference on Camel Production and Imporovement, 10-13 December 1990, Tobruk, Libya*:49-53.

Krug W (1971) A survey of trypanosomiasis with particular reference to livestock in the South West province of Ethiopia. *Bulletin of Epizootic Diseases in Africa 19*:243-255.

Krzyzaniak L (1978) New light on early food production in the Central Sudan. *Journal of African History 19(2)*: 159-72.

Krzyzaniak L and Kobusiewicz M (1984) *Origin and Early Development of Food-producing Cultures in North Eastern Africa.* Poznan. Poznan Archaeological Museum.

Krzyzaniak L and Kobusiewicz M (1989) *Late Prehistory of the Nile Basin and the Sahara.* Poznan. Poznan Archaeological Museum.

Krzyzaniak L, Kroeper K and Kobusiewicz M (1996) *Interegional contacts in the Later Prehistory of North East Africa.* Poznan.

Kuls W (1958) *Beitrage zur Kulturgeographie der Suedathiopischen Seenregion.* Frankfurter geographische Hefte 32, University of Frankfurt, Federal Republic of Germany.

Kurashina H (1978) *An Examination of the Prehistoric Lithic Technology in East Central Ethiopia.* PhD thesis, University of California, Berkely, Ann Arbor.

Kurimoto E (1995) Trade relations between Ethiopia and the Nile Valley during the 19th century. *Journal of Ethiopian Studies 28(1)*:53-68.

Kutzbach J E (1987) The changing pulse of the Monsoon. In Fein and Stephens (eds):247-268.

Ladizinsky G (1975) Oats in Ethiopia. *Economic Botany 29(3)*:238-241.

Lamb H H (1977) *Climate: Past, Present and Future. Volume 2: Climate History and the Future.* Methuen, London.

Lamb H F, Gasse F, Benkaddour A, Hamouti El, van der Kaars S, Perkins W T, Pearce N J and Roberts C N (1995) Relation between century-scale Holocene arid intervals in tropical and temperate zones. *Nature 373*:134-137.

Lamb H F and van der Karrs S (1995) Vegetational response to Holocene climate change: pollen and palaeolimnolgical data from the Middle Atlas, Morocco. *The Holocene 5(4)*:400-408.

Laurent-Tackholm V (1951) The plant of the Naqada. *Annales du Service des Antiquities de l'Egypte 51*:299-312.

Leakey L S B (1931) *The Stone Age Cultures of Kenya Colony.* London, CUP.

Leakey L S B (1943) Industries of the Gogora Rock Shelter, Lake Tana. *Journal of East Africa and Uganda Natural History Society 17*: 199-203.

Leakey L S B and Ogot B A (1980) *Proceedings of the 8th PanAfrican Congress of Prehistoric and Quaternary Studies,* Nairobi.

Leakey M D (1943) Notes on the ground and polished stone axes of East Africa. *Journal of the East African and Ugandan Natural History Society 17*:182-195.

Leakey M D (1945) Report on the excavations at Hyrax Hill, Nakuru, Kenya Colony 1937-8. *Transactions of the Royal Society of South AFrica 30(4)*:271-409.

Leakey M D (1966) Excavation of burial mounds in Ngorongoro Crater. *Tanzania Notes and Records 66*:123-125.

Leakey M D and Leakey L S B (1950) *Excavations at Njoro River Cave.* Oxford, Clarendon.

Leclant J (1973) Gravures rupestres de Nubie, Maghreb et Sahara: etudes geographiques apportees a Jean Despois. *Societe de Geographie, Paris*:239-246.

Leclant J and Miquel A (1959) Reconaissances dans l'Agame. Gaulo Makeda et Sabea. *Annales d'Ethiopie 3*:107-114.

Lee R B (1965) *Subsistence Ecology of !Kung Bushmen.* Ann Arbor, Michigan PhD Thesis, University of California, Berkley

Lee R B (1969) !Kung Bushman subsistence: an input-output analysis. In Vayda A P (ed) *Environment and Cultural Behaviour: Ecological Studies in Cultural Anthropology.* Natural History Press, New York:47-79.

Lee R B (1972) Population Growth and the Beginnings of Sedentary Life Among the !Kung Bushmen. In *Population Growth:Anthropological Implications.* Spooner T (ed) Cambridge, Massachusetts: 329-42.

Lee R B (1979) *The !KungSan: Men and Women in a Foraging Society.* Cambridge University Press, Cambridge.

Lee R B and DeVore I (1968) *Man the Hunter.* Aldine, Chicago.

Lee R B and DeVore I (1975) *Kalahari Hunter-Gatherers. Studies of the !Kung San and their Neighbours.* University of Harvard Press, Cambridge, Massachucetts.

Legge T (1996) The beginnings of caprine domestication in South West Asia. In Harris (ed):238-262.

Lemordant D (1971) Contribution a l'ethnobotanique Ethiopienne. *Journal of Argicultural and Tropical Botany Application 18(1-6)*: 1-35; *18(4-6)*: 142-179.

Lemordant D (1972) The history and ethnobotanmy of the Konso. *Journal d'Agriculture Tropicale et de Botanique Appliquee 19(12)*:560-582.

Le Quellec J-L (1987) *L'art rupestre du Fezzan sepentrional (Libye) Widyan Zreda et Tarut (Wadi esh-Shati).* Cambridge Monographs in Africa Archaeology 22.

Levine D N (1974) *Greater Ethiopia; the Evolution of a Multi-Ethnic Society.* University of Chicago Press, London.

Lewis I M (1982) *A Pastoral Democracy.* Africana Publishing Co., Nairobi.

Lezine A M and Bonnefille R (1982) Diagramme pollinique Holocene d'un sondage du Lac Abijata (Ethiopie, 7°42' Nord). *Pollen et Spores 24(3-4)*:463-480.

Lhote H (1959) *The Search for the Tassili Frescoes, the Story of the Prehistoric Rock Paintings of the Sahara.* Hutchinson, London.

Liban A (1955) The Ethiopian plough. *Ethnographic Society University College of Addis Ababa Bulletin 4*:23-24.

Livingstone D A (1967) Postglacial vegetation of the Ruwenzori Mountains in Equatorial Africa. *Ecological Monographs 37*:25-52.

Livingstone D A (1980) Environmental changes in the Nile headwaters. In Williams and Faure (eds):339-360.

Loftus R T, MacHugh D E, Bradley D G, Sharp P M and Cunningham P (1994) Evidence for two independent domestications of cattle. *Proceedings of the National Academy of Sciences, USA 91*:2757-2761.

Lynch T F (1983) Camelid pastoralism and the emergence of Tiwanaku civilisation in the South Central Andes. *World Archaeology 15(1)*:1-13.

Mackel R, Schultka W and Walther D (1989) Vegetation and land use changes of Mount Kulal, North Kenya, during the Holocene. In Mahaney (ed):405-420.

MacNeish R S (1950) A synopsis of the archaeological sequence in the Sierra de Tamaulipas. *Revista Mexico de Instituto Antropologia 11*:79-96.

MacNeish R S (1971) Speculation about how and why food production and village life developed in the Tehuacan Valley, Mexico. *Archaeology 24(4)*:307-315.

Mahaney W C (1989) Agriculture and its effect on montane forest soils, Naro Moru area, Kenya, East Africa. In Mahaney (ed):421-434.

Mahaney W C (1989) *Quaternary Environmental research on East African Mountains*. A A Balkema, Rotterdam.

Maire R (1955*) Flore de l'Afrique du Nord. Vol 3*. Lechevalier, Paris.

Maley J (1989) Late Quaternary climatic changes in the African rainforest - forest refugia and the major role of sea surface temperature variations. In Leinen M and Sarnthein M (eds*) Palaeoclimatology and Palaeometerology: Modern and Past Patterns of Global Atmospheric Transport*. Kluwer Academic Publicatione, Dordrecht:585-616.

Maley J (1993) Climatic and vegetational history of the Equatorial regions. In Shaw et al (eds):43-52.

Mann J A, Kimber C T and Miller F R(1983) The origin and early cultivation of sorghums in Africa. *Bulletin no.1434 Texas A and M University, College Station, Texas.*

Marco Polo *The Travels of Marco Polo*. Translated by Benedetto L F and Ricci A (1931).

Marks A E, Mohammed-Ali A and Fattovich R (1986) The archaeology of the Eastern Sudan: a first look. *Archaeology 39(5)*: 44-50.

Marks A E and Sadr K (1988) Holocene environments and occupations in the Southern Atbai, Sudan: a preliminary formulation. In Bower and Lubell (eds): 69-90.

Marquardt W H (1985) Complexity and scale in the study of fisher-gatherer-hunters: an example from the Eastern United States. In Price and Brown (eds):59-98.

Marshall F (1989) Rethinking the role of *Bos indicus* in Africa. *Current Anthropology 30*:235-240.

Martin R D (1972) Concepts of human territoriality. In Ucko et al (eds).

McBurney C B M (1967) *The Haua Fteah (Cyrenica) and the Stone Age of the South East Mediterranean*. Cambridge. C.U.P.

McCarthy F D and McArthur M (1960)The food quest and the time factor in Aboriginal economic life. in Mountford C P (ed) *Records of the America-Australian Scientific Expedition to Arnhem Land II: Anthroplogy and Nutrition*. Melbourne University Press, Melbourne:145-194.

McCorriston J (1992) *The early development of agriculture in the ancient Near East: an ecological-evolutionary study*. PhD Dissertation, Department of Anthropology, Yale University.

McCorriston J and Hole F (1991) The ecology and seasonal stress and the origins of agriculture in the Near East. *American Anthropologist 93*: 46-69.

McIntosh R J and McIntosh S K (1981) The inland Niger delta before the empire of Mali: evidence from Jenne-jeno. *Journal of African History 22*:1-22.

McIntosh S K and McIntosh R J (1980) *Prehistoric investigations at Jenne, Mali*. Oxford BAR International Series 89.

Meadow R H (1989) Osetological evidence for the process of animal domestication. In Clutton-Brock (ed):80-90.

Meadow R H (1996) Agriculture and pastoralism in northwest South Asia. In Harris (ed):390-412.

Megaw J V S (1977) *Hunters, Gatherers and the First Farmers Beyond Europe*. Leicester University Press, Surrey.

Mehra K L (1963a) Considerations on the African origin of *Eleusine coracana* (L.)Gaertn. *Current Science 32*: 300-301.

Mehra K L (1963b) Differentiation of the cultivated and wild *Eleusine* species. *Phyton 20*: 189-198.

Mehra K L (1991) Prehistoric Ethiopia and India: contacts through sorghum and millet genetic resources. In Hawkes et al (eds):160-168.

Mekbib H, Demissie A and Tulu A (1991) Pulse crops of Ethiopia: genetic resources and their utilisation. In Hawkes et al (eds): 329-343.

Mellaart J (1975) *The Neolithic of the Near East*. London. Thames and Hudson.

Mengesha M H (1975) Crop germplasm diversity and resources in Ethiopia. In Frankel O H and Hawkes J G (eds*) Genetic Resources for Today and Tomorrow*:449-453.

Messerli B and Winiger M (1980) The Saharan and East African uplands during the Quaternary. In Williams and Faure (eds):87-132.

Meyer C, Todd J M and Beck C (1991) From Zanzibar to Zagros: a copal pendant from Eshunna. *Journal of Near Eastern Studies 50*:289-297.

Meyers J T (1971) The origins of agriculture: an evaluation of three hypotheses. In *Prehistoric Agriculture*. Struever S (ed) The Natural History Press, New York:101-121.

Michels J W (1988) Regional political organization in the Axum-Yeha area during the Pre-Axumite and Axumite era. *Paper presented at the 10th conference of Ethiopian Studies, Paris 1988.*

Michels J W (1988) The Axumite kingdom: a settlement archaeological perspective. *Proceedings of the 9th International Conference of Ethiopian Studies 5, Moscow*:173-183.

Miller L F and Makonnen T (1965) Organisation and operation of three Ethiopian case farmers. *Experimental Station Bulletin 35*, Agricultural College, Addis Ababa.

Mloszewski M J (1983) *The Behaviour and Ecology of the African Buffalo*. Cambridge University Press, New York.

Mloszewski M J and Mahaney W C (1989) A check-list of wild Africa Bovidae related to montane habitats, with an expanded note on geographic behaviour of buffalo on Mount Kenya. In Mahaney (ed):309-326.

Mohammed Ali A S (1984) Evidence of early food production in North East Africa: an alternative model. In Krzyzanizak and Kobusiewicz (eds):65-72.

Mohammed M U, Bonnefille R, and Johnson T C (1995) Pollen and isotopic records in Late Holocene sediments from Lake Turkana, Kenya. *Palaeogeography, Palaeoclimatology, Palaeoecology 119*:371-383.

Mohammed M U (1992) *Palaeoenvironment et palaeoclimatologie des derniers millenaires en Ethiopie, contribution palynologique.* Thesis de Universite d'Aix Marseille 3.

Mohammed M V and Bonnefille R (1991) The recent history of vegetation and climate around Lake Langano (Ethiopia). *Palaeoecology of Africa and the Surrounding Islands 22*:275-286.

Mohammed-Ali A S (1982) *The Neolithic Period in the Sudan, c.6000-2500 BC.* Cambridge Monographs in African Archaeology 6. BAR International Series 139.

Mohammed I M (1986) *The Archaeology of Central Darfur (Sudan) in the 1st Millennium AD.* Cambridge Monographs in African Archaeology 14.

Mooney H (1961) The natural vegetation of Ethiopia. *Ethiopian Observer 5(3):* 203-206.

Moore A M T (1982) Agricultural origins in the Near East - a model for the 1980s. *World Archaeology 14:*224-236.

Morner N A and Wallin B (1977) 10 000 year temperature record from Gotland, Sweden. *Palaeogeography, Palaeoclimatology, Palaeoecology 21*:113-138.

Mortensen P (1986) The Bahrain Temple - its chronology and foreign relations reconsidered. *Bahrain Through the Ages. Archaeological Proceedings of the 1983 Bahrain Historical Conference.*

Moysey F (1943) Excavation of a rock shelter at Gogora, Lake Tana. *Journal of East African and Ugandan Natural History Society 17*: 196-8.

Munro-Hay S (1989) *Excavations at Axum.* British Institute in Eastern Africa, London.

Munson P J (1977) Africa's prehistoric past. In *Africa.* O'Meara P (ed) Bloomington, Indiana University Press: 62-82.

Murdock G P (1959) *Africa, its peoples and their Culture History.* New York.

Murdock (1967) *Ethnographic Atlas.* University of Pittsburgh Press, Pittsburgh.

Murphy H F (1963) Fertility and other data on some Ethiopian soils. *Experimental Station Bulletin No 4.* Imperial College of Agricultural and Mechanical Arts, Alamya.

Murphy P J, Witcombe J R, Shewry P R and Mifln B J (1982) The origin of 6-rowed 'wild' barely from the Western Himalaya. *Euphytica 31:*183-192.

Murray J (1970) *The First European Agriculture.* Edinburgh.

Muzzolini A (1983) *L'art rupestre du Sahara central: classification et chronologie. Le boeuf dans la prehistorire africaine.* These 3eme cycle, Universite de Provence, Aix-en-Provence.

Muzzolini A (1986) *L'art Rupestre Prehistorique des Massifs Centraux Sahariens.* Oxford. British Archaeological Reports.

Muzzolini A (1993) The emergence of a food-producing economy in the Sahara. In Shaw T et al (eds): 227-39.

Muzzolini A (1995) *Les Images Rupestres du Sahara.* Prehistorique du Sahara 1, France.

National Atlas of Ethiopia (1988) 1:4 000 000 Ethiopian Mapping Agency, Addis Ababa.

Naville E (1898) *The Temple of Deir el Bahari. Part III.* Egypt Exploration Fund, London.

Ndiiri W (1996) The impact of man on the environment of the Kenya Coast during the Holocene period. *Paper presented at the 13th Biennial Congress of SAFA, Posnan, September 1996.*

Negash A (1990) Distribution of rock paintings and engravings in Ethiopia. *Proceedings of the 1st National Conference of Ethiopian Studies, Addis Ababa, April 11-12:*289-302.

Negassa M (1985) Patterns of phenotypic diversity in an Ethiopian barley collection, the Arussi-Bale Highlands as a centre of origin of barley. *Hereditas 102*:139-150.

Nelson C M (1980) The Elmenteitan lithic industry. In Leakey and Ogot (eds):275-278.

Nelson C M (1995) The work of the Koobi Fora Field School at the Jarigole Pillar Site. *Kenya Past and Present 27*:49-63.

Nelson C M (n.d.) *A Comparative Analysis of 29 Later Stone Age Occurrences from East Africa.* Unpublished PhD dissertation, University of California, Berkeley.

Nesbitt M (1993) Archaeobotanical evidence for early Dilmun Diret at Saar, Bahrain. *Arabian Archaeology and Epigraphy 4(1)*:20-47.

Neumann (1996) (abstract) *Proceedings for the PanAfrican Conference* . Harare, June 1995.

Newcomer M, Grace R and Unger-Hamilton R (1986) Investigating microwear polish with blind tests. *Journal of Archaeological Science 13*:203-217.

Nicholson G E (1960) The production, history, uses and relationships of cotton (*Gossypium* spp) in Ethiopia. *Economic Botany 14*:3-36.

Nicolaisen J (1963) *Ecology and Culture of the Pastoral Tuareg with Particular Reference to the Tuareg of Ahaggar and Air.* National Museum of Copenhagen, Ethnogratiste Raekke IX.

Niklewski J and van Zeist W (1970) A late Quaternary pollen diagram from North West Syria. *Acta Botanica Neerlandica 19:*737-754.

O'Brien S R, Mayewski P A, Meeker L D, Meese D A, Twickler M S and Whitlow S I (1995) Complexity of Holocene climate as reconstructed from a Greenland ice core. *Science 270*:1962-1964.

O'Connor D (1982) Egypt 1552-664 BC. In Clark (ed):830-940.

Onyango-Abuje J C (1977) Crescent Island: a preliminary report on excavations at an East African Neolithic site. *Azania 12*:147-159.

Onyango-Abuje J C (1980) Temporal and spatial distribution of Neolithic cultures in East Africa. In Leakey and Ogot (eds):288-292.

Orlov A A (1929) The barleys of Abyssinia and Eritrea. *Bulletin of Applied Botany 20*: 283-342.

Owen R B, Barthelme J W, Renault R W and Vincens A (1982) Palaeolimnology and archaeology of Holocene deposits North East of Lake Turkana, Kenya. *Nature 298*:523-529.

Pankhurst R (1961) *An Introduction to the Economic History of Ethiopia (from early times to 1800)*. Lalibela House, London.

Pankhurst R (1965) *Travellers in Ethiopia*. Oxford University Press, London.

Pankhurst R (1966) Some factors depressing the standard of living of peasants in traditional Ethiopia. *Journal of Ethiopian Studies 4(2)*: 45-98.

Pankhurst R (1968) *Economic History of Ethiopia 1800-1935*. Haile Selassie I University Press, Addis Ababa.

Pankhurst R (1984) *The History of Famines and Epidemics Prior to the Twentieth Century*. Relief and Rehabilitation Commission, Addis Ababa.

Pastouret L, Chamley H, DelibriasG, Duplessy J C and Thiede J (1978) Late Quaternary climatic changes in the Western tropical Africa deduced from deep sea sedimentation off the Niger Delta. *Oceanologica Acta 1*:217-232.

Pearsall D M (1989) Adaptation of prehistoric hunter-gatherers to the high Andes: the changing role of plant resources. In Harris and Hillman (eds):318-332.

Pearsall D M (1989) *Palaeoethnobotany: a Handbook of Procedures*. Academic Press, London.

Periplus (1912*) The Periplus of the Erythraean Sea*. Translated by Schoff W H.

Perkins D (1964) Prehistoric fauna from Shanidar, Iraq. *Science 144*:1565-1566.

Perrott R A and Street-Perrott F A (1982) New evidence for a late Pleistocene wet phase in Northern inter-tropical Africa. *Palaeoecology of Africa 14*:57-75.

Peters J (1985-6) *Bijdrage tot de Archeozoologie van Soedan en Egypte*. Unpublished PhD thesis, Rijksuniversiteit Gent, Fakulteit des Watenschappen.

Petit Maire N, Fontugne M and Rouland C (1991) Atmospheric methane ratio and environmental changes in the Sahara and Sahel during the past 130 000 years. *Palaeogeography, Palaeoclimatology, Palaeoecology 86*:197-204.

Phillips J (1997) Punt and Aksum: Egypt and the Horn of Africa. *Journal of African History 38*: 423-457.

Phillipson D W (1977a) *The Later Prehistory of Eastern and Southern Africa*. Heinemann, London.

Phillipson D W (1977b) The excavation of Gobedra rock shelter, Axum: an early occurrence of cultivated finger millet in Northern Ethiopia. *Azania 12*: 53-82.

Phillipson D W (1977c) Lowasera. *Azania 12:*1-32.

Phillipson D W (1984) Aspects of food production in Northern Kenya. In Krzyzaniak and Kobusiewicz (eds):489-495.

Phillipson D W (1992) *African Archaeology*. Cambridge University Press, Cambridge.

Phillipson D W (1993) The antiquity of cultivation and herding in Ethiopia. In Shaw et al (eds): 344-57.

Phillipson D W (1994) The significance and symbolism of the Axumite stelae. *Cambridge Archaeological Journal 4*: 189-210.

Phillipson D W and Reynolds A (1996) BIEA excavations at Aksum, Northern Ethiopia. *Azania 31*:99-148.

Pichi-Sermolli R E G (1951) Missione di Studio al Lago Tana. *Ricerche Botanica 7*:1-39.

Pichi-Sermolli R E G (1955) Tropical East Africa (Ethiopia, Somaliland, Kenya and Tanganyika). *Arid Zone Research 6: Plant Ecology, Reviews of Research*. UNESCO:302-360.

Pichi-Sermolli R E G (1957) Una carte geobotanica dell'Africa Orientale (Eritrea, Ethiopia, Somalia). *Webbia 13(1):*15-132+map.

Pickersgill B (1989) Cytological and genetical evidence on the domestication and diffusion of crops within the Americas. In Harris and Hillman (eds):426-439.

Porter B and Moss R (1952) *Topographical Bibliography of Ancient Egyptian Hieroglyphic Texts, Reliefs and Paintings VII: Nubia, Deserts and Outside Egypt*. Oxford University Press, Oxford.

Porteres R (1958) *Journal of Agriculture and Tropical Botany Appl. 5*:454-463.

Porteres R (1976) African cereals: *Eleusine*, fonio, black fonio, tef, *Brachiaria, Paspalum, Pennisetum* and African rice. In Harlan et al (eds):156-172.

Potts D T (1993) The late Prehistoric, Protohistoric and early Historic periods in Eastern Arabia (c.5000-1200 BC). *Journal of World Prehistory 7:*163-212.

Potts D T (1994) Contribution to the agrarian history of Eastern Arabia. II The Cultivators. *Arabian Archaeology and Epigraphy 5(4)*:236-275.

Prance G T (1987) Biogeography of Neotropical plants. In Whitmore T C and Prance G T (eds) *Biogeography and Quaternary History in Tropical America*. Clarendon Press, Oxford Monographs in Biogeography 3, Oxford:46-65.

Price T D and Brown J A (1985) *Prehistoric Hunter-Gatherers: the Emergence of Social Complexity*. Academic Press, London.

Purseglove J W (1968) *Tropical Crops: Dicotyledons. I and II*. Longman, London.

Purseglove J W (1972) *Tropical Cereals*: Monocotyledons. London.

Purseglove J W (1976) The origins and migration of crops in Tropical Africa. In Harlan et al:*-309.

Qualset C O (1975) Sampling germplasm in a centre of diversity: an example of disease resistance in Ethiopian barley. In Frankel O H and Hawkes J G (eds) *Crop Genetic Resources for Today and Tomorrow*. Cambridge University Press, Cambridge:81-98.

Rafferty J E (1985) The archaeological record on sedentariness, recognition, development and implications. *Advances in Archaeological Method and Theory 8*: 113-56.

Rahmato D (1988) *Peasant Survival Strategies*. International Institute for Relief and Development/Food for the Hungry International, Geneva.

Rappaport R A (1984) *Pigs for the Ancestors: Ritual in the Ecology of a New Guinea People*. Yale University Press, New Haven.

Rao S R (1963) Excavations at Rangpur and other explorations in Gujarat. *Ancient India 18-19*. New Delhi.

Reed C A (1977) *Origins of Agriculture*. World Anthropology.

Reihl H and Meitin J (1979) Discharge of the Nile River: a barometer of short period climatic variation. *Science 206*:1178-1179.

Reinolds J (1994) Le Neolitique au Soudan. *La Nubie 196*:6-11.

Renaut R W and Owen R B (1980) Late Quaternary fluvio-lacustrine sedimentation and lake levels in the Baringo basin, northern Kenya Rift Valley. *Recherches Geologiques en Afrique 5*: 130-33.

Renfrew C (1987) *Archaeology and Language: the Puzzle of Indo-European Origins*. Jonathan Cape, London.

Renfrew C (1994) World linguistic diversity. *Scientific American 261(4)*:82-90.

Renfrew C (1996) Language families and the spread of farming. In Harris (ed):70-92.

Renfrew J M (1969) The archaeological evidence for the domestication of plants: methods and problems. In Ucko and Dimbleby (eds): 149-72.

Renfrew J M (1973) *Palaeoethnobotany*. Columbia University Press, New York.

Renfrew J M (1991) *New Light on Early Farming: Recent Developments in Palaeoethnobotany*. Edinburgh University Press, Edinburgh.

Renvoize S A, Cope T A, Cook F E M, Clayton W D and Wickens G E (1992) Distribution and utilisation of grasses in arid and semi-arid regions. In Chapman T (ed) *Desertified Grasslands: their Biology and Management*. Academic Press, London:3-16.

Rhindos D (1984*) The Origins of Agriculture: an Evolutionary Perspective*. Academic Press, New York.

Rhindos D (1985) Darwinian selection, symbolic variation and the evolution of culture. *Current Anthropology 26*: 65-77.

Richardson J L and Dussinger R A (1986) Palaeolimnology of mid-elevation lakes in the Kenya Rift Valley. *Hydrobiologia 147*: 167-74.

Rightmire G P (1975) New studies of post-Pleistocene human skeletal remains from the Rift Valley, Kenya. *American Journal of Physical Anthropology 42*:351-369.

Riley K W, Mateo N, Hawtin G C and Yadav R (1990) *Mountain Agriculture and Crop Genetic Resources*. International Development Research Centre, Nepal.

Robbins L H (1972) Archaeology in the Turkana district, Kenya. *Science 176*:359-366.

Robbins M C (1974) *The Lothagam site: a Later Stone Age fishing settlement in the Lake Rudolf basin, Kenya*. East Lansing, Michigan State University Museum Anthropological Series 1(2).

Robbins M C (1980*) Lopoy: a Late Stone Age fishing and pastoral settlement west of Lake Turkana, Kenya*. Michigan State University Museum Anthropological Series 3(1).

Roberts N, Taieb M, Barker P, Damnati B, Kole M and Williamson D (1993) The timing of the Younger Dryas event in East Africa from lake level changes. *Nature 366*:146-148.

Roberts N (1980) *Late Quaternary Geomorphology and Palaeoecology of the Konya Basin, Turkey*. Unpublished PhD thesis, London University.

Roberts N (1983) Age, palaeoenvironments and climatic significance of late Pleistocene Konya Lake, Turkey. *Quaternary Research 19:*154-171.

Roberts N (1984) Reply to comment by Ann P El-Moslimany. *Quaternary Research 21*:117-120.

Roberts N (1990) Ups and downs of African lakes. *Nature 346*:107.

Roberts N and Barker P (1993) Landscape stability and biogeomorphic response to past and future climatic shifts in intertropical Africa. In Thomas D S G and Allison R J (eds) *Landscape Sensitivity*. John Wiley and Sons Ltd, London.

Roberts N and Wright J E Jr (1993) Vegetational and lake-level climatic history of the Near East and Southwest Asia. In Wright J E Jr, Kutzbach J E, Webb III T, Ruddiman W F, Street-Perrott F A and Bartlein P J (eds) *Global Climates since the Last Glacial Maximum*. University of Minnesota Press, Minneapolis:194-220.

Robertshaw P J and Collett D P (1983) The identification of pastoral peoples in the archaeological record: an example from East Africa. *World Archaeology 15*: 67-78.

Rosenberg M (1990) The Mother of Invention: evolutionary theory, territoriality and the origins of agriculture. *American Anthropologist 92*: 399-415.

Rossel G (1991) The diffusion of plantain (*Musa* sp AAB) and banana (*Musa* sp AAA) in Africa: a case for linguists, taxonomists and historians, focused on Nigerian crop names. In Leakey R E and Slikkerveer L J (eds) *Origins and Development of Agriculture in East Africa: the Ethnosystems Approach to the Study of Early Food Production in Kenya*. Iowa State University, Iowa:129-158.

Rossel G (1996) *Musa* and *Ensete* in Africa: taxonomy, nomenclature and uses. In Sutton J E G (ed):130-145.

Rossignol-Strick M (1983) African monsoons an immediate climate response to orbital insolation. *Nature 304*:46-49.

Rossignol-Strick M and Duzer D (1979) West African vegetation and climate since 22 500 BP from deep sea cores palynology. *Pollen et Spores 21*:105-134.

Rostovzeft M (1941) *The Social and Economic History of the Hellenistic World*. **Roth H L** (1887) On the Origins of Agriculture. Journal of the Royal Anthropological Institute of Great Britain and Ireland 16:102-136.

Rosen S A (1988) Notes on the origins of pastoral nomadism: a case study from the Negev and Sinai. *Current Anthropology 29*:498-506.

Roubet C (1971) La prehistoire dans la IV Mission Internationale en Afar (Ethiopie). *L'Anthropologie 75*:324-327.

Roubet C (1978) Une economie pastorale, pre-agricole en Algerie orientale: le Neolithique de tradition capsienne. *L'Anthropologie 82*:583-586.

Rouk H F and Mengesha M H (1963) *An introduction to tef (*Eragrostis abyssinica *Schad.), a nutritious cereal grain of Ethiopia*. Debre Zeit Agricultural Research Station Bulletin 26.

Rowley-Conwy P (1983) Sedentary hunters: the Ertebolle example. In Bailey G (ed) *Hunter-Gatherer Economy in Prehistory*. Cambridge University Press, Cambridge:111-126.

Rowley-Conwy P (1988) The camel in the Nile Valley. New radiocarbon accelerator (AMS) dates from Qasr Ibrim. *Journal of Egyptian Archaeology 74*:245-248.

Rowley-Conwy P (1991) Sorghum from Qasr Ibrim, Egyptian Nubia, c.800 BC-AD 1811: a preliminary study. In Renfrew J (ed) *New Light on Early Farming*. Edinburgh University Press, Edinburgh:191-212.

Sadr K (1986) A Pan-Grave related culture. *Nyame Akuma 27*:48-49.

Sadr K (1988) Settlement Patterns and Land-use in the Late Prehistoric Southern Atbai, East Central Sudan. *Journal of Field Archaeology 15(4)*: 381-403.

Sadr K (1988) *The development of Nomadism: the view from ancient North Africa*. PhD dissertation, Southern Methodist University, Ann Arbor.

Sadr K (1991) *The Development of Nomadism in Ancient North East Africa*. University of Pennsylvania Press.

Sahlins M (1972) *Stone Age Economics*. Chicago, Aldine.

Sahlins M and Service E R (1960) *Evolution and Culture*. University of Michigan Press, Ann Arbor.

Saied A M A H (1977) Discovery of the site of the 12th Dynasty port at Wadi Gawasis on the Red Sea shore. *Revue d'Egyptologie 29*:140-176.

Salt H (1814) *A Voyage to Abyssinia*. London.

Saraswat K S (1986) Ancient plant economy of Harappans from Rohira, Punjab (c.200-1 700 BC). *The Palaeobotanist 35*: 32-38.

Saraswat K S (1993) Plant economy of late Harappans at Hulas. *Pauratattra 23*:1-22.

Sauer C O (1952) *Agricultural Origins and Dispersals*. New York, American Geographical Society.

Scheiwnfurth G (1868) *Pflanzen Geographische Skizze des Gesammten Nil-Gebiets und der Uferlander des Rothen Meeres*. Petermanns Mitt.

Schiemann E (1951) New results of the history of cultivated cereals. *Heredity 5*:305-320.

Schiffer M B (1987) *Formation Processes of the Archaeological Record*. University of New Mexico. Albuquerque.

Schoeninger M J and De Niro M J (1982) Carbon isotope ratios on apatite from fossil bone cannot be used to reconstruct diets of animals. *Nature 297*:577-578.

Schon W (1992) The late Neolithic of Wadi el Akhdar (Gilf Kebir) and the Eastern Sahara. *Archaeology du Nil Moyen 6*:131-175.

Schulz E (1987) Holocene vegetation in the Tadrart Acacus: the pollen record of two early ceramic sites. In Barich B E (ed) *Archaeology and Environment in the Libyan Sahara. The Excavations in the Tadrart Acacus 1978-1983*. BAR International Series 368:313-326.

Schulz E (1996) Holocene vegetation history of the Sahara. The predominance of climate or man? *Paper presented at the 13th Biennial Congress of SAFA, Posnan, September 1996.*

Sebald V O (1968) Breicht Botanisches studien und sammlungen an Tana-see und im Senyen-gerbirge (Aethiopien). *Stuttgart Beitr Naturkd 194*:1-40.

Seegeler C (1983) *Oil Plants in Ethiopia: their Taxonomy and Agricultural Significance*. PUDOC, Washington.

Selinus R (1971) Dietary Studies in Ethiopia: Dietary pattern among the Rift Valley Arsi Galla. *American Journal of Clinical Nutrition 24*:365-377.

Selinus R, Awalom G and Gobezie A (1971) Dietary studies in Ethiopia II. Dietary pattern in two rural communities in Northern Ethiopia. *Acta Soc Med Uppsala 76:* 17-38.

Semenov S A (1964) *Prehistoric Technology*. London.

Servant M and Servant-Vildary S (1980) L'environment quaternaire du bassin du Tchad. In Williams M A J and Faure H (eds): 133-62.

Service E R (1962) *Primitive Social Organisation: an Evolutionary Perspective*. Random House, New York.

Service E R (1975) *Origins of the State and Civilisation: the Process of Cultural Evolution*. Norton, New York.

Sethe K (1908-9) *Urkunden der 18 Dynastie*. Hinnichs, Leipzig.

Shao Q, Chang-Sen L and Chiren B (1982) Origin and evolution of cultivated barley: wild barley from west Szechuan and Tibet, China. *Barley Genetics Newsletter 12*: 37-42.

Shao Q, Jeqiz Z and Ansheng L (1990) Wild and cultivated barleys in the Himalayas. In Riley K W, Mateo N, Hawtin G C and Yadav R (eds) *Mountain Agriculture and Crop Genetic Resources*. International Development Research Centre:305-310.

Shao Q (1982) *Wild Barley from Xizhang (Tibet)*. Science Press.

Shaw T (1977) Hunters, gatherers and first farmers in West Africa. In Megaw (ed): 69-125.

Shaw T and Daniel S G H (1984) Excavations at Iwo Eleru, Ondo State, Nigeria. *West African Journal of Archaeology 14*:1-269.

Shaw T, Sinclair P, Andah B and Okpoko A (1993) *The Archaeology of Africa: Food, Metals and Towns*. Routledge. London.

Shennan S (1985) *Experiments in the Collection and Analysis of Archaeological Survey Data: East Hampshire Survey.*

Sherratt A (1980) Water, soil and seasonality in early cereal cultivation. *World Archaeology 11(3)*:313-330.

Shinnie P L (1984) The mainlines of socio-economic development in the Sudan in post-Neolithic times. In Krzyzaniak and Kobusiewicz (eds):109-115.

Shipek F (1989) An example of intensive plant husbandry: the Kumeyaoy of South California. In Harris and Hillman (eds):159-170.

Siegenthaler L E (1960) Useful plants of Ethiopia. *Experimental Station Bulletin 14(1)*. Agricultural College, Addis Ababa.

Simmonds N W (1976) *Evolution of Crop Plants*. Longman. London.

Simoons F J (1958) The agricultural implements and cutting tools of Begemder and Semyen, Ethiopia. *South West Journal of Anthropology 14*:388.

Simoons F J (1960) *Northwest Ethiopia: Peoples and Economy*. University of Wisconsin Press, Madison.

Simoons F J (1965) Some Questions on the Economic Prehistory of Ethiopia. *Journal of African History 6(1)*: 1-12.

Skinner N (1984) Afroasiatic vocabulary: evidence for some culturally important items. *Africanan Marburgensia special issue*.

Smith A B (1984) Origins of the Neolithic in the Sahara. In Clark and Brandt (eds): 84-92.

Smith A B (1989) Cattle domestication in North Africa. *African Archaeological Review 4*:197-203.

Smith A B (1992) *Pastoralism in Africa*. Hurst and Co., London.

Smith A M (1992) Holocene palaeoclimatic trends from palaeoflood analysis. *Palaeogeography, Palaeoclimatology, Palaeoecology 97*:235-240.

Smith B D (1995) *The Emergence of Agriculture*. Scientific American Library, New York.

Smith J S C and Lester R N (1980) Biochemical systematics and evolution of Zea, Tripsacum and related genera. *Economic Botany 34*:201-218.

Smith S E (1980) The environmental adaptation of Nomads on the West African sahel: a key to understanding prehistoric pastoralists. In Williams and Faure (eds): 467-488.

Snowden J D (1936) *J D Snowden; the Cultivated Races of Sorghum*. London.

Sperling L (1984) The adoption of camels by Samburu cattle herders. *Paper for seminar on Significance and Prospects for Cattle Pastoralism in Kenya*. Marsabit 1984.

Srejovic D (1972) *Europe's First Monumental Sculpture: New Discoveries at Lepenski Vir*. Thames and Hudson, London.

Stager J C (1984) The diatom record of Lake Victoria (East Africa): the last 17 000 years. In *Proceedings of the 7th Diatom Symposium, Philadelphia*, 22-27 August 1982. Mann D G (ed), Koenigstein, Otto Koeltz Science Publishers: 455-76.

Stager J C and Mayewski P A (1997) Abrupt early to mid-Holocene climatic transition registered at the Equator and the poles. *Science 276:* 1834-1836.

Stahl A B (1984) Hominid dietary selection before fire. *Current Anthropology 25*:151-168.

Stahl A B (1989) Plant-food processing: implications for dietary quality. In Harris and Hillman (eds):171-196.

Stanley S (1966) *Some Prehistoric Aspects of the Ethiopia Rural Economy*. Addis Ababa Haile Selassie 1 University.

Stanely R S and Rose E K (1980) Diet, nutrition and population dynamics in the Basin of Mexico. *World Archaeology 11:*185-207.

Staudt G (1961) the origin of cultivated barleys, a discussion. *Economic Botany 15*: 203-212.

Stemler A B L (1980) Origins of plant domestication in the Sahara and the Nile Valley. In Williams and Faure (eds): 503-27.

Stemler A B L (1990) A scanning electron microscope analysis of plant impressions in pottery from the sites of Kadero, el Zakiab, Um Direiwa and el Kadada. *Archeologie du Nil Moyen 4*:87-105.

Stemler A B L, Harlan J R and de Wet J M J (1977) The sorghums of Ethiopia. *Economic Botany 31*: 446-460.

Stemler A B L and Falk R H (1981) SEM of archaeological plant specimens. *Scanning Electron Microscopy P3*:191-196.

Steward J H (1955) *Theory of Culture Change*. Urbana, University of Illinois Press.

Stiehler W (1948) Studien zur landwirtschafts und siedlungsgeographic Athiopians. *Erdkunde 2*: 257-282.

Stiles D (1980) A diachronic study of the demography, human ecology and history of pastoral groups in North Kenya. *Report to the Ford Foundation and UNESCO*.

Stiles D (1988) Historical inter-relationships between the Boni with pastoral peoples of Somalia and Kenya. *Kenya Past and Present 20*:38-45.

Stiles D (1993) The past and present of hunter-gatherers in Kenya. *Kenya Past and Present 25*:39-45.

Street F A (1975-77) Late Quaternary precipitation estimates for the Ziway-Shala Basin, South Ethiopia. *Palaeoecology of Africa and Surrounding Islands 2*:135-143.

Street F A (1979) *Late Quaternary lakes in the Ziway-Shala basin, Southern Ethiopia*. Unpublished PhD thesis, University of Cambridge.

Street F A and Grove A T (1976) Environmental and climatic implications of late Quaternary lake-level fluctuations in Africa. *Nature 261*: 385-90.

Street Perrott F A, Roberts N and Metcalfe S (1985) Geomorphic implications of Late Quaternary hydrological and climate changes in the Northern hemisphere tropics. In Douglas I and Spence T (eds) *Environmental Changes and Tropical Geomorphology*. London, Allen and Unwin:165-183.

Street-Perrott F A, Mitchell J F B, Marchand D Sand Brunner J S (1990) Milankovitch and albedo forcing of the tropical monsoons: a comparison of geological evidence and numerical simulations for 9000 BP. *Transaction of the Royal Society of Edinburgh, Earth Science 81:*407-427.

Stryer L (1995) *Biochemistry*. 4th Edition. Freeman, New York.

Stuiver M, Grootes P M and Braziunas T F (1995) The GISP $\delta^{18}O$ climate record of the past 16 500 years and the role of the sun, ocean and volcanoes. *Quaternary Research 44:*341-354.

Sturtevant W C (1969) History and ethnography of some West Indian starches. In Ucko and Dimbleby (eds):177-199.

Sutton J E G (1973) *The Archaeology of the Western Highlands of Kenya*. BIEA Memoir 3, Nairobi.

Sutton J E G (1974) Aquatic civilisations of Middle Africa. *Journal of African History 15(4)*: 527-46.

Sutton J E G (1977) The African Aqualithic. *Antiquity 51*: 25-34.

Sutton J E G (1996) *The Growth of Farming Communities in Africa from the Equator Southwards*. Proceedings of the Conference of the British Institute in Eastern Africa, 4-8 July 1994, Cambridge.

Tackholm V (1956) *Student's Flora of Egypt.* Anglo-Egyptian Bookshop, Cairo.

Tadesse D (1975) Tef *(Eragrostis tef)* cultivars: morphology and classification. Part II. *Debre Zeit Agricultural Research Centre Bulletin No.66.* Alemeya University of Agriculture, Ethiopia:1-73.

Talbot M R, Livingstone D A, Palmer P G, Maley J, Melack J M, Delibrias G and Gulliksen S (1984) Preliminary results from sediment cores from Lake Bosumtwi, Ghana. *Palaeoecology of Africa and the Surrounding Islands 16:* 173-92.

Talbot M R (1980) Environmental responses to climate change in the West African Sahel over the past 20 000 years. In Williams and Faure (eds):37-62.

Tesfaye G (1979) Decouverte de grauvres prehistoriquss dans la Vallee de Gunda-Gundi (Agame, Tigray). *Abbay 10*:75-76.

Thomas H (1938) *Discovery of Abyssinia by the Portuguese in 1520.* London.

Tiercelin J J, Renaut W R, Delibrias G, LeFournier J and Beida S (1981) Late Pleistocene and Holocene lake levels fluctuations in the Lake Borgoria basin, Northern Kenya Rift valley. *Palaeoecology of Africa and the Surrounding Islands 13:* 105-20.

Trigger B (1972) Determinants of urban growth in pre-Industrial societies. In Ucko et al (eds):575-599.

Trimingham J S (1952) *Islam in Africa.* Oxford University Press, London.

Tringham R (1971) *Hunters, Fishers and Farmers of Eastern Europe 6000-3000 BC.* Hutchinson, London.

Tringham R, Cooper G, Odell G, Voyrek B and Whitman A (1974) Experimentation in the formation of edge damage: a new approach to lithic analysis. *Journal of Field Archaeology 1*:171-196.

Tykot R H (1996) Obsidian procurement and distribution in the Central and Western Mediterranean. *Journal of Mediterranean Archaeology 91*:39-82.

Tylock W (1975) The evidence of the proto-lexicon for the cultural background of the Semitic peoples. In Bynon J and T (eds) *Hamito-Semitica.* Mouton, the Hague:55-61.

Ucko P J and Dimbleby G W (1969) *The Domestication and Exploitation of Plants and Animals.* Duckworth, London.

Ucko P J, Tringham R and Dimbleby G W (1972) *Man, Settlement and Urbanism.* Cambridge, Massachusetts.

Ullendorf E M (1955) *The Semitic Languages of Ethiopia.* Taylor's Foreign Press, London.

Ullendorf E M (1960) *The Ethiopians. An Introduction to Country and People.* London.

Unger A (1866) Botanische streifzuge auf dem geibet der kulturgeschichte. VII. *Inhalte an organischen Einschulssen. Akademie Wissenchaften Wien Sitzungsber. Math.-Naturwiss. Kl. 54*:33-62.

Unger A (1866) Notiz uber fossile holtzer aus Abyssinien. *Sitzungsberichte der k Akademie der Wissenchaften zu Berlin.*

Unger-Hamilton R (1988) *Method in Microwear Analysis: Prehistoric Sickles and other tools from Arjoune, Syria.* Oxford

van Beek G W (1969) *Hajir bin-Humeid: Investigations at a Pre-Islamic Site in Southern Arabia.* Baltimore, John Hopkins University Press.

van Campo E, Duplessy J C and Rossignol-Strick M (1982) Climatic conditions deduced from 150 kyr oxygen isotope-pollen record from the Arabia Sea. *Nature 296*:56-59.

van der Merwe N J and Vogel J C (1983) Recent carbon isotope research and its implications for African archaeology. *African Archaeological Review 1*:33-56.

van der Veen M (1995) Ancient agriculture in Libya: a review of the evidence. *Acta Palaeobotanica 35(1)*:85-98.

van Zeist W (1972) Palaeobotanical results in the 1970 season at Cayonu, Turkey. *Helinium 12*:3-19.

van Zeist W (1980) Holocene vegetation of North West Syria. *Palaeohistoria 19*:19-85.

van Zeist W and Bakker-Heeres J A H (1975) Evidence for linseed cultivation before 6000 BC. *Journal of Archaeological Science 2*:215-219.

van Zeist W and Bottema S (1977) Palynological investigations in Western Iran. *Palaeohistoria 19*:19-85.

van Zeist W and Bottema S (1982) Vegetational history of the Eastern Mediterranean and the Near East during the last 20 000 years. In Bintliff J L and van Zeist W (eds) *Palaeoclimates, Palaeoenvironments and Human Communities in the Eastern Mediterranean Region.* BAR International Series S133:277-321.

van Zeist W and Bottema S (1991) Late Quaternary vegetation of the Near East. *Beihefte zum Tubinger Atlas des Vorderen Orients, Reihe A (Naturwissenschaften) Nr.18.* Reichert, Weisbaden.

van Zeist W and Wright H E Jr (1963) Preliminary pollen studies at Lake Zeriber, Zagros Mountains, South West Iran. *Science 140*:65-67.

van Zinderen Baker E M (1951) Archaeology and Palynology. *South African Archaeology Bulletin 23*:80-87.

van Zinderen Baker E M (1976) Tentative vegetation maps of Africa south of the Sahara during glacial and interglacial maximum. *Palaeoecology of Africa 9.*

Vansina J T (1979) Bantu in the crystal ball. *History in Africa 6*: 287-333.

Vansina J T (1980) Bantu in the crystal ball II. *History in Africa 7*: 293-325.

Vavilov N I (1926) *Studies on the Origins of Cultivated Plants.* Leningrad Institute of Applied Botany and Plant Breeding, Leningrad.

Vavilov N I (1951) The origin, variation, immunity and breeding of cultivated plants. *Chronica Botanica 13.*

Vavilov N I (1960) *World Resources of Cereals, Leguminous Seed Crops and Flax, and their Utilisation in Plant Breeding.* Academy of Science USSR, Moscow 1957.

Vavilov N I (1992) *Origin and Geography of Cultivated Plants.* (Lore D. Transl.) Cambridge University Press, Cambridge.

Vermeersch P M, Paulissen E and van Neer W (1989) The late Palaeolithic Makhadma site (Egypt): environment and subsistence. In Krzyzaniak and Kobusiewicz (eds): 277-86.

Vickers W T (1989) Patterns of foraging and gardening in a semi-sedentary Amazonian community. In Kent S (ed) *Farmers as Hunters; the Implications of Sedentism.* Cambridge University Press, Cambridge:46-59.

Vincens A (1989) Early Holocene pollen data from an arid East African region, Lake Turkana (Kenya): botanical and climatic implications. *Palaeoecology of Africa and the Surrounding Islands 20:*87-97.

Vishnu-Mittre (1968) Protohistoric records of agriculture in India. *Trans Bose Res Inst Calcutta 31*: 87-106.

Vishnu-Mittre (1969) Remains of rice and millet. In Sankalia HD, Deo S B and Ansai Z D (eds) *Excavations at Ahar (Tiambavati), Poona.* Decca College Postgraduate and Research Institute, Poona.

Vishnu-Mittre (1974) The beginnings of agriculture: palaeobotanical evidence in India. In Hutchinson (ed).

Vishnu-Mittre (1977) Changing economy in India. In Reed (ed):569-588.

Vita-Finzi C (1969) *The Mediterranean Valley.* Cambridge.

Vita-Finzi C (1970) Prehistoric economy in the Mount Carmel area of Palestine: site catchment analysis. *Proceedings of the Prehistoric Society 36*:1-37.

von Bothmer R, Jacobsen N, Baden C, Jorgensen R B and Linde-Laursen I (1995) *An Ecological Study of the Genus Hordeum.* 2nd Edition. IPGRI.

von Breitenbach F (1963) *The Indigenous Trees of Ethiopia.* 2nd Edition. Addis Ababa.

Wandibba S (1980) The application of attribute analysis to the study of Later Stone Age/Neolithic pottery ceramics in Kenya. In Leakey and Ogot (eds):283-285.

Wandibba S (1990) Ancient and modern ceramic traditions in the Lake Victoria Basin of Kenya. *Azania 25*:69-78.

Ward D J (1962) Some evolutionary aspects of certain morphologic characters in a world collection of barleys. *US Department of Agriculture Technical Bulletin 1276.*

Wasylikowa K, Harlan J R, Evans J, Wendorf F, Schild R, Close A E, Krolik H and Housley R A (1993) Examination of Botanical Remains of Early Neolithic Houses at Nabta Playa, Western Desert, Egypt, with Special Reference to Sorghum Grains. In Shaw et al (eds): 154-64.

Wasylikowa K, Schild R, Wendorf F, Krolik H, Kubiak-Martens L and Harlan J R (1995) Archaeobotany of the early Neolithic site E-75-6 at Nabta Playa, Western Desert, South Egypt. *Acta Palaeobotanica 35(1):* 133-155.

Wasylikowa K (1992) Holocene flora of the Tadrart Acaucus area, South West Libya, based on plant macrofossils from Uan Muhuggiag and Ti-n-Torha/Two Caves archaeological sites. *Origini 16:*125-159.

Wasylikowa K and Kubiak-Martens L (1995) Wild sorghum from the early Neolithic site of Nabta Playa, southern Egypt. *Res Archaeobotanicae 9th Symposium IWGP.* Kroll H and Pasternak R (eds): 345-358.

Watanabe H (1972) The Ainu. In Biccieri M (ed) *Hunters and Gatherers Today.* Holt, Reinhardt and Winston, New York:451-484.

Webb P, von Braun J and Yohannes Y (1992) *Famine in Ethiopia: policy implications of coping failure at national and household levels.* IFPRI Research Report 92, Washington.

Weber S A (1991) *Plants and Harappan Subsistence.* New Delhi, Oxford and IBH.

Webster J B (1979) *Chronology, Migration and Drought in Interlacustrine Africa.* Longman, London.

Weedman K (1996) The ethnoarchaeology of hide working and scraper use in Southern Ethiopia. *Paper presented at the 13th Biennial Congress of SAFA, Posnan, September 1996.*

Wendorf F (1968) *The Prehistory of Nubia.* Dallas, Southern Methodist University Press.

Wendorf F, Said R and Schild R (1970) Egyptian prehistory; some new concepts. *Science 169*: 1161-71.

Wendorf F and Schild R (1974) *A Middle Stone Age Sequence from the Central Rift Valley, Ethiopia.* Polish Academy of Science, Warsaw.

Wendorf F and Schild R (1976a) *Prehistory of the Nile Valley.* New York Academic Press.

Wendorf F and Schild R (1976b) The use of ground grain during the late Palaeolithic of the Lower Nile Valley, Egypt. In Harlan et al (eds): 269-88.

Wendorf F and Schild R (1980) *Prehistory of the Eastern Sahara.* New York Academic Press.

Wendorf F and Hassan F A (1980) Holocene ecology and prehistory in the Egyptian Sahara. In Williams and Faure (eds): 407-20.

Wendorf F and Schild R (1984a) Some implications of late Palaeolithic cereal exploitation at Wadi Kubbanyia (Upper Egypt). In Krzyzaniak and Kobusiewicz (eds): 117-28.

Wendorf F and Schild R (1984b) The emergence of food production in the Sahara. In Clark and Brandt (eds):93-101.

Wendorf F, Schild R and Close A E (1984a) *Loaves and Fishes: The Prehistory of Wadi Kubbanyia.* Dallas, Southern Methodist University Press.

Wendorf F, Schild R and Close A E (1984b) *Cattle Keepers of the Eastern Sahara. The Neolithic of Bir Kiseiba.* Dallas. Southern Methodist University Press.

Wendorf F, Schild R and Close A E (1989) *The Prehistory of Wadi Kubbanyia. Volume 2. Palaeoeconomy, Environment and Stratigraphy.* Dallas. Southern Methodist University Press.

Wendorf F, Close A E, Schild R, Wasylikowa K, Housley R A, Harlan J R and Krolik H (1992) Saharan exploitation of plants 8000 years ago. *Nature 359*:721-724.

Wenke R J and Casini M (1979) The Epi-palaeolithic - Neolithic Transition in Egypt's Fayum Depression. In Krzyzaniak and Kobusiewicz (eds): 193-204.

Werth E (1939) Emmer und Gerste aus dem 5. Jarrtausend v. Chr. und andere vorgeschichtliche Kulturpflanzenfunde. *Botanischen Gesellschaft 57*:453-462.

Westphal E (1974) *Pulses in Ethiopia, their Taxonomy and Agricultural Significance.* PUDOC, Wageningen.

Westphal E (1975) *Agriculture in Ethiopia*. Agricultural Research Reports 826. Wageningen.

Wetterstrom W (1993) Foraging and farming in Egypt: the transition from hunting and gathering to horticulture in the Nile Valley. In Shaw et al (eds): 165-226.

White F (1983) *The Vegetation of Africa*. UNESCO Natural Resources Research 20, Paris.

Whitehead G O (1926) Meroitic remains. *Sudan Notes and Records 9(2)*:51-58.

Whiteway R S (1902) *The Portuguese Expedition to Abyssinia in 1541-1543*. London.

Whitmore T C and Prance G T (1987) *Biogeography and Quaternary History in Tropical America*. Clarendon Press, Oxford.

Wilding R (1980) The desert trade of Eastern Ethiopia. In Leakey and Ogot (eds):379-380.

Wilding R (1984) A preliminary check list of plants used in the Borana economy. *JEPSS Research Memo 12*.

Wilding R (1985a) The pastoralists of the Lowlands of the Horn of Africa. Historical Studies in the Service of Development. *EPSS Research Memo 22*.

Wilding R (1985b) The history of pastoralism and the emergence of the Borana Oromo. A review of issues. *Joint Ethiopian Pastoral Systems Study Research Report 15*.

Wilke P J, Bettinger R, King T F and O'Connel J F (1972) Harvest selection and domestication in seed plants. *Antiquity 46*:203-209.

Wilkes G (1984) Maize: domestication, racial evolution and spread. In Harris and Hillman (eds):440-455.

Williams M A J (1984) Late Quaternary prehistoric environments in the Sahara. In Clark and Brandt (eds):74-83.

Williams M A J (1988) After the deluge: the Neolithic landscape in North Africa. In Bower and Lubell (eds):43-60.

Williams M A J, Bishop P M, Dakin F M and Gillespie R (1977) Late Quaternary lake levels in Southern Afar and the adjacent Ethiopian Rift. *Nature 267:* 690-93.

Williams M A J and Adamson D A (1980) Late Quaternary depositional history of the Blue and White Nile rivers in central Sudan. In Williams and Faure (eds):281-304.

Williams M A J and Faure H (1980) *The Sahara and the Nile*. A A Balkema. Rotterdam.

Williamson K (1989) Linguistic evidence for the prehistory of the Niger Delta. In *The Early History of the Niger Delta*. Alagoa E J, Anozie F N and Nzewunwa N (eds) Hamburg, Buske: 65-119.

Wilson R T (1984) *The Camel*. Longman, Harlow, England.

Wilson R T and Gebre Mariam W (1979) Medicine and magic in Central Tigray: a contribution to the ethnobotany of the Ethiopian Plateau. *Economic Botany 33*: 29-34.

Winkler H A (1938) *Rock Engravings of Southern Upper Egypt I and II*. Egypt Exploration Society, London.

Witcombe J (1978) *Euphytica 27*:601-604.

Woldetatios T (1973) Agroclimatology of tef. *Eragrostis* in Ethiopia. In *Agroclimatology of the Highlands of East Africa. Proceedings of the Technical Conference:*152-160.

Wolde-Michael H (1985) *The History of Famine in Ethiopia*. Mimeographed.

Wood C A (1976) Farmer's response to drought in Ethiopia. In Hussein (ed):67-88.

Wood C A (1977) A preliminary chronology of Ethiopian droughts. In Dalby D, Harrison Church R J and Bezzaz F (eds) *Drought in Africa*. International African Institute, London:68-73.

Woodburn J (1970) *Hunters and Gatherers: the Material Culture of the Nomadic Hadza*. British Museum, London.

Worster D (1990) The ecology of order and chaos. *Environmental History Review 14:*1-18.

Wright H E Jr (1977) Environmental change and the origins of agriculture in the Old and New worlds. In Reed (ed):281-318.

Wright H E Jr (1993) Environmental determinism in Near Eastern prehistory. *Current Anthropology 34*:459-469.

Wright H E Jr (1976) The environmental setting for plant domestication in the Near East. *Science 194*:385-89.

Wright R (1961) A painted rock shelter on Mount Elgon, Kenya. *Proceedings of the Prehistoric Society 27*:28-34.

Wylde A B (1901) *Modern Abyssinia*. London.

Xolocotzi E H (1987) Experiences leading to greater emphasis on man in ethnobotanical studies. *Economic Botany 4*:6-11.

Yan Z and Petit-Maire N (1994) The last 140 ka in the Afro-Asian arid/semi-arid transitional zone. *Palaeogeography, Palaeoclimatology, Palaeoecology 110*:217-233.

Yellen J E, Brooks A S, Cornelissen E, Mehlmam M J and Stewart K (1995) Middle Stone Age worked bone industry from Katanda, Upper Semliki Valley, Zaire. *Science 268*:553-536.

Yemane K, Robert C and Bonnefille R (1987) Pollen and clay mineral assemblages of a late Miocene lacustrine sequence from the North West Ethiopian Highlands. *Palaeogeography, Palaeoclimatology, Palaeoecology 60*:123-141.

Yen D E (1989) The domestication of the environment. In Harris and Hillman (eds):55-78.

Young R L (1996) Missing plant foods? Where is the archaeobotanical evidence for sorghum and millet. *Paper presented at the 13th Biennial Congress of SAFA, Posnan, September 1996.*

Young T C, Smith P E and Mortensen (1982) *The Hilly Flanks and Beyond: Essays on the Prehistory of South West Asia*. Oriental Institute of the University of Chicago, Chicago.

Zahn R (1994) Fast flickers in the Tropics. *Nature 372*:621-622.

Zamara A and Fields M L (1979) Nutritive quality of fermented cowpeas (*Vigna sinesis*) and chick peas (*Cicer arietinum*). *Journal of Food Science 44*:234-236.

Zarins J (1988) Prehistoric trade in the South Red Sea - the obsidian evidence. *Paper presented at the Conference on The Indian Ocean in Antiquity. London 1988.*

Zarins J and al-Badr H (1985) Archaeological investigations in the Southern Tihama Plain II: *Atlal 10:*

Zarins J, Murad A and Al-Yaish Kh (1981) The 2nd preliminary report on the South West province. *Atlal 5*:9-42.

Zarins J and Zahrani A (1986) Recent archaeological investigations in the Southern Tihama Plain. The sites of Athar and Sihi. *Atlal* 9:55-64.

Zeven A C and Zhukovsky P M (1975) *Dictionary of Cultivated Plants and their Centres of Diversity*. PUDOC, Washington.

Zhimin A (1989) Prehistoric agriculture in China. In *Harris and Hillman* (eds):643-649.

Zhukovsky P M (1975) *World Gene Pool of Plants for Breeding. Mega-centres and Endemic Micro-genecentres.* (English Translation), N I Vavilov Institute of Plant Industry, Leningrad.

Zimmerer K S (1994) Human ecology and the 'new ecology': the prospect and promise of integration. *Annals of the Association of American Geographers 84:*108-125.

Zohary D (1970) Centre of diversity and centre of origin. In *Genetic Resources in Plants - Their Exploration and Conservation.* Frankell O H and Bennet E (eds). Blackwell Scientific Publications, Oxford: 33-42.

Zohary D (1971) Origin of South West Asian cereals: wheats, barley, oats and rye. In Davies P H et al (eds*) Plant Life of South West Asia*. Botanical Society of Edinburgh:235-265.

Zohary D and Hopf M (1993) *Domestication of Plants in the Old World. The Origin and Spread of Cultivated Plants in West Africa, Europe and the Nile Valley.* 2nd Edition. Oxford University Press, Oxford.

CAMBRIDGE MONOGRAPHS IN AFRICAN ARCHAEOLOGY

* No 1 - BAR S75, 1980 **The Niger Delta** *Aspects of its Prehistoric Economy and Culture* by Nwanna Nzewunwa. ISBN 0 86054 083 9
* No 2 - BAR S89, 1980 **Prehistoric Investigations in the Region of Jenne, Mali** *A Study in the Development of Urbanism in the Sahel* by Susan Keech McIntosh and Roderick J. McIntosh ISBN 0 86054 103 7
* No 3 - BAR S97, 1981 **Off-Site Archaeology and Human Adaptation in Eastern Africa** *An Analysis of Regional Artefact Density in the Amboseli, Southern Kenya* by Robert Foley. ISBN 0 86054 114 2
* No 4 - BAR S114, 1981 **Later Pleistocene Cultural Adaptations in Sudanese Nubia** by Yousif Mukhtar el Amin. ISBN 0 86054 134 7
* No 5 - BAR S119, 1981 **Settlement Patterns in the Iron Age of Zululand** *An Ecological Interpretation* by Martin Hall. ISBN 0 86054 143 6
* No 6 - BAR S139, 1982 **The Neolithic Period in the Sudan, c. 6000-2500 B.C.** by Abbas S. Mohammed-Ali. ISBN 0 86054 170 3
* No 7 - BAR S195, 1984 **History and Ethnoarchaeology in Eastern Nigeria** *A Study of Igbo-Igala relations with special reference to the Anambra Valley* by Philip Adigwe Oguagha and Alex Ikechukwu Okpoko. ISBN 0 86054 249 1
* No 8 - BAR S197, 1984 **Meroitic Settlement in the Central Sudan** *An Analysis of Sites in the Nile Valley and the Western Butana* by Khidir Abdelkarim Ahmed. ISBN 0 86054 252 1
* No 9 - BAR S201, 1984 **Economy and Technology in the Late Stone Age of Southern Natal** by Charles Cable. ISBN 0 86054 258 0

No 10 - BAR S207, 1984 **Frontiers Southern African Archaeology Today** edited by M. Hall, G. Avery, D.M. Avery, M.L. Wilson and A.J.B. Humphreys. ISBN 0 86054 268 8. £23.00.

* No 11 - BAR S215, 1984 **Archaeology and History in Southern Nigeria** *The ancient linear earthworks of Benin and Ishan* by P.J. Darling. ISBN 0 86054 275 0
* No 12 - BAR S213, 1984 **The Later Stone Age of Southernmost Africa** by Janette Deacon. ISBN 0 86054 276 9
* No 13 - BAR S254, 1985 **Fisher-Hunters and Neolithic Pastoralists in East Turkana, Kenya** by John Webster Barthelme. ISBN 0 86054 325 0

No 14 - BAR S285, 1986 **The Archaeology of Central Darfur (Sudan) in the 1st Millennium A.D.** by Ibrahim Musa Mohammed. ISBN 0 86054 367 6.

No 15 - BAR S293, 1986 **Stable Carbon Isotopes and Prehistoric Diets in the South-Western Cape Province, South Africa** by Judith Sealy. ISBN 0 86054 376 5.

* No 16 - BAR S318, 1986 **L'art rupestre préhistorique des massifs centraux sahariens** by Alfred Muzzolini.. ISBN 0 86054 406 0

No 17 - BAR S321, 1987 **Spheriods and Battered Stones in the African Early and Middle Stone Age** by Pamela R. Willoughby. ISBN 0 86054 410 9

* No 18 - BAR S338, 1987 **The Royal Crowns of Kush** *A study in Middle Nile Valley regalia and iconography in the 1st millennia B.C. and A.D.* by Lázló Török.. ISBN 0 86054 432 X

No 19 - BAR S339, 1987 **The Later Stone Age of the Drakensberg Range and its Foothills** by H. Opperman. ISBN 0 86054 437 0

* No 20 - BAR S350, 1987 **Socio-Economic Differentiation in the Neolithic Sudan** by Randi Haaland. ISBN 0 86054 453 2

No 21 - BAR S351, 1987 **Later Stone Age Settlement Patterns in the Sandveld of the South-Western Cape Province, South Africa** by Anthony Manhire. ISBN 0 86054 454 0

No 22 - BAR S365, 1987 **L'art rupestre du Fezzan septentrional (Libye) Widyan Zreda et Tarut (Wadi esh-Shati)** by Jean-Loïc Le Quellec. ISBN 0 86054 473 7

* No 23 - BAR S368, 1987 **Archaeology and Environment in the Libyan Sahara** *The excavations in the Tadrart Acacus*, 1978-1983 edited by Barbara E. Barich. ISBN 0 86054 474 5

No 24 - BAR S378, 1987 **The Early Farmers of Transkei, Southern Africa Before A.D. 1870** by J.M. Feely. ISBN 0 86054 486 9

No 25 - BAR S380, 1987 **Later Stone Age Hunters and Gatherers of the Southern Transvaal** *Social and ecological interpretation* by Lyn Wadley. ISBN 0 86054 492 3

No 26 - BAR S405, 1988 **Prehistoric Cultures and Environments in the Late Quaternary of Africa** edited by John Bower and David Lubell. ISBN 0 86054 520 2

No 27 - BAR S418, 1988 **Zooarchaeology in the Middle Nile Valley** *A Study of four Neolithic Sites near Khartoum* by Ali Tigani El Mahi. ISBN 0 86054 539 3

No 28 - BAR S422, 1988 **L'Ancienne Métallurgie du Fer à Madagascar** by Chantal Radimilahy. ISBN 0 86054 544 X

No 29 - BAR S424, 1988 **El Geili The History of a Middle Nile Environment, 7000 B.C.-A.D. 1500** edited by I. Caneva. ISBN 0 86054 548 2

No 30 - BAR S445, 1988 **The Ethnoarchaeology of the Zaghawa of Darfur (Sudan) Settlement and Transcience** by Natalie Tobert. ISBN 0 86054 574 1

No 31 - BAR S455, 1988 **Shellfish in Prehistoric Diet Elands Bay, S.W. Cape Coast, South Africa** by W.F. Buchanan. ISBN 0 86054 584 9

No 32 - BAR S456, 1988 **Houlouf I** *Archéologie des sociétés protohistoriques du Nord-Cameroun* by Augustin Holl. ISBN 0 86054 586 5

* No 33 - BAR S469, 1989 **The Predynastic Lithic Industries of Upper Egypt** by Liane L. Holmes. ISBN 0 86054 601 2 (two volumes)

No 34 - BAR S521, 1989 **Fishing Sites of North and East Africa in the Late Pleistocene and Holocene** *Environmental Change and Human Adaptation* by Kathlyn Moore Stewart. ISBN 0 86054 662 4

* No 35 - BAR S523, 1989 **Plant Domestication in the Middle Nile Basin** *An Archaeoethnobotanical Case Study* by Anwar Abdel-Magid. ISBN 0 86054 664 0

No 36 - BAR S537, 1989 **Archaeology and Settlement in Upper Nubia in the 1st Millennium A.D.** by David N. Edwards. ISBN 0 86054 682 9

No 37 - BAR S541, 1989 **Prehistoric Settlement and Subsistence in the Kaduna Valley, Nigeria** by Kolawole David Aiyedun and Thurstan Shaw. ISBN 0 86054 684 5

No 38 - BAR S640, 1996 **The Archaeology of the Meroitic State** *New perspectives on its social and political organisation* by David N. Edwards. ISBN 0 86054 825 2

No 39 - BAR S647, 1996 **Islam, Archaeology and History** *Gao Region (Mali) ca. AD 900 - 1250* by Timothy Insoll. ISBN 0 86054 832 5

No 40 – BAR S651, 1996 **State Formation in Egypt**: *Chronology and society* by Toby A.H. Wilkinson. ISBN 0 86054 838 4

No 41 - BAR S680, 1997 **Recherches archéologiques sur la capitale de l'empire de Ghana** *Etude d'un secteur d'habitat à Koumbi Saleh, Mauritanie. Campagnes II-III-IV-V (1975-1976)-(1980-1981)* by S. Berthier. ISBN 0 86054 868 6

No 42 - BAR S689, 1998 **The Lower Palaeolithic of the Maghreb** *Excavations and analyses at Ain Hanech, Algeria* by Mohamed Sahnouni. ISBN0 86954 875 9

No 43 - BAR S715, 1998 **The Waterberg Plateau in the Northern Province, Republic of South Africa, in the Later Stone Age** by Maria M. Van der Ryst. ISBN 0 86054 893 7

No 44 - BAR S734, 1998 **Cultural Succession and Continuity in S.E. Nigeria** *Excavations in Afikpo* by V. Emenike Chikwendu. ISBN 0 86054 921 6

* out of print

www.ingramcontent.com/pod-product-compliance
Ingram Content Group UK Ltd.
Pitfield, Milton Keynes, MK11 3LW, UK
UKHW061847041026
15591UKWH00015B/955

9 780860 549710